Enzymatic Analysis

Biological Methods

Enzymatic Analysis: ***A Practical Guide***
by ***Janet V. Passonneau*** and ***Oliver H. Lowry,*** 1993
Characterization of Proteins,
edited by ***Felix Franks,*** 1988
Methods of Protein Microcharacterization: ***A Practical Handbook***
edited by ***John E. Shively,*** 1986
Techniques in Cell Cycle Analysis,
edited by ***Joe W. Gray*** and ***Zbigniew Darzynkiewicz,*** 1986
Experiments in Molecular Biology,
edited by ***Robert J. Slater,*** 1986
Liquid Chromatography in Clinical Analysis,
edited by ***Pokar M. Kabra*** and ***Laurence J. Marton,*** 1981
Metal Carcinogenesis Testing: ***Principles and In Vitro Methods,***
by ***Max Costa,*** 1980

Enzymatic Analysis

A Practical Guide

by

Janet V. Passonneau
Washington, DC

and

Oliver H. Lowry
Washington University, St. Louis, Missouri

Humana Press **Totowa, NJ**

999 Riverview Drive, Suite 208
Totowa, New Jersey 07512

Printed in the United States of America. 10 9 8 7 6 5 4 3 2 1

Library of Congress Cataloging-in-Publication Data

Main entry under title:

Enzymatic analysis : a practical guide / edited by Janet V. Passonneau
and Oliver H. Lowry.
p. cm. —— (Biological methods)
Includes index.
ISBN 0-89603-238-8
1. Enzymes——Analysis——Laboratory manuals. I. Passonneau, Janet
V. II. Lowry, Oliver H. III. Series.
QP601.E5152 1993
574.19'285——dc20 92-28742
CIP

Preface

Enzymatic Analysis: A Practical Guide is a multipurpose manual of laboratory methods. It offers a systematic scheme for the analysis of biological materials from the level of the whole organ down to the single cell and beyond. It is intended as a guide to the development of new methods, to the refinement of old ones, and to the adaptation in general of methods to almost any scale of sensitivity.

As some may realize, the book is a sequel to *A Flexible System of Enzymatic Analysis,* originally published in 1972. The major changes, other than an appropriate interchange of authors, consist of a wholly new chapter of methods and protocols for measuring enzymes, the addition of 13 new entries in the metabolite chapter, and a much superior chapter on enzymatic cycling. With considerable nostalgia, we have switched from DPN and TPN to NAD and NADP nomenclature, which no doubt will make Otto Warburg turn over in his grave.

The incentives for the methodology in this book came from the rigorous demands of quantitative histochemistry and cytochemistry. These demands are specificity, simplicity, flexibility, and, of course, sensitivity—all likewise desirable attributes of methods for other purposes.

The specificity is provided by the use of enzyme methods. Simplicity is achieved by leading all reactions to a final pyridine nucleotide step. Flexibility and part of the sensitivity result from the use of fluorometry for measuring the oxidation or reduction of the pyridine nucleotide. Greater sensitivity is provided by chemical amplification by means of "enzymatic cycling."

Enzymatic Analysis is in a sense very sharply focused. It does not have the comprehensive character of either Bergmeyer's *Methods of Enzymatic Analysis* or of Glick's *Quantitative Histochemistry.* Instead, with a few exceptions, only methods and tools actually used by the authors and their long-suffering colleagues and trainees are presented. It is hoped, however, that the principles and techniques presented here can be exploited for new analyses when needed and therefore provide a useful *de facto* breadth.

The book is intended for novices as well as experts who are familiar with other analytical styles. It therefore contains rudimentary information, perhaps oversimplified, on such matters as enzyme kinetics and the use of buffers. It is divided into four parts: *Part I,* the general section is comprised of five chapters, the first three of which deal with the properties of the pyridine nucleotides, kinetics, and constriction pipets. Chapter 4 describes methods and problems in the preparation of tissues for analyses. Chapter 5, "Improvement, Modification, Adaptation, Troubleshooting, and the Development of New Methods," is meant to be the "heart" of the book. We hope most users will read this chapter.

Part II consists of three chapters. The first gives detailed protocols for 47 metabolites, the second for 44 enzymes, and the last presents new or improved enzymatic cycling methods for NADP, NAD, ATP + ADP, and GTP + GDP.

Part III, the quantitative histochemistry section, is comprised of four chapters that include information on the preparation of frozen-dried material, dissection of samples for analysis, the fishpole balance for weighing samples, and finally, the generalities of analysis with emphasis on the "oil well technique."

Part IV is an appendix containing information that may be useful to have available at the bench.

It will be seen that no presentation is made of methods dependent on radioactive tracers. No implication should be taken from this that radioactive methods are any less important than direct chemical methods. On the contrary, we would argue that one set of techniques complements the other. One common misconception worth dispelling is that radioactive methods are inherently more sensitive than chemical ones. For example, to count carrier-free ^{14}C in 2 h with an error of not over 3% requires about 5×10^{-14} moles. Direct chemical assays with a single enzymatic cycling step can provide 50 times this sensitivity and, with double cycling, 50,000 times this sensitivity.

Unless noted otherwise, abbreviations for chemical compounds will follow the recommendation of the *Journal of Biological Chemistry.* The symbols, *M*, m*M*, μ*M*, n*M*, n*M*, etc, will be used exclusively to indicate concentration, i.e., molar, millimolar, micromolar, nanomolar, 10^{-8} molar, etc. *Amounts* of material will be indicated by the terms mol, mmol, μmol, 10^{-12} mol, etc. For example, an enzyme velocity given as 20 μ*M*/ /min would mean 20 μmol/L/min (20 μmol/min). Wavelengths will be given as nm (nanometers or mμ). Enzyme activities will be given in international units (U), i.e., μmol/min.

We wish to acknowledge the ideas and criticism generously given by many colleagues. Special thanks must go to Drs. Helen B. Burch, David B. McDougal, Jr., Franz M. Matschinsky, Nelson D. Goldberg, Takahiko Kato, Sosamma J. Berger, William H. Outlaw, Jr., Elizabeth K. Barbehenn, W. David Lust, D. A. Rottenberg, F. A. M. deAzeredo; and to Demoy W. Schulz, Joyce G. Carter, Joseph G. Brown, V. R. Lauderdale, Gretchen K. Feussner, Nira R. Roberts, Elizabeth J. Crawford, Jeanne A. Lopez, Mary Jane Brock, Maggie M.-Y, Chi, Carol S. Hintz, Mary Ellen Pusateri, Jane Pirolo, and Jill K. Manchester; and to a long series of patient postdoctoral fellows who have struggled with us to develop and use this methodology.

Finally we are indebted to the American Cancer Society and the National Institutes of Health for a great many years of generous financial support that made this all possible.

Janet V. Passonneau
Oliver H. Lowry

Contents

Part 1

General Principles

Chapter 1

Pyridine Nucleotides

All of the methods of enzymatic analysis to be considered depend on pyridine nucleotides. In 1935 Negelein and Hass *(1)* in Warburg's laboratory described a method for the determination of glucose-6-P dehydrogenase activity based on the increase in absorption in the near UV as NADPH was produced. This is apparently the first publication of a method of this type. Greengard *(2)* was the first to describe fluorometric pyridine nucleotide methods for measuring metabolites. Since 1935, an enormous number of enzymes and metabolites have been measured with the aid of NAD and NADP. In fact, with the use of auxiliary enzymes, almost every substance of biological interest could be measured with a pyridine nucleotide system.

The usefulness of NAD and NADP for analytical purposes depends on some of their unusual properties.

1. They serve as the natural oxidizing and reducing agents in a wide variety of specific enzyme systems. With the appropriate enzyme as catalyst, they can selectively oxidize or reduce a particular substrate in the presence of innumerable other compounds. If one or more axuiliary enzymes are used, it is a rare substance that cannot be specifically oxidized or reduced in this way.
2. The reduced forms of the nucleotides NADH and NADPH, in addition to absorbing near UV light, are fluorescent, whereas the oxidized forms are not. Because the fluorescence can be measured accurately at concentrations down to $10^{-7}M$, oxidation or reduction of a pyridine nucleotide can be measured with great sensitivity.
3. The reduced forms can be completely destroyed in acid without affecting the oxidized forms. Conversely, the oxidized forms can be destroyed in alkali without affecting the reduced forms. This means that, at the end of a reaction, the excess pyridine nucleotide of the reagent can be destroyed, leaving the fraction that has been oxidized or reduced intact for subsequent measurement by procedures that provide even greater sensitivity than the native fluorescence.

4. At this point, two possibilities are available. Both the oxidized and reduced forms can be converted to highly fluorescent forms in strong alkali. This permits accurate meaurements at concentrations down to $10^{-8}M$. Alternatively, much greater sensitivity, by orders of magnitude, can be provided by "enzymatic cycling" in which the pyridine nucleotide acts as the catalytic intermediate for a two enzyme system (Chapter 5).

To make the fullest analytical use of the pyridine nucleotides, it is desirable to be thoroughly familiar with those properties that have analytical relevance. In this chapter, these properties will be described, together with a discussion of fluorometry and its limitations. In addition, the preparation, standardization, and storage of pyridine nucleotides will be detailed.

Spectral Absorption and Fluorescence of Reduced Pyridine Nucleotides

NADH and NADPH have identical absorption bands in the near UV light, which peak at 340 nm. NAD^+ and $NADP^+$ do not absorb at this wavelength. Therefore, changes in oxidation or reduction can be measured in the spectrophotometer.

Some of the light that is absorbed is reemitted as fluorescence. Although only a small fraction of the absorbed light is emitted, this can be measured to give much greater sensitivity than can be easily attained from the measurement of the absorption itself. In the spectrophotometer (i.e., with absorption measurements), it is the nonabsorbed light that is registered on the phototube, and very small decreases in this transmitted light are hard to measure. The ordinary spectrophotometer requires for precision a decrease in the transmitted light of 5% or more, and there is little to be gained by increasing the intensity of the incident light or sensitivity of the phototube. In the fluorometer, the photocell sees only the emitted fluorescence, and this is directly proportional to the intensity of the incident light. Thus, the sensitivity can be increased almost without limit by increasing the intensity of the light source, the amplification, or the phototube sensitivity.

Although spectrophotometry does not readily offer the sensitivity of fluorometry, it is exceedingly useful for standardization purposes and for measurements with concentrations of NADH and NADPH above the range suitable for fluorometry. The molar extinction coefficient (ε) for NADH or NADPH at 340 nm is 6270. This means that a 0.1-m*M* (100-μ*M*) solution would have an optical density (OD) of 0.627. This corresponds to 23.6% transmission (I) through a 1-cm light path. (With a 1-cm light path, log [I_o/I] = OD, or for a 100-μ*M* solution, log [100/23.6] = 0.627.)

With ordinary equipment, accurate measurements can be made with NADH concentrations of 10–200 μ*M*, or 5–100 nmol in 0.5 mL. With special equipment (microcells or special differential amplifiers), much smaller amounts of NADH can be measured, but fluorometry offers a sinpler and easier way to achieve high sensitivity.

In the fluorometer, the analytically useful range of concentration with NADH (NADPH) is from 0.1 to 10 μ*M*, equivalant to 0.1–10 nmol in 1 mL. Thus, the upper limit for the fluorometry is about the lower limit for accurate spectrophotometry (with conventional equipment). Fluorometric sensitivity can be further increased, and extended to oxidized as well as reduced pyridine nucleotides, by treatment with strong alkali *(see page 10).*

Limitations of Fluorometry

For the great advantages of fluorometry, a certain price has to be paid. It has a few peculiarities and limitations that should be clearly recognized.

Quenching

Any substance that absorbs the exciting light, including the pyridine nucleotide itself, will reduce the intensity of the exciting light as it passes through the solution. This will thereby diminish the emitted light (i.e., will "quench" the fluorescence). In consequence, the emitted light is proportional to the concentration of the substance measured only so long as the absorption of the exciting light is negligible. It is for this reason that the upper concentration limit for NADH measurement is set at about 10 μ*M* (corresponding to an OD of 0.063 or a 5% absorption of light at the middle of a 1-mL fluorometer tube).

A second source of error is absorption of part of the exciting light or the emitted light by colored substances that might be present. Because the fluorescence is in the visible region, the presence of materials absorbing the emitted light can be detected by eye. Substances absorbing the exciting light might go unnoticed, but can be detected by seeing whether an added amount of NAD(P)H gives the expected increment of fluorescence.

Fluorescence Enhancement

A number of cations can enhance the fluorescence of pyridine nucleotides in alkaline solutions. For example, the presence of 1 m*M* Mg^{2+} at pH 11.5 enhances the native fluorescence of NADH by 50% and that of NADPH by 300%. The enhancement with cations is not analytically useful, since it is affected by the order of mixing and the fluorescence is not stable with time. Enhancement by cations should therefore be avoided by either adding an excess of EDTA or keeping the pH at 10 or below for reading.

Effect of Enzymes on Native Fluorescence

The fluorescence of NADH or NADPH may be affected (usually enhanced) by the enzymes they serve as coenzyme. In most situations, this is not a problem. It can become significant when high levels of enzyme are used or if the enzyme has a very strong effect. The difficulty is that the enhancement is not the same at all coenzyme levels, since only that fraction of the coenzyme that is in combination with the enzyme is affected. Consequently, proportionality is lost at some point, usually at a low level, because high molar concentrations of enzyme are seldom used. The safe procedure is to test the system with several levels of NADH or NADPH in the working range and, if necessary, calculate analytical results from a standard curve.

Effect of pH

There appears to be no effect of pH *per se* on the fluorescence of NADH (NADPH) from pH 6 to 13. However, pHs below 7 are ordinarily avoided because of instability of the reduced forms. At pHs above 10.5, NAD^+ ($NADP^+$) begins to be converted into a fluorescent product. Therefore, high pH is avoided when the oxidized nucleotides are present.

Temperature Effects

Another potential source of error is the large negative temperature coefficient that is characteristic of fluorescence in general. In the case of pyridine nucleotides, this amounts to 1.6% per degree. If standards and samples are all read at the same temperature, there is of course no problem, but differences in temperature must be avoided.

Fluorescence Blanks

An ultimate limitation of fluorescence is set by blank readings. Most reagents are either slightly fluorescent or contain fluorescent impurities. With H_2O alone, it is difficult to reduce the instrumental reading below a value equivalent to the fluorescence of 0.05 μ*M* NADH. Part of this is owing to failure of the optical system to exclude all scattered light from the phototube; part may be owing to slight fluorescence of the light filters, and part is possibly owing to the Raman spectrum of the H_2O itself.

The use of disposable glassware has become commonplace. Although rigorous cleaning seemed essential in the past, our experience with disposable tubes indicates that they are adequate for all except perhaps the most sensitive procedures. This applies at least to the larger tubes (10 × 75 mm, 7 × 70 mm, and 6 × 50 mm). When direct assays in the fluorometer require sensitivity of 0.1 μ*M* or less, it may be useful to rinse the tubes in distilled or deionized water a few times before use and allow them to drain dry, without heat.

To rinse, the tubes are placed open end up in 600 mL beakers, and the beaker is filled completely to maintain alignment. The tubes are filled with water, taking care that all air bubbles are removed. This can be done by gentle shaking or by inserting a clean glass rod in the tube to displace the air. The tubes are then inverted against a stainless-steel screen and allowed to drain. After several rinses, the tubes are drained by inverting over the screen, with a rod to elevate the beaker from the surface of the bench. The outside of the tubes should be wiped dry with a lint-free cloth before use. The blank fluorescence reading of the tubes filled with clean water need not exceed the equivalent of 0.1 μ*M* NAD(P)H. Occasionally, there are tubes in which the glass appears to be fluorescent. However, there are commercially available tubes that are adequate for all but the most sensitive determinations.

An increase in scattered light, whether from a scratched tube or from turbidity in the solution, will increase the blank even in the best fluorometer. In a poorly designed instrument or with improper filters, the effect will be much greater.

It is unnecessary to tolerate a blank with distilled H_2O greater than the equivalent of 0.1 μ*M* NADH. If the H_2O blank is higher than this, the fault can be with the fluorometer, the filters, the H_2O, the vessel used to store the H_2O, the fluorometer tubes (scratched, dirty, or dusty), or even in some cases tubes that are made of glass that is fluorescent. In any event, it is strongly recommended that the fluorometer be zeroed with the shutter closed, i.e., with the instrument balanced against the dark current alone. This means that the reagent blank will be directly monitored, and serious blank problems detected and eliminated. If the sample generates only a 0.2 μM signal and the reagent and tubes contribute a 20 μM signal, precision becomes out of the question.

Limits of Useful Sensitivity

In many instances, the useful sensitivity of a fluorescence measurement is set by factors outside of the instrumentation. As sensitivity is increased, contamination becomes an increasing problem. Certain substances, such as inorganic phosphate, ammonia, or lactate, are so common that it becomes impracticable to measure them at concentrations below $10^{-6}M$. In general, at the highest sensitivity, there is certain to be danger of contamination with traces of fluorescent material or specks of dust. The greatest useful direct sensitivity is achievable if the assay is based on the difference between two readings made before and after the addition of the specific reagent. In this case, it is helpful if the time lapse is not too great and essential that the instrument be highly stable.

Fluorescence Reference Standards

All fluorescence readings have to be made by comparison with some reference solution. A stable working standard, such as quinine sulfate, is ordinarily used, and all readings made against it. Such standards ought not to serve as a basis for calculation, since they may change with time and may not have the same temperature coefficient as NADH. In each analytical procedure, a true standard is carried through the entire process.

Some problems have been encountered in preparing stable quinine standards of the concentrations desired. The standards are routinely prepared in test tubes like those used for fluorometric analyses. If the tubes are sealed with corks, the solutions evaporate, and require constant rechecking. If the tubes are sealed by heating in the flame, the quinine solutions frequently do not fluoresce as much as expected. The following procedure is recommended to avoid these problems.

The Pyrex™ tubes (10×75 mm) are heated and pulled in an oxygen-gas flame, so that there is a long, narrow neck. At this stage, the tubes are filled with 50% HNO_3 and heated for 5 or 10 min in a boiling water bath. The tubes are then rinsed thoroughly three or four times with distilled water, filled with distilled water, and again heated. The water is shaken out, and quinine sulfate (0.02–0.2 µg/mL) prepared in 0.01*N* H_2SO_4 is added to the tubes in a vol of 1 mL. The narrow neck of the tube can then be easily sealed in the flame without risk to the quinine sulfate solution.

Standards prepared in such a manner yield the expected fluorescence and are relatively stable with time *if they are kept in the dark when not in use.* Although quinine sulfate is not very light sensitive, if these working standards are continually exposed to laboratory light, they will fade markedly over weeks and months.

The ratio between the fluorescence readings with quinine and NADH varies somewhat with the wavelength of the exciting light and the transmission characteristics of the secondary light filter. With a tungsten light source and the secondary filter recommended, quinine is about 30 times more fluorescent than NADH or NADPH on a molar basis. This means that a 1-µ*M* NADH solution is equivalent to approx 0.013 µg/mL of quinine sulfate (0.03 µ*M*). (Quinine fluorescence is strongly quenched by halide ion; therefore, standards must be prepared in H_2SO_4 rather than HCl).

Excitation and Emission Wavelengths and Light Filters

The fluorescence spectra of NADH and NADPH appear to be identical in position and magnitude in pure solution. Their fluorescence is not light sensitive. Because the absorption maximum is 340 nm, this is ordinarily the optimal wavelength for excitation. There may be situations, however, in which the ratio of blank-to-sample fluorescence can be reduced by using a somewhat longer wavelength. The usual fluorometer actually provides excitation with the 365 nm Hg line. This is no disadvantage as far as sensitivity is concerned, since the useful sensitivity limit is determined by the blank fluorescence and not by the absolute output of fluorescent light. With a mercury light source, a very satisfactory primary filter is CS filter No. 7-37, glass No. 5860, which has peak transmission (about 25%) at 360 nm. With the weaker UV light provided by a tungsten light source CS No. 7-60, glass No. 5840 with peak transmission (about 70%) at 355 nm may be preferable, especially for weak solutions.

The fluorescence emission maximum is at 460 nm. Here again the best wavelength for measurement may not always be at the peak. The actual choice can be made empirically by finding the optimal ratio between sample and blank fluorescences. For general purposes, the best filter combination we have found consists of a combination of filters CS No. 4-72, glass No. 4303 and CS No. 3-72, glass No. 3387. Because No. 3387 is slightly fluorescent, it is placed nearest the photocell, where it is shielded from most of the scattered light by the other filters. Otherwise, somewhat higher blank readings will be obtained.

Conversion of Pyridine Nucleotides in Alkali into Fluorescent Products

As discovered by Kaplan et al. *(3),* the destruction of NAD^+ and $NADP^+$ in strong alkali results in the formation of highly fluorescent products. There is no difference between the products of the two nucleotides in regard to absorption spectrum, fluorescence spectrum, or fluorescence intensity. NADH and NADPH, which are very stable in alkali, can also be converted into these products if they are first oxidized to NAD^+ and $NADP^+$.

The absorption spectrum has its peak at 360 nm (i.e., it is shifted only 20 nm from the peak for native reduced nucleotides). The fluorescence spectrum is narrower than in the case of NADH and NADPH, but the peak is the same position (460 nm *[4]*). The fluorescence developed in 6*M* NaOH, as recommended, is about 10 times greater than the native fluorescence of the reduced nucleotides. The same fluorometer filters used for the native fluorescence can ordinarily be used to measure the fluorescence induced in alkali.

The yield of the induced fluorescence product increases almost linearily with NaOH concentration from 0.05 to 6*M*, even though the nucleotides are destroyed completely over the whole range. This appears to be the consequence of two competing reactions. One results in destruction of NAD^+ ($NADP^+$) without producing fluorescence and is *not* accelerated by increasing NaOH above 0.05*M*. The other reaction converts NAD^+ ($NADP^+$) to a fluorescent product. This second reaction is accelerated in proportion to NaOH concentration. The competing reactions must have almost equal temperature coefficients, since the same fluorescence yield is obtained to within 1%, at 38 and 60°C.

In 6*M* NaOH, maximum fluorescence is developed in 1 h at 25°C, 30 min at 38°C, and 10 min at 60°C. If tubes are heated to hasten development of fluorescence, they must be cooled to exactly room temperature before reading, because the fluorescence has a negative temperature coefficient of 1.3%/degree. Once the fluorescence has been developed, the intensity is the same between pH 10.5–15 (10*M* NaOH).

An important disadvantage of the original strong alkali method is that the fluorescence is light sensitive. This can make reading difficult, or imprecise, especially with lowest levels, for which high light intensity may be required. Fortunately, light sensitivity is almost abolished by incorporating 10 m*M* imidazole in the strong alkali *(5)*. It is preferable to include the imidazole before the development of fluorescence to protect against room light. Imidazole added to 6*M* NaOH has been found to retain its efficacy for at least 3 mo. The rate of fluorescence development is not affected by the imidazole.

One other precaution should be mentioned. Strong NaOH is viscous and has a high specific gravity. If 6*M* NaOH is run slowly into a tube containing the sample, there will be incomplete mixing at the interface. This means that some of the nucleotide will be exposed to weaker alkali than intended and will start to be destroyed with a low-fluorescence yield. Therefore, mixing should be as prompt as possible, and because of the viscosity of the NaOH, it is necessary to mix especially well. A second mixing after heating may be indicated. Because the fluorescence developed is proportional to NaOH concentration, the volumes of the samples and standards should be as nearly alike as possible.

Analytically Useful Range with Strong Alkali Enhancement

Reproducible measurements can be made down to 0.01-μ*M* concentrations and up to nearly 10 μ*M*. The upper limit is set by the point of departure from linearity resulting from substantial absorption of the exciting light by the pyridine nuleotide product. The lower limit is set by the fluorescent blank of the NaOH solution. This can easily be kept down to the equivalent of 0.01 μ*M* NAD^+. Fresh NaOH solutions usually have a much higher blank than this, especially if they have come into contact with organic matter (or rubber stoppers!). This blank fluorescence is light sensitive and can be eliminated by leaving the bottle in direct sunlight for a few hours.

When working at low levels, the absolute value of the blank is less important than its reproducibility. The effect of the blank can often be minimized if a reading can be made on individual tubes of 6*M* NaOH before the sample addition.

Interfering Substances

Because of the high sensitivity of the strong alkali method, possible interfering substances can usually be diluted out. However, certain reactive substances must be kept at rather low concentrations during fluorescence development. These include pyruvate and α-ketoglutarate, which decrease the fluorescence yield at concentrations above 0.1 m*M*, and glucose, which inhibits at concentrations much above 1 m*M*. The presence of 6 m*M* (0.02%) H_2O_2 in the strong alkali will reduce interference from all of these carbonyl compounds. In any event, it is desirable to measure standards under conditions identical to those for the unknown samples.

Increased Fluorescence from NADH and NADPH with Strong Alkali

When reduced pyridine nucleotides are treated with H_2O_2 in 6*M* NaOH-imidazole, they are oxidized and yield the same highly fluorescent products that are obtained without H_2O_2 from the oxidized forms. This tenfold increase in fluorescence may be of advantage not only when the concentration is too low to measure accurately by the native fluorescence, but also when a blank is present that would be disturbing without the enhancement.

Oxidation of the NADH (NADPH) by H_2O_2 is inappreciable at pHs between 8–12, but proceeds quite rapidly in 6*M* NaOH. There even appears to be oxidation in 6*M* NaOH without H_2O_2. The time to produce full fluorescence is somewhat longer than in the case of NAD^+ and $NADP^+$, because of the extra time needed for the oxidation. With 3 m*M* H_2O_2, full fluorescence is obtained in 2 h at 25°C, 1 h at 38°C, and 15 min at 60°C. With higher levels of peroxide, partial destruction occurs (e.g., there is 4% loss with 10 m*M* H_2O_2 and 40% with 100 m*M*). As in the case of NAD^+ and $NADP^+$, tubes that have been heated must be cooled exactly to room temperature before readings are made.

The H_2O_2 is added to the 6*M* NaOH-imidazole from a 3% (0.9*M*) solution, that is prepared within a week from 30% H_2O_2. If the stronger H_2O_2 is added directly to the 6*M* NaOH, a precipitate will form that may be difficult to dissolve. The 6*M* NaOH-H_2O_2 solution is not very stable, and should be prepared within an hour of use if kept at room temperature, and within a few hours if kept in ice water.

Anything that reacts with H_2O_2 in alkali could interfere if the level is too high. Increasing the H_2O_2 concentration can often compensate.

Stability and Selective Destruction of Reduced and Oxidized Forms of Pyridine Nucleotides

It is of extreme analytical importance that oxidized and reduced pyridine nucleotides are differentially sensitive to destruction by acid and alkali. NADH and NADPH are rapidly destroyed in acid under conditions that leave NAD^+ and $NADP^+$ completely intact. Conversely, NAD^+ and $NADP^+$ are readily destroyed in alkali without discernible loss of NADH or NADPH. This makes it possible at the end of a reaction to eliminate the excess pyridine nucleotide, while preserving the fraction that has been oxidized or reduced for subsequent measurement.

The following describes the influence of pH, temperature, and certain other factors on the rates of destruction. This permits selection of conditions appropriate to a particular analytical situation.

Destruction of NADH and NADPH in Acid and Their Stability in Alkali

The rate of destruction of reduced pyridine nucleotides is a linear function of hydrogen ion concentration over a wide pH range (Fig. 1). The time calculated for 99% destruction of NADH at 23°C is 1.2 min at pH 2 (0.01*M* HCl) and 2 h at pH 4. The rate is three or four times faster at 38°C and 20 times faster at 60°C. NADPH has not been studied so thoroughly, but it is somewhat more easily destroyed than NADH. At 30°C, for example, it is destroyed 80% faster than NADH.

At pH 2, NAD^+ is roughly 100,000 times more stable than NADH. Thus, an enormous excess of NADH (NADPH) can be eliminated with no detectable loss of NAD^+ ($NADP^+$). The destruction, of course, follows first-order kinetics. Thus, if 99% is destroyed in 1.2 min, 99.99% should be destroyed in 2.4 min, and so on.

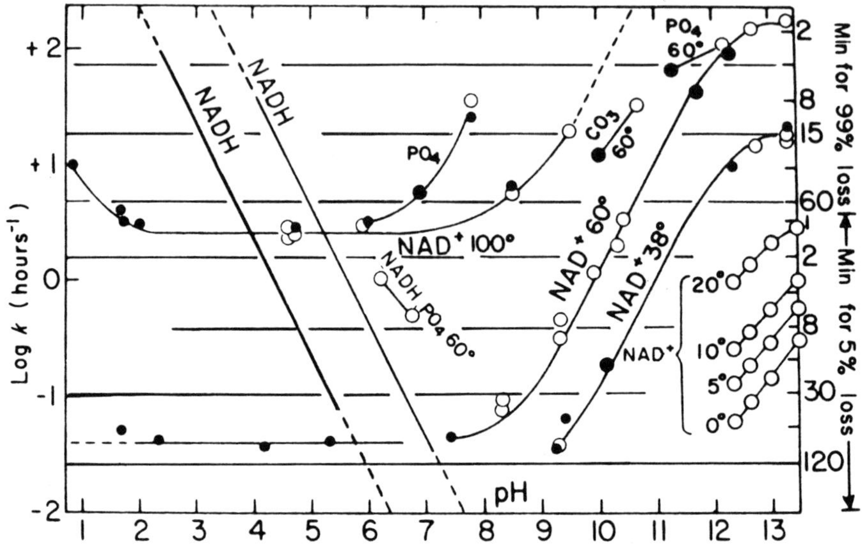

Fig. 1. Rates of destruction of NAD^+ and NADH as a function of pH value and temperature. The pH values were all measured at 25°. For the NAD^+ observations, solutions with pH values below 3 contained sulfuric acid or sulfuric acid plus sodium sulfate. Unless otherwise indicated, buffers for the rest of the samples were acetate, 0.04–0.2 M, at pH 4.0–5.2; Tris, 0.1 M, at pH 7.3–8.3; 2-NH_2-2-methyl-1,3-propanediol, at pH 8.3–9.3; and 2-NH_2-2-methylpropanol at pH 9.3–10.5. NaOH was used above pH 11.5. Adapted from Lowry et al. (*6*,1961) with additional data from Burch et al. (7,1967).

Except at low concentrations in small volumes *(see p. 16)*, NADH becomes exceedingly stable as the pH increases. It can, in fact, be heated at 100°C for 60 min in 0.1*M* NaOH without significant loss. The chief danger of loss above pH 8 is from oxidation. This is accelerated by impurities in the solution and is an important factor in storage of NADH solutions *(see next section)*.

At pH 7, in most solutions, the rate of destruction at 25°C is about 0.2%/h, and this would increase to 2%/h at pH 6. Phosphate accelerates the destruction of NADH. For example, a concentration of 0.1*M* phosphate increases the destruction rate 3.5-fold at pH 6.2 and sevenfold at pH 6.8, whether the temperature is 25, 38, or 60°C. EDTA has little effect on the phosphate acceleration. The loss is only partially owing to oxidation of NADH to NAD^+. This phenomenon may

Table 1
Percentage of NADH Remaining After Storage for 80 Days

	NADH, 0.4 m*M*		NADH, 40 m*M*	
pH	4°	–20°	4°	–20°
9.1	92	90	75	86
10.5	89	74	66	93
11.2	91	57	35	91
12.7	74	8	29	88

[a]The media at the pHs given were, respectively, 100 m*M* 2-amino-2-methyl-1, 3-propanediol-HCl buffer; 75 m*M* Na_2CO_3; 25 m*M* $NaHCO_3$; 40 m*M* Na_2PO_4; 40 m*M* K_2HPO_4; and 50 m*M* NaOH.

be of practical importance in the choice of buffer for analytical reactions in the pH range.

Stability of NADH and NADPH During Storage

Table 1 represents a protocol of a storage test conducted with 0.4 and 40 m*M* NADH at various alkaline pHs at two temperatures. It is seen that strong solutions of NADH appear to keep reasonably well at –20°C at all pH values tested, but undergo large losses at 4°C, the loss increasing with increasing pH. Conversely, weak NADH solutions are fairly stable at 4°C from pH 9–11, but were destroyed at –20°C at pH 10.5 and above. At –85°C (not shown in the table), no loss was observed at any of the four pHs.

The results can be explained if one assumes that impurities in the NADH preparation, or conceivably the NADH itself, catalyze the oxidation, especially at more alkaline pH values. This would account for greater loss in stronger samples at 4°C. At –20°C impurities would be very concentrated in the residual liquid phase and would accelerate oxidation of weak NADH samples, but strong samples would be protected by precipitation of much of the NADH. At –85°C, there should be no liquid phase at all and, consequently, no chance for catalyzed oxidation. If this is the correct explanation, stability could probably be achieved at any temperature below –40°C.

NADPH has not been tested as extensively as NADH, but appears to possess similar stability characteristics. Thus, a 0.5 m*M* solution of NADPH in 0.02*M* NaOH (pH 12.3) showed no loss in a week at 4 or –85°C, but a 13% loss at –20°C.

NADH Oxidation in Small Volumes

For some unknown reason, when NADH is present in small volumes at high dilution, it becomes sensitive to oxidation even at neutral pH. The sensitivity increases approx as the inverse square root of the volume. Thus, it is proportionately 10 times faster in 10 µL than in 1 mL. The percentage oxidation also appears to increase roughly as the inverse cube root of the concentration. Thus, it was found to be four times faster at 1 µ*M* than at 100 µ*M* concentration. As an example, in one experiment, a 5-µL vol of 2 µ*M* NADH solution was 10% oxidized to NAD^+ in an hour at room temperature.

The oxidation does not appear to be the result of metal catalysis, at least it is not prevented by either EDTA or dithiothreitol. Fortunately, it is almost completely prevented by ascorbic acid at 1 or 2 m*M* concentration.

Enzyme Inhibition by an NADH Degradation Product

When NADH is destroyed by oxidation during storage, either in solution or in the dry state, a compound accumulates that is inhibitory for a number of dehydrogenases for which NADH serves as a coenzyme. This can cause serious underestimation of the activity of such enzymes. Consequently, original bottles of NADH should be stored at –15°C or below, and NADH solutions in which loss has occurred should be discarded.

Destruction of NAD^+ and $NADP^+$ in Alkali and Stability in Acid

The effects of pH on the oxidized pyridine nucleotides are almost the mirror image of those on the reduced nucleotides (Fig. 1). The stability is high and independent of pH from 2 to 6.5. Above pH 7, the rate increases, until it is 5000-fold faster at pH 12.5–13. The results are those expected if NAD^+ is converted from a relatively stable form to a much more unstable form by removal of one proton from an acid group with pK_a of about 10. At 60°C, the rate reaches a plateau at about pH 12.5. At lower temperatures, the maximum rate occurs above 13.

$NADP^+$ has not been studied as extensively as NAD^+, but it appears to be slightly more stable (15%) than NAD^+ at alkaline pHs *(6)*. This may be contrasted with the somewhat lower stability of NADPH as compared to NADH noted above.

Effect of Temperature

The destruction of NAD^+ is strongly affected by temperature. In the temperature range of 38–100°C, the rate of NAD^+ destruction increases 2.8-fold for each 10° rise in temperature (11%/degree). At least in the pH 12–13 range, the temperature coefficient is even larger below 20°C (Fig. 1). The rate increases more than fourfold between 0°C and 10° at pH 12.5. This is of analytical importance since it makes it possible to stop an enzyme reaction with alkali at 0° without suffering serious losses of NAD^+ or $NADP^+$. This is of particular significance for the preparation of tissue extracts without loss of either the reduced or oxidized pyridine nucleotides (Burch et al., 1967) *(7)*.

Effects of Salts

The rate of destruction of NAD^+ is increased by a number of salts. At pH 10 and 60°C, for example, the rate is increased 40% by 0.2*M* NaCl and 10% by 0.2*M* sodium maleate. Carbonate has a greater effect. At 60°C, the rate was found to be increased 380, 500, and 630% by 0.1, 0.2, and 0.5*M* carbonate buffer. This acceleration is not affected by 1 m*M* EDTA. Phosphate accelerated NAD^+ destruction in the neighborhood of pH 7 (Fig. 1).

Destruction of NAD^+ and $NADP^+$ as an Analytical Step

When NADH or NADPH is to be measured at the end of an analysis by the fluorescence induced in strong alkali or by cycling, it is necessary to destroy the excess NAD^+ or $NADP^+$. If the alkaline fluorescence method is to be used, the destruction needs to be done with minimal fluorescence production. A pH between 11.7–12.3 is optimal. In this range, destruction can be accomplished with fluorescence of the order of 1% of that produced in 6*M* NaOH. The proper pH can be obtained with NaOH (an excess of 0.005–0.02*M*). However, it is usually easier to use a buffer, that permits a wide latitude in adjusting the pH of well-buffered solutions. For this purpose a mixture of K_2HPO_4 and Na_3PO_4 is recommended. During heating, a final ratio of $HPO^{2-}{}_4$:$PO^{3-}{}_4$ between 2:1 and 1:3 is satisfactory. Na_3PO_4 may precipitate out of 6*M* NaOH if the concentration is too high. The concentration of P_i in the strong alkali should therefore be kept below 0.05*M*. In adjusting the pH it is not infrequently overlooked that $NH^+{}_4$ is an acid, with a pK_a of 9.4, and that $(NH_4)_2SO_4$, perhaps added with an enzyme solution, must be taken into account in calculating the amount of NaOH or buffer to be added.

If the NADH (NADPH) is to be measured by cycling, the amount of fluorescence produced during destruction is ordinarily of little importance, and for safety's sake, stronger alkali (e.g., 0.1 or 0.2*M* NaOH) can be used in the destruction. Appropriate half-times for destruction of NAD^+ at 60°C are 2 and 0.5 min at pH 11.7 and pH 12.3, respectively. From these figures, the time necessary to destroy 99.9% would be 20 min at pH 11.7 and 5 min at pH 12.3.

Preparation, Standardization, and Storage of Pyridine Nucleotide Solutions

Pyridine nucleotides, whether oxidized or reduced forms, are stored desiccated at –20°C. In spite of statements to the contrary, it is not safe to store NADH at 0–25°C even if it is kept in the dark.

Primary standardization of stock solutions is made in the spectrophotometer. A procedure for this purpose is given in the case of each of the nucleotides.

NAD^+ and $NADP^+$ solutions can be stored frozen for months without significant change in concentration. NADH and NADPH solutions ordinarily need to be restandardized more frequently unless stored at very low temperature *(see below)*.

General Standardization Protocol

It is convenient to pipet an exactly measured volume of the appropriate reagent into each of three cuvets. One of the three cuvets serves as a blank, the other two provide duplicate assays. In cells of 4 or 5 mm inner width, a convenient vol is 400 or 500 µL. This permits easy mixing without introducing a rod or inverting the cell. The nucleotide is added in a small volume to give a concentration of 50–100 µ*M* (e.g., 5 µL of 10 m*M* nucleotide added to 500 µL of reagent). The cuvets are all read at 340 nm, the enzyme is added, and the solutions are well mixed. Readings are made at suitable intervals until there is no further change in OD. Under the conditions recommended, each reaction should be complete within 10 min. The calculations are based on molar extinction coefficient of 6270 at 340 nm for NADH and NADPH. The concentration (m*M*) of the standard solutions is equal to:

$$\text{OD} \times \frac{1000}{6270} \times \frac{\text{total volume}}{\text{volume of nucleotide solution}} \qquad \text{[Eq. 1-1]}$$

NAD^+ and $NADP^+$ Solutions

These keep well dissolved in H_2O. To prepare stock solutions of about 100 m*M* concentration, dissolve 70 mg of NAD^+ or 80 mg of $NADP^+$/mL. It is convenient for frozen storage to prepare 1 or 2 mL in an 8-mL screw cap tube. Portions of the stock solution are diluted to approx 10 and 1m*M* with water. For purposes of further calculation, these dilutions should be made quantitatively, e.g., 200- and 20-µL vol measured with calibrated constriction pipets are each diluted with 2 mL of H_2O, measured with accurate volumetric pipets. All three dilutions can be stored at –20°C and are indefinitely stable. However, since there may be some evaporation on long storage, the solutions should be restandardized at intervals when used as a basis of calculation in fluorometric tests.

Standardization of NAD^+

Reagent: Tris-HCl buffer, pH 8.7 (80 m*M* Tris base, 20 m*M* Tris HCl); ethanol, 1*M* (about 5% by vol); EDTA, 1 m*M*. Dilute yeast alcohol dehydrogenase to a concentration of 900 U/mL (3 mg/mL) in 20 m*M* Tris-HCl buffer (pH 8.1). If the assay is conducted with 400 or 500 µL of reagent, a vol of 2 µL of this enzyme solution should complete the reaction in 10 min or less.

Standardization of $NADP^+$

Reagent: Tris-HCl buffer, pH 8.1 (50 m*M* Tris base, 50 m*M* Tris HCl); glucose-6-P, 1 m*M*. Baker's yeast glucose-6-P dehydrogenase is diluted to a concentration of about 8 U/mL (20 µg/mL) in 20 m*M* Tris-HCl, pH 8.1. Of this, 2 µL are used in the assay when the reagent vol is 500 µL. NAD^+ can also be standardized in this reagent if glucose-6-P dehydrogenase from *Leuconostoc mesenteroides* is substituted for the baker's yeast enzyme, but 20 times more units (NADP) must be used.

NADH and NADPH Solutions

Although, as seen earlier, NADH and NADPH are extremely stable toward direct destruction in alkali, they are susceptible to oxidation even at low temperature. Therefore, storage of reduced nucleotide solutions without serious loss presents much greater problems than with NAD^+ and $NADP^+$. The storage properties described above suggest that, unless a freezer at –40°C or below is available, NADH and

NADPH solutions should be prepared at concentrations no greater than 5 m*M*, at a pH of 9–11, and be stored at 4°C. If a low-temperature freezer is available, stronger solutions can be prepared and probably stored indefinitely without loss.

If it is desirable to keep NAD^+ and $NADP^+$ to a minimum in reduced nucleotide solutions, it is recommended that these be prepared in a carbonate at pH 10.6 (80 m*M* Na_2CO_3:20 m*M* $NaHCO_3$). In this case, NAD^+ or $NADP^+$ initially present, or that which may accumulate during storage, can be destroyed by heating 3 min at 100°C. Heating is not otherwise recommended for reactions to be followed fluorometrically, since there is some increase in blank fluorescence in the process. To prepare an approx 5 m*M* NADH or NADPH solution, use 4 or 5 mg of powder/mL of the carbonate buffer.

Standardization of NADH

Reagent: Phosphate buffer, pH 7.0 (60 m*M* Na_2HPO_4, 40 m*M* NaH_2PO_4); pyruvate, 1 m*M*. Lactate dehydrogenase (heart or skeletal muscle) is diluted to a concentration of 125 U/mL (0.5 mg/mL) in 20 m*M* phosphate buffer (pH 7.0). For 500 µL of reagent, 1 µL of this enzyme solution should suffice.

Standardization of NADPH

Reagent: Tris-HCl buffer, pH 8.1 (50 m*M* Tris base, 50 m*M* Tris-HCl); ADP, 100 µ*M;* α-ketoglutarate, 5 m*M;* ammonium acetate, 5 m*M*. Glutamate dehydrogenase (beef liver) is diluted to a concentration of 600 U/mL (5 mg/mL) with 20 m*M* Tris-HCl (pH 8.1). Of this, 5 µL should be sufficient for an assay with 500 µL of reagent. NADH can also be standardized in this reagent, rather than in the pyruvate-lactate dehydrogenase system, since glutamate dehydrogenase reacts with either coenzyme.

Fluorometers

The two most desirable features of a fluorometer are stability and low background reading. If the readings generated are stable and reproducible, small increments in reading can be measured with precision. If the background reading is low, high sensitivity can be used when needed. Any fluorometer provided with a photomultiplier tube is likely to have more than enough sensitivity, but this will not be usable to full capacity if the instrumental blank is too high.

There are two basic types of fluorometers, those that isolate the light wavelengths for excitation and fluorescence with filters, and those that do so spectroscopically. The latter are much more expensive, and, because they require a very intense light source (usually a xenon arc lamp), are likely to be unstable. They are also ordinarily designed to use square cuvets rather than test tubes, and are therefore far less convenient for analyzing many samples that require multiple readings. For all these reasons, for the present purpose, filter-type instruments are much to be preferred. These are either direct reading instruments, such as that produced by Optical Technology Devices, Elmsford, NY (formerly by the Farrand Optical Company), or else they operate on a null point principle, in which the fluorescence light is balanced against a small fraction of the exciting light itself. This type is available from the Turner Company, Sequoia-Turner Corp., Mountain View, CA. The null point principle has the advantage that variations in the light source are canceled out. Almost all of our experience, however, has been with the direct reading Farrand instrument, which has the advantage that sensitivity is continuously variable over the whole range, instead of in steps only (null point instruments could, of course, be made with continuously variable sensitivity).

The fluorometer may be furnished with a turntable to permit insertion of several tubes into the instrument at once. Unless this device is exceedingly well designed, it will increase the variability of the readings. This is because a very slight difference in tube position will have a big effect on the emitted light that reaches the sensitive area of the phototube. It is recommended that the turntable be fixed rigidly in position and the tubes inserted and read one at a time.

CHAPTER 2

A Kinetics Primer for the Tissue Analyst

A rudimentary working knowledge of enzyme kinetics is of great value in the use of enzymes as analytical tools. An analysis may fail if too little enzyme is used, because of incomplete reaction, but it may also fail if too much enzyme is used. This might not be true if absolutely pure, single-action, enzymes were available.

In this case, a tremendous excess of enzyme could be employed and no harm would be done. Even crystalline enzymes, however, are frequently contaminated with interfering activities, or the enzyme itself may have a disturbing side action. Fortunately, if just the "right amount" is employed, enzyme preparations that are only partially purified can often be used successfully. The "right amount" of enzyme (the least amount that will do the job) can be rather exactly defined. In order to determine how much enzyme should be used, kinetic information is needed. The subject of enzyme kinetics often fills the neophyte with fear. It is the hope here to present simple kinetics in such a way that even the relatively inexperienced tissue analyst can proceed with confidence.

Nonenzyme Kinetics

Before discussing enzyme kinetics, it may be useful to present some fundamental concepts applicable to all chemical reactions.

"Order" of the Reaction

Chemical reactions are defined as "first order," "second order," or "higher order" if there are one, two, or more reactants. The decay of a radioactive element is a first-order reaction; the combination of an alcohol with an acid to form an ester is a second-order reaction. The hydrolysis of an ester is also second order, but because one of the reactants (H_2O) is present in enormous excess, the reaction has the mathematical form of a first-order reaction. It is called "pseudo first order" or usually simply "first order." There is also a (pseudo) "zero order" reaction, which may sound impossible. Any reaction that proceeds with constant velocity over the period of observation fits the definition of (pseudo) zero order. For example, the rate of decay of radium over a 24-h period is, for practical purposes, zero order, and when measuring the activity of an enzyme it is desirable to make the reaction as nearly zero order as possible.

Only reactions following zero-, first-, or second-order kinetics are of interest here. Of these we shall mainly emphasize those of first order.

First-Order Reactions

Consider the first-order reaction:

$$A \xrightarrow{k} B \qquad \text{[Re. 2-1]}$$

Its velocity (v) can be described by the equation:

$$v = k[A] \qquad \text{[Eq. 2-1]}$$

As the reaction proceeds, the velocity will decrease exactly as [A] decreases (Fig. 1). When A is half-gone, the velocity will be half of the initial velocity, and so forth. [A] is said to diminish by a "die-away" curve or by "logarithmic decrement"; k is a "first-order rate constant." Note its meaning. It is equal to the fraction of A, which would be converted to B in unit time *if the velocity did not fall off.* For example, if $k = 0.3$/min, the velocity *at any instant* is always 0.3[A]/per min (i.e., 30%/min of the amount of A present at that moment).

Equation (2-1) describes the situation at any moment in Fig. 1, but we need an equation that sums up the total change in [A] in an interval of time (t). This (integrated) equation is:

$$2.3 \log[A]_o/[A] = kt \qquad \text{[Eq. 2-2]}$$

where $[A]_o$ is the initial amount of A. When A is half-gone, 2.3 log(1/0.5) = kt = 0.7 (actually 0.69). The time is the "half-time" or $t_{1/2}$, i.e.:

$$t_{1/2} = 0.7/k \qquad \text{[Eq. 2-3]}$$

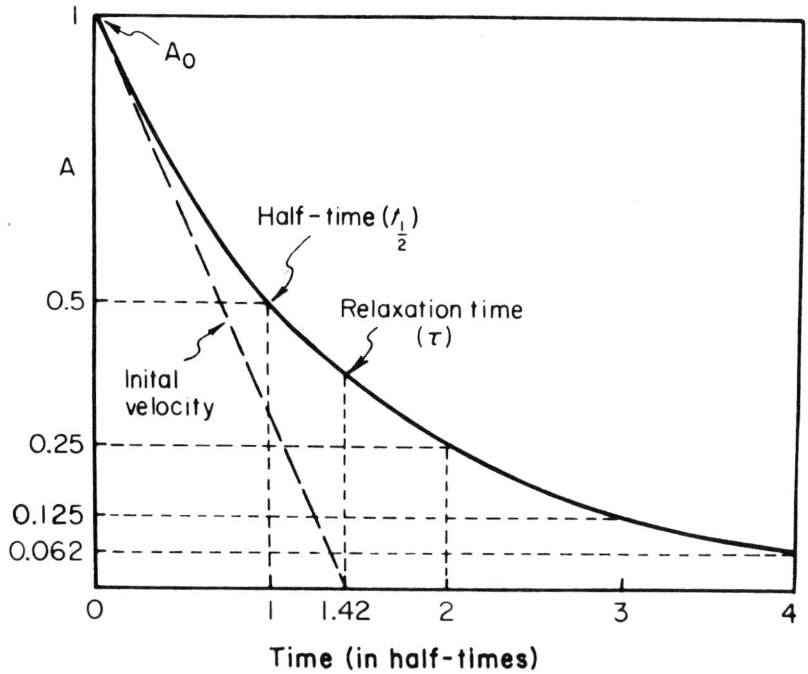

Fig. 1. Plot of a first-order reaction.

For example, if $k = 0.3/\text{min}$, $t_{1/2} = 2.3$ min. Most people find the half-time of a reaction easier to visualize than the rate constant. (A somewhat more erudite term than the half-time is the "relaxation time," t, which is simply the reciprocal of the first-order rate constant [Fig. 1].)

Note that after 1, 2, 4, and 6 half-times, A will be 50, 75, 94, and 98.5% gone (i.e., *a simple irreversible first-order reaction will be complete, for most practical purposes, in 5 or 6 half-times*). This applies to both true and pseudo-first-order reactions, and to enzymatic as well as nonenzymatic reactions.

Reversible First-Order Reactions

If the reaction is reversible, the situation is more complicated:

$$A \underset{k_2}{\overset{k_1}{\rightleftarrows}} B \qquad \text{[Re. 2 - 2]}$$

The net velocity is the difference between the two velocities (v_1 and v_2) in the forward and reverse direction. The velocity equation is:

$$v = v_1 - v_2 = k_1[A] - k_2[B] \qquad \text{[Eq. 2-4]}$$

Thus, the initial velocity (i.e., when [B] = 0) is the same as for an irreversible reaction, but the net velocity will diminish more rapidly and will become zero when $k_1[A] = k_2[B]$. When this happens, the system is in equilibrium ($[A] = [A]_{eq}$, $[B] = [B]_{eq}$). It is clear that:

$$\frac{[A]_{eq}}{[B]_{eq}} = \frac{k_2}{k_1} = K_{eq} \qquad \text{[Eq. 2-5]}$$

This describes the fundamental fact that the equilibrium constant is determined by the two rate constants. Since $[B] = [A]_o - [A]$, [B] can be eliminated from Eq. (2-4) to give:

$$v = (k_1 + k_2)[A] - k_2[A]_o \qquad \text{[Eq. 2-6]}$$

This can be integrated, and by taking advantage of Eq. (2-5) the integrated equation can be put into a simple form:

$$2.3 \log \frac{[A]_o - [A]_{eq}}{[A] - [A]_{eq}} = (k_1 + k_2)t \qquad \text{[Eq. 2-7]}$$

This is exactly analogous to Eq. (2-2). The half-time of *approach to equilibrium* is:

$$t_{1/2} = \frac{0.7}{k_1 + k_2} \qquad \text{[Eq. 2-8]}$$

and the reaction will be 50, 75, and 98.5% as far as it will ever go in 1, 2, and 6 half-times. If, for example, $k_1 = 0.3$/min, and $k_2 = 0.075$/min, $K_{eq} = 4$ (from Eq. [2-5]), i.e., equilibrium will occur when [A] has fallen to 20% of $[A]_o$. The half-time (i.e., when [A] = 60% of $[A]_o$) will be 0.7/0.375 = 1.9 min. Note that the approach to equilibrium is faster than if the reaction had not been reversible and [A] had been approaching zero instead of 20%.

Second-Order Reactions

Most analytical reactions have the form of first- or zero-order reactions, or something in between. However, we will encounter some analytical reactions that follow a second-order time-course, and, moreover, a brief presentation of certain aspects of the second-order reaction may be helpful in understanding enzyme kinetics. Consider the irreversible reaction:

$$A + B \xrightarrow{k} C \qquad \text{[Re. 2-3]}$$

The velocity equation is:

$$v = k[A][B] \qquad \text{[Eq. 2-9]}$$

Thus, the concentrations of both A and B affect the velocity, and the constant k has a more complex meaning than in the case of a first-order reaction. It is numerically equal to the *instantaneous rate* expressed as the fraction of one reactant, which is converted to product in *unit time* when the other reactant is present at *unit concentration.* For example, suppose $k = 0.3$ L/mol/min ($0.3M^{-1}\,\text{min}^{-1}$). This would mean that if [A] is 1*M*, the instantaneous velocity will correspond to the disappearance of B at the rate of 30%/min. If [A] is only 0.01*M*, the rate of disappearance of B will be only 0.3%/min. Thus, a second-order rate constant has both a time and a concentration dimension.

The integrated form of Eq. (2-9), which would describe the time-course of a second-order reaction, will not be presented. However, in the enzyme section below, a graphic representation will be given for the time-course when [A] is varied relative to [B].

If Re. (2-3) is reversible:

$$A + B \underset{k_2}{\overset{k_1}{\rightleftarrows}} C \qquad \text{[Re. 2-4]}$$

i.e., it is second order in one direction and first order in the other. The velocity equation is:

$$v = k_1[A][B] - k_2[C] \quad \text{[Eq. 2-10]}$$

At equilibrium, $k_1[A][B] = k_2[C]$ and:

$$K_{eq} = \frac{[A][B]}{[C]} = \frac{k_2}{k_1} \quad \text{[Eq. 2-11]}$$

It is important to realize the difference between Eq. (2-11) and Eq. (2-5). K_{eq} has a concentration dimension, and the position of equilibrium is no longer independent of absolute concentrations. For example, let $k_1 = 10M^{-1}$ min^{-1} and $k_2 = 0.1$ min^{-1}. From Eq. (2-11), $K_{eq} = 0.01M$ = 10 mM. At equilibrium, if [A] = [B] = 0.1M, [C] = 1M, i.e., 91% of the total; if [A] = [B] = 0.1 mM, [C] = 0.001 mM, i.e., only 1% of the total. (If the reaction is A + B ↔ C + D, the equilibrium constant has no concentration term, and the position of equilibrium is, of course, independent of absolute concentrations.)

A shift in equilibrium with concentration occasionally has analytical significance. For example, glutamate can be oxidized enzymatically with NAD^+ to form α-ketoglutarate and NH_3. Two components react to form three; therefore, the equilibrium position is concentration-dependent. In the millimolar concentration range the equilibrium is unfavorable, but in the micromolar range, the reaction can easily be driven to essential completion.

Enzyme Kinetics

The kinetics of catalyzed reactions are more complicated than those of spontaneous reactions, because the catalyst must of necessity enter into the reaction, usually by forming a transient complex with the reactant(s). The velocity of an enzymatic reaction will be determined by the amount of enzyme and by some function of the reactant (substrate), but in contrast to a spontaneous reaction, the velocity will usually not be directly proportional to reactant concentration.

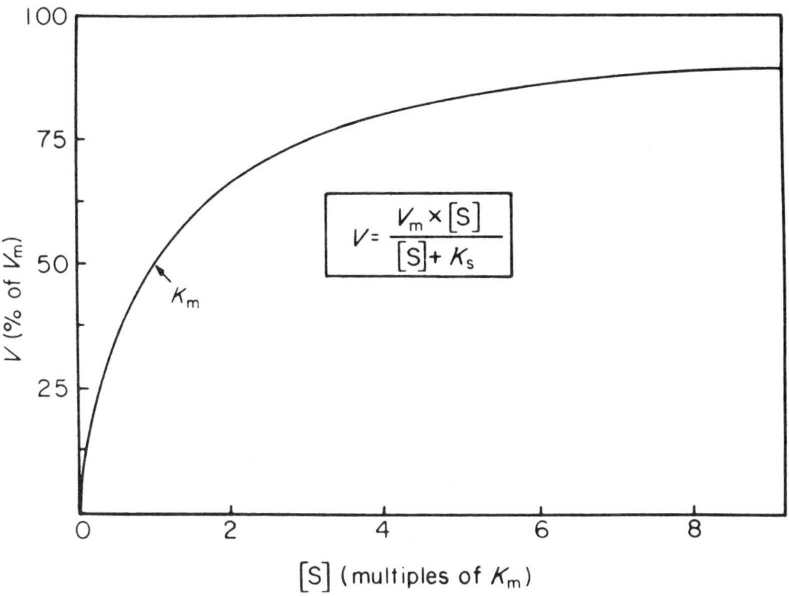

Fig. 2. Velocity of an enzyme reaction as a fraction of substrate concentration. This describes the situation for an enzyme with "normal" kinetics.

One-Step Reactions with One Substrate

In simple uncomplicated cases, the initial velocity of an enzyme reaction increases with substrate concentration in the manner shown in Fig. 2. At low substrate concentrations, the rate rises almost linearly with concentration. At higher concentrations, the increase gradually lessens, until the reaction is hardly accelerated by increasing substrate. Under such conditions, the substrate is said to have "saturated" the enzyme, and the rate observed is the "maximum velocity" or V_{max} for that concentration of enzyme. The actual chain of events in an enzyme reaction may be very complex, but in the case of a single substrate, S, and an irreversible reaction that forms product, P, the situation can usually be explained for practical purposes by three reactions:

$$E + S \underset{k_2}{\overset{k_1}{\rightleftarrows}} ES \xrightarrow{k_3} P + E \qquad \text{[Re. 2-5]}$$

where E is free enzyme and ES is a transient complex between enzyme and substrate. (It would be equally satisfactory to write "EP" instead of "ES.") From what has been said earlier, it will be evident that of the three reactions taking place, one is second order and two are first order. The third reaction (ES → P) is the one ordinarily measured as the overall enzyme velocity (represented below as either v or v_3).

Usually, at the start of such an enzyme reaction, S greatly exceeds the enzyme on a molar basis. In this case, Re. (2-5) can be considered to take place in three phases. In the first phase, ES is formed faster by reaction 1 than it breaks down by reactions 2 and 3; therefore, ES accumulates and E falls. This is the "presteady-state" phase, which usually lasts only a matter of milliseconds. In the second phase, a "steady state" has been established in which reaction 1 is balanced by reactions 2 and 3, but S has not diminished appreciably. ES is therefore at its peak level and the net velocity (decrease in S or increase in P) is called "initial velocity." Finally, in the third (and usually longest) phase, S falls toward zero, during which time a steady state is maintained between E, S, and ES.

At all times the net velocity is proportional to [ES], i.e.:

$$v = k_3\,[\mathrm{ES}] \qquad \text{[Eq. 2-12]}$$

It will be seen from Re. (2-5) that as [S] is made larger and larger, it will finally diminish [E] to a negligible value (i.e., the total enzyme, E_T, will be in the form ES, and therefore, the velocity will be maximal) thus:

$$V_{\mathrm{max}} = k_3\,[\mathrm{E_T}] \qquad \text{[Eq. 2-13]}$$

When velocity is maximal, it is independent of modest changes in S and is called "zero" order in S. (It is, in reality, at least a second-order reaction, and should properly be called pseudo zero order. It is difficult to conceive of a *true* zero order reaction.)

At lower substrate levels, an equation is needed relating v to S and V_{max}. The three separate velocities of Re. (2-5) are:

$$\begin{aligned} v_1 &= k_1[\mathrm{E}][\mathrm{S}], \\ v_2 &= k_2[\mathrm{ES}], \\ v_3 &= k_3[\mathrm{ES}] \end{aligned} \qquad \text{[Eq. 2-14]}$$

When the steady state is reached, $v_1 = v_2 + v_3$, or:

$$k_1[\text{E}][\text{S}] = k_2[\text{ES}] + k_3[\text{ES}] = (k_2+k_3)[\text{ES}] \qquad \text{[Eq. 2-15]}$$

Rearranging:

$$\frac{[\text{E}][\text{S}]}{[\text{ES}]} = \frac{k_2+k_3}{k_1} = K_\text{m}(\text{or } K_\text{S}) \quad \text{or} \quad \frac{[\text{E}]}{[\text{ES}]} = \frac{K_\text{m}}{[\text{S}]} \qquad \begin{matrix}\text{[Eq. 2-16]}\\ \text{[Eq. 2-17]}\end{matrix}$$

K_m (the "Michaelis constant") has the form of an equilibrium constant (cf. Eq. [2-11]) . It is a true equilibrium constant if k_2 is much larger than k_3. K_m is a readily determinable constant, whereas, k_1 and k_2 are usually very difficult to evaluate. From Eq. (2-17), it is seen that when [S] = K_m, [E] = [ES], i.e., half of the total enzyme will be present as ES. Since v is proportional to [ES] (Eq. [2-17]) , when [S] = K_m, v is half-maximal. Thus, *K_m is numerically equal to the substrate concentration, which gives half-maximal velocity* (*see* Fig. 2).

The desired equation can now be formulated. Rearranging Eq. (2-17), [ES]K_m = [E][S]. Substituting [E_T] – [ES] for [E] and rearranging (E_T = total enzyme):

$$[\text{ES}] = \frac{[\text{E}_\text{T}][\text{S}]}{[\text{S}]+K_\text{m}} \qquad \text{[Eq. 2-18]}$$

If both sides are multiplied by k_3, this becomes (from Eqs. [2-12] and [2-13]):

$$v = \frac{V_\text{m}[\text{S}]}{[\text{S}] + \text{K}_\text{m}} = \frac{V_\text{m}}{1 + K_\text{m}/[\text{S}]} \qquad \text{[Eq. 2-19]}$$

which is the Michaelis-Menten equation. This describes the hyperbolic curve plotted in Fig. 2. The reciprocal form of Eq. (2-19) is often more convenient because it describes a straight line:

$$\frac{1}{v} = \frac{1}{V_\text{m}}\left(1 + \frac{K_\text{m}}{[\text{S}]}\right) \qquad \text{[Eq. 2-20a]}$$

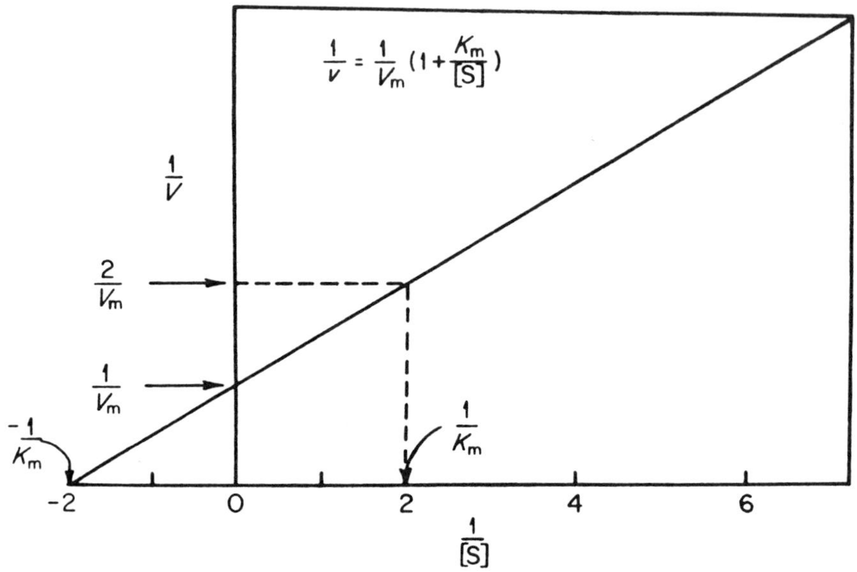

Fig. 3. Reciprocal Lineweaver-Burk plot of velocity against substrate concentration for an enzyme with "normal" kinetics.

Equation (2-20a) is the basis of the widely used Lineweaver-Burk plot, or reciprocal plot, of 1/v against 1/[S] (Fig. 3). The intercept on the vertical axis is $1/V_m$, and the intercept on the horizontal axis is $-1/K_m$.

First-Order Enzyme Reactions

When [S] is small compared to K_m, Eq. (2-19) reduces to:

$$v = V_m[S]/K_m \qquad \text{[Eq. 2-20b]}$$

Since, with a given level of enzyme, V_m and K_m are both constants, they can be replaced by their ratio, which we can denote by k and call the "apparent first-order rate constant":

$$k = V_m/K_m \qquad \text{[Eq. 2-21a]}$$

from which:

$$v = k[S] \qquad \text{[Eq. 2-21b]}$$

Table 1
Influence of Maximum Velocity and Michaelis Constant on Apparent First-Order Rate Constant and Reaction Times[a]

Enzyme	V_{max}[b] μ*M*/min	K_m μ*M*	k min^{-1}	$t_{1/2}$ min	$t_{0.75}$ min	$t_{0.98}$ min
A	700	1000	0.7	1	2	5.6
A	200	1000	0.2	3.5	7	20
B	50	50	1	0.7	1.4	3.9

[a]The reaction times apply to initial substrate levels well below the K_m.
[b]At the enzyme concentration used.

This has the form of a first-order equation, and we speak in this case of the substrate concentration being at a "first-order level." Everything said above about nonenzymatic first-order reactions is applicable in this case. Equation (2-2) can be rewritten (S replacing A) as:

$$t = \frac{2.3}{k} \log \frac{[S]_o}{[S]} \qquad \text{[Eq. 2-22]}$$

For the majority of analytical situations, *k* is more important than either V_m or K_m alone. For example, in the hypothetical cases of Table 1, enzyme B at an activity (V_{max}) of 50 μ*M*/min is more active toward low levels of substrate than enzyme A at an activity of 700 μ*M*/min.

If a substance is to be measured enzymatically, and the concentration is well below the K_m, the time required, as shown earlier, is 5 or 6 half-times. The same time would be required whether $[S]_o$ was 1/10th or 1/1000 of K_m.

Note that, in calculating *k*, it is necessary to put V_m and K_m into rational compatible terms. It is often convenient to express K_m as μ*M* (μmol/L), and to express V_m as μ*M*/min (i.e., μmol/L/min). In this case *k* will be expressed as *per minute*. The advantage of this is that the international enzyme unit (U) is 1 μmol/L/min, therefore, U/L = μ*M*/min. Stock enzyme solutions can be conveniently expressed as m*M*/min (U/mL).

Mixed Zero-Order and First-Order Enzyme Reactions

If $[S]_o$ approaches K_m or exceeds it, more time is required to complete the reaction than when $[S]_o$ is at a first-order level. The time required can be calculated from the integrated form of Eq. (2-19):

$$t = \frac{2.3K_m}{V_m}\log\frac{[S]_o}{[S]} + \frac{[P]}{V_m} \quad \text{[Eq. 2-23]}$$

where P is the product formed (i.e., S_o-S). If k is again substituted for V_m/K_m:

$$t = \frac{2.3}{k}\log\frac{[S]_o}{[S]} + \frac{[P]}{V_m} \quad \text{[Eq. 2-24]}$$

This may be seen to differ from the equation for a first-order reaction (Eq. [2-22]) only by the addition of the term $[P]V_m$.

In other words, *to complete a one-step enzyme reaction, with any level of substrate, requires the time calculated for a first-order reaction plus the time that would be required if the entire reaction were to proceed at the* V_m.

For the reaction to be 98% complete (and ignoring the 2% difference between $[S]_o$ and [P]), Eq. (2-23) becomes (with sufficiently close approximation):

$$t_{0.98} = \frac{4}{k} + \frac{[S]_o}{V_m} \quad \text{[Eq. 2-25]}$$

or since $k = 0.7/t_{1/2}$:

$$t_{0.98} = 6\,t_{1/2} + \frac{[S]_o}{V_m} \quad \text{[Eq. 2-26]}$$

Consider as an example an enzyme with $K_m = 200\ \mu M$ and V_m (at the level of enzyme used) = 50 μM/min. This gives $k = 0.25$/min and $t_{1/2} = 3$ min. To give 98% conversion to product with initial substrate levels of 1 μM, 100 μM, and 1 mM would require 18 min, (18 + 2) min, and (18 + 20) min, respectively. (*See* Table 2 for this and other examples.)

In a metabolite assay, the problem is usually the converse of the above (i.e., how much enzyme should be used to complete a reaction

Table 2
Effect of Substrate Concentration on Reaction Time[a]

V_{max} μ*M*/min	K_m μ*M*	$[S]_o$ μ*M*	*k*, min^{-1}	$t_{1/2}$ min	$t_{0.98}$ min
50	200	1	0.25	3	18
50	200	100	0.25	4	20
50	200	1000	0.25	13	38
10	10	1	1	0.7	4
10	10	100	1	5.7	14
10	10	1000	1	51	104

[a]The half-times were calculated from Eq. (2-23); for example, in the third line, $t_{1/2} = 0.7/k + [P]/50 = 3 + 10 = 13$. The 98% times ($t_{0.98}$) were calculated in a similar manner (*see* text).

in a given time). In this case, it is simpler to replace *k* by V_m/K_m in Eq. (2-25), which then becomes:

$$t_{0.98} = \frac{4K_m + [S]_o}{V_m} \quad \text{or} \quad V_m = \frac{4K_m + [S]_o}{t_{0.98}} \qquad \text{[Eq. 2-27]}$$

For example, it is desired to have a reaction (98%) complete in 10 min. The K_m is 10 μ*M*, and the highest concentration to be measured will be 100 μ*M* in the assay system. The enzyme stock is a 1% solution with activity of 8 U/mg (i.e., 80 μmoles/min/mL, or 80,000 μ*M*/min). How much of this stock will be needed? From Eq. (2-27), $V_m = (4 \times 10 + 100)/10 = 14$ μ*M*/min. Therefore, an enzyme concentration of 1:5000 should suffice. Figure 4 provides a graphic calculation of the enzyme requirement with different substrate levels and with enzymes having different Michaelis constants.

When K_m is substantially larger than $[S]_o$, Eq. (2-27) reduces to $V_m = 4\,K_m/t$. Thus, if the K_m is 200 μ*M* to complete the reaction (98%) in 20 min would require enzyme activity equal to 40 μ*M*/min or 40 U/L.

One-Step Reactions with Two Substrates

The majority of analytical reactions to be considered will involve two substrates (A, B). In many cases, however, one of the reactants (B) will be present in substantial excess and will therefore not be greatly reduced during the reaction. This means that for practical purposes the

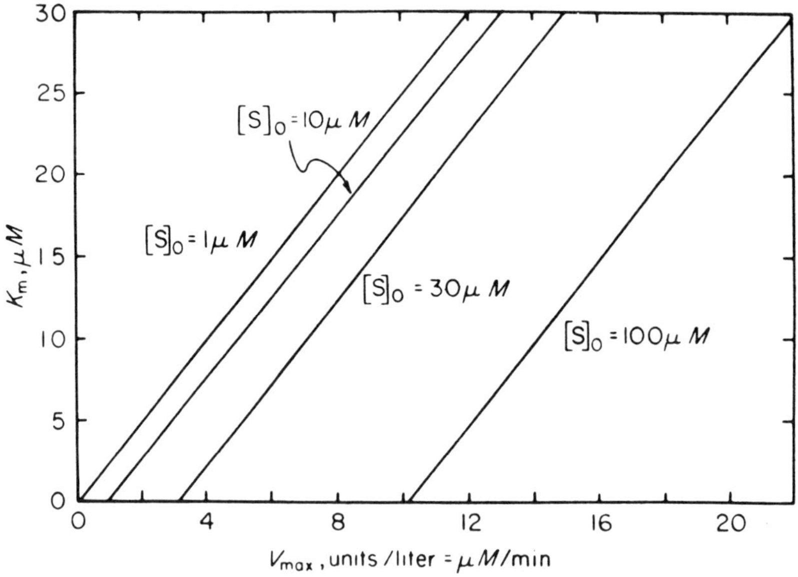

Fig. 4. Enzyme required for an enzyme reaction with normal kinetics to be 98% complete in 10 min. The lines are calculated from Eq. (2-27). For enzymes with K_m's of 50, 100, 200, and 500 μ*M*, the minimal enzyme requirements for $[S_o] = 100$ μ*M* would be 30, 50, 90, and 210 U/L, respectively, and only 10 U/L less if $[S]_o = 1$ μ*M*.

kinetic situation will reduce to that for one substrate, except that the level of B will influence one or both of the kinetic parameters for A(V_m and K_A). (For the case in which B as well as A is substantially reduced during the reaction, *see* the next subsection.)

The mutual effects of two or more cosubstrates can be complex; nevertheless, if [B] is held constant, the reaction will usually obey Eq. (2-20) (and therefore Eqs. [2-22]–[26]), i.e.:

$$v = \frac{V_m'[A]}{[A] + K_m'} \qquad \text{[Re. 2-6]}$$

where V_m' and K_m' only apply to the particular level of B. In designing or modifying an analytical method, it is useful to know how the concentration of B affects the kinetics of A in the analytical range, but it should not be necessary to go into a complete and perhaps complicated kinetic analysis. Without any attempt to be comprehensive, some of the common types of kinetic interactions between cosubstrates can be mentioned.

CASE 1: "RANDOM ORDER OF ADDITION"

In the simplest case, each substrate acts as though it combines independently with the enzyme.

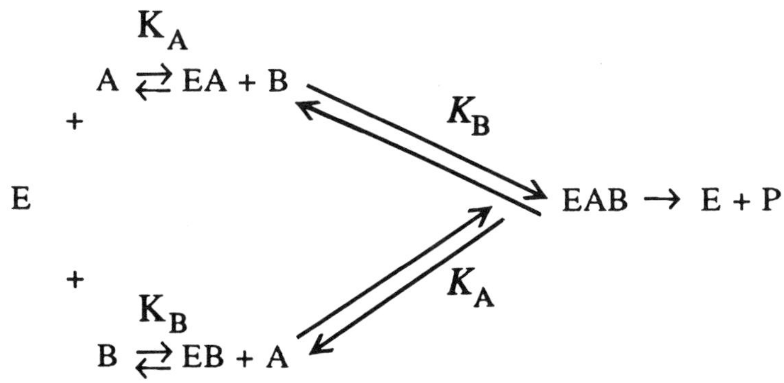

[Re. 2-7]

The net velocity is proportional to the amount of enzyme that is combined simultaneously with both substrates (EAB). The K_m for each substrate (K_A and K_B) is independent of the concentration of the other substrate, and the velocity equation is:

$$v = \left(\frac{V_m[A]}{[A] + K_A}\right)\left(\frac{[B]}{[B] + K_B}\right) \quad \text{[Eq. 2-28]}$$

which is simply Eq. (2-19) multiplied by a term applying to B. If [A] is much less than K_A, Eq. (2-28) reduces to:

$$v = \frac{V_m[A]}{K_A}\left(\frac{[B]}{[B] + K_B}\right) = k_a[A]\left(\frac{[B]}{[B] + K_B}\right) \quad \text{[Eq. 2-29]}$$

If [B] is also much less than K_B, this becomes:

$$v = \frac{V_m[A][B]}{K_A K_B} = k_{ab}[A][B] \quad \text{[Eq. 2-30]}$$

which is the same as the equation for a second-order nonenzymatic reaction (Eq. [2-11]). Note that as in the nonenzymatic case, k_{ab} is a second-order rate constant, and therefore, has both a time and concentration dimension (*see* examples in Table 3).

If [B] is constant, Eq. (2-28) reduces to:

$$v = \frac{V_m'[A]}{[A] + K_A} \qquad \text{[Re. 2-8]}$$

and Eqs. (2-29) and (2-30) reduce to v = k_a' [A], where:

$$k_a' = k_a \frac{[B]}{[B] + K_B} \qquad \text{[Eq. 2-31]}$$

Table 3 gives examples of the infuence of Michaelis constants and concentrations of substrate B on the apparent first-order rate constant for substrate A. This table would be relevant for enzymatic analysis of substrate A.

Case 2: "Cooperative Addition"

In a second two-substrate case, each substrate influences the K_m for the other in a manner that can be represented by:

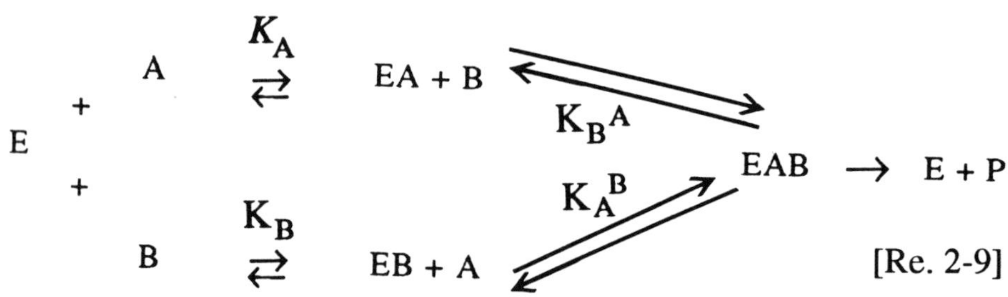

[Re. 2-9]

Here there are two Michaelis constants for each substrate, one (K_A or K_B) that applies to the free enzyme and another (K_A^B or K_B^A) that applies to the enzyme-second substrate combination. There are many instances in which the presence of one substrate will favor combination with the other (positive cooperation) (i.e., $K_A^B < K_A$). If so, the effect must be mutual (i.e., $K_B^A < K_B$). In fact, whether the cooperation is positive or negative:

Table 3
Effect of Michaelis Constants and Levels of One Substrate on the Apparent First-Order Rate Constant of the Other Substance[a]

Enzyme	V_{max} µ*M*/min	K_A µ*M*	K_B µ*M*	k_a min^{-1}	k_{ab} µM^{-1} min^{-1}	[B], µ*M*	V^{1}_{max} µ*M*/min	k_a' min^{-1}	$t_{0.98}$ min
I	1000	500	200	2	0.01	5000	960	1.9	2.1
						200	500	1	3.9
						10	48	0.1	39.
II	1000	500	20	2	0.1	5000	996	2.0	2.0
						10	333	0.67	5.8
III	1000	50	200	20	0.1	5000	960	19.2	0.2
						10	48	0.96	4.1

[a]This table applies to the case of a two-substrate reaction showing "random order of addition." k_a is the apparent first-order rate constant (V_{max}/K_A) at infinite concentration of B; k_a' is the apparent first-order rate constant at the particular concentration of B, and is defined by Eq. (2-31). The time required for 98% disappearance of A is designated "$t_{0.98}$."

$$K_A/K_A^B = K_B/K_B^A \quad \text{[Eq. 2-32]}$$

The velocity equation is:

$$v = \frac{V_m[A][B]}{[A]([B] + K_B^A) + K_A^B([B] + K_B)} \quad \text{[Eq. 2-33]}$$

If [B] is held constant, this reduces to Eq. (2-19):

$$v = \frac{V_m'[A]}{[A] + K_A'} \quad \text{[Re. 2-10]}$$

where:

$$V_m' = V_m\left(\frac{[B]}{[B] + K_B^A}\right) \quad \text{[Eq. 2-34]}$$

and:

$$K_A' = K_A^B\left(\frac{[B] + K_B}{[B] + K_B^A}\right) \quad \text{[Eq. 2-35]}$$

At low levels of A, the apparent first-order equation is:

$$k_a' = \frac{V_m}{K_A^B}\left(\frac{[B]}{[B] + K_B}\right) \quad \text{[Eq. 2-36]}$$

in analogy to Eq. (2-31) for the simpler case.

In the above equations, when [B] is much larger than either K_A or K_B^A, $V_m' = V_m$, $K_A' = K_A^B$, and $k_a' = V_m/K_A^B$.

Consider an example in which $K_A = K_B = 1$ mM, $K_A^B = K_B^A = 0.1$ mM, and [B] is held constant at the levels shown in Table 4.

Note that [B] has a favorable effect on both V_m', and K_A' and therefore an even more favorable effect on k_a'.

Table 4

[B], m*M*	V_m', m*M*/min	K_A', m*M*	k_a', min^{-1}
Infinite	1.00	0.10	10.0
1.00	0.91	0.18	5.0
0.10	0.50	0.55	0.9
0.01	0.09	0.92	0.1

CASE 3: "ORDERED ADDITION"

A third type of two-substrate case is in reality an extreme example of the case just discussed in which $K_A{}^B$ is zero and K_B is infinitely large.

$$E + A \underset{}{\overset{K_A}{\rightleftarrows}} EA + B \overset{K_B{}^A}{\rightleftarrows} EAB \overset{k_5}{\longrightarrow} E + P \qquad \text{[Re. 2-11]}$$

This is called "ordered addition" of substrates, since it describes the situation in which the second substrate can only combine with the enzyme if the first substrate is in place. For strictly ordered addition, "rapid equilibrium" kinetics, on which Eq. (2-33) is based, lead to erroneous results. Thus, according to rapid equilibrium kinetics, K_A' becomes infinitely small as the concentration of B approaches infinity, whereas K_A' in fact approaches K_A. The reason for the discrepancy is that the assumption of rapid equilibrium does not properly take into account the contribution of k_5 to the kinetic situation. Further consideration of this special case is beyond the scope of this book.

Enzyme Reactions with Second-Order Kinetics

There are a number of analytical situations involving two substrates in which both substrates are initially below their respective Michaelis constants and in which the auxiliary substrate B falls substantially during the analytical reaction. This is often the case when B is NADH or NADPH, and the disappearance is measured directly in the fluorometer. It is not possible to use a large excess of the pyridine nucleotide, because during the reaction, the percentage change would be too small to measure accurately. In this situation, the reaction is no longer pseudo first order. The practical consequence is that the reaction velocity falls

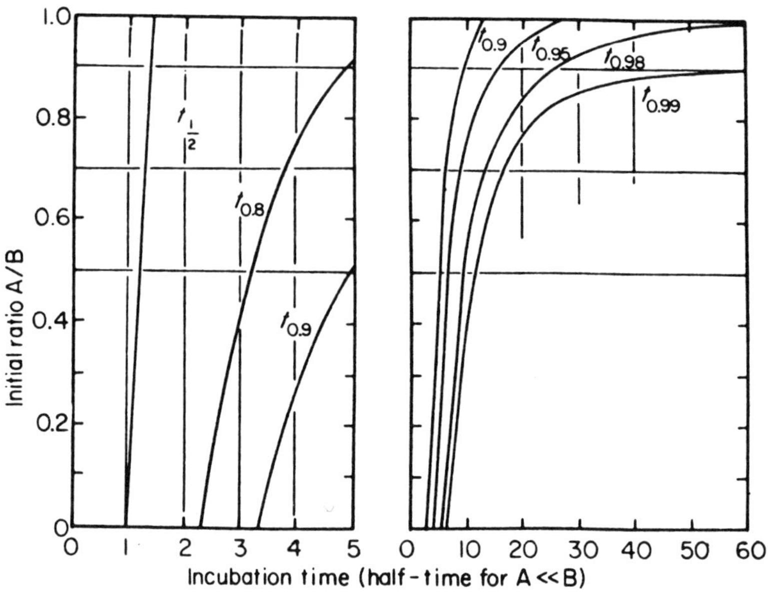

Fig. 5. Incubation time required for an enzyme reaction between two substrates when both are present well below their respective Michaelis constants. The ratio of A to B is given on the vertical axis. The incubation time required to convert different fractions of A to product are indicated by the curves. Time is recorded on the horizontal axis as multiples of the half-time for the case when A is much smaller than B.

off faster than in a first-order reaction. As shown above, the velocity is described by the equation for a nonenzymatic second-order reaction:

$$v = k_{ab}[A][B] \quad \text{[Eq. 2-30a]}$$

The integrated form is somewhat complicated and awkward to use. Therefore, a graphic representation is provided (Fig. 5). From this can be calculated the time required to use up a given percentage of substrate A when A is varied relative to B.

The graph is constructed with time recorded in multiples of the half-time that would be found when A is much smaller than B. When A is 20% of B, it requires 5.5 and 7 such half-times, respectively, to convert 95 and 98% of A to product (i.e., only about 25% longer than if A were much smaller than B), but when A is 90% of B, the graph shows it requires 15 and 25 half-times to convert 95 and 98% of A to product.

When A equals B, it would require 70 such half-times (off the graph) to remove 98% of A!

This graph applies strictly only to the case of random addition with no cooperative effects between the two substrates. However, it will in fact apply well enough in most cases, since by definition both substrates are present at levels substantially below the Michaelis constants, and cooperative effects will therefore usually be negligible.

Two-Step Reactions

Many analytical enzyme reactions involve two or more steps. Usually all but the first step will be first order. Therefore what follows will be limited to situations where both steps are first order, or where the first step is zero order and subsequent steps are first order.

Two-Step Reaction with First Step Zero Order, Second Step First Order

This is the usual situation in a two-step assay for enzyme activity. The enzyme to be measured catalyzes the first step and forms product B at a steady rate.

$$A \xrightarrow{v_1} B \xrightarrow{k_2} C \qquad \text{[Eq. 2-37]}$$

The problem is to determine what discrepancy there may be between the disappearance of A and the appearance of C (i.e., how fast and to what extent B accumulates [Fig. 6]).

Since step 2 by definition is first order, $v_2 = k_2[\mathrm{B}]$, where, according to Eq. (2-21), $k_2 = V_{m2}/K_\mathrm{B}$. Ultimately, a steady state will be established, at which time $v_1 = v_2$. When this occurs:

$$v_1 = k_2[\mathrm{B}]_{max} \quad \text{or} \quad [\mathrm{B}]_{max} = v_1/k_2 \qquad \text{[Eq. 2-38]}$$

where $[\mathrm{B}]_{max}$ is the steady-state level of B.

Thus, for the case of a zero order reaction followed by a first-order reaction, $[\mathrm{B}]_{max}$ will be numerically equal to the velocity of the first reaction divided by the rate constant for the second reaction. For example, if the velocity of the first reaction is 1 μ*M*/min and the rate constant for the second reaction is 3/min, $[\mathrm{B}]_{max}$ will be 1 μ*M*/3 = 0.33 μ*M*.

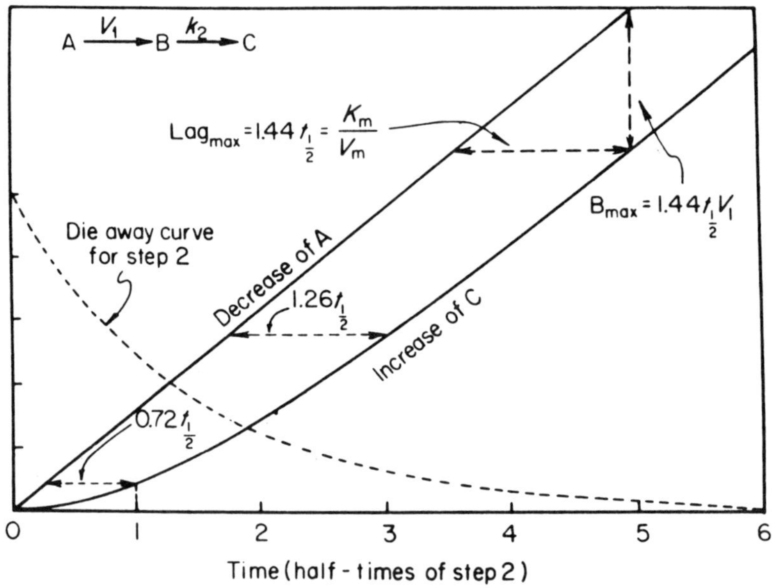

Fig. 6. Lag in a two-step reaction in which the first step is zero order; and the second step is first order. The vertical distance between the lines equals the amount of B accumulated, and the horizontal distance between the lines equals the time lag between disappearance of A and appearance of C.

Since $k = 0.69/t_{1/2}$ (Eq. [2-3)], Eq. (2-38) can also be written as:

$$[\mathrm{B}]_{\mathrm{max}} = 1.44\,t_{1/2}v_1 \qquad \text{[Eq. 2-39]}$$

where $t_{1/2}$ refers to the second step. This says that $[\mathrm{B}]_{\mathrm{max}}$ represents the amount of A converted to B in 1.44 half-times for the second step.

It is often more useful to consider the lag time between the disappearance of A and the appearance of C. The maximum lag time (Fig. 6) is the time it takes to accumulate $[\mathrm{B}]_{\mathrm{max}}$ or:

$$\mathrm{lag}_{\mathrm{max}} = \frac{[\mathrm{B}]_{\mathrm{max}}}{v_1} = \frac{1}{k_2} = \frac{K_{\mathrm{B}}}{V_{\mathrm{m2}}} = 1.44t_{1/2} \qquad \text{[Eq. 2-39a]}$$

Notice that the lag time depends solely on the first-order rate constant of the *second step*.

The amount of B present before the steady state is reached is described by the integrated equation:

$$[B] = (v_1/k_2)(1 - 0.5^{t/t_{1/2}}) \qquad \text{[Eq. 2-40]}$$

Accordingly, when *t* is equal to 1, 2, 3, and so on, half-times (for step 2), [B] will be 1/2, 3/4, 7/8, and so on, of the maximum; or what amounts to the same thing, the lag time will be 1/2, 3/4, 7/8, and so on, of the maximum lag time (Fig. 6).

Example: $v_1 = 0.5\ \mu M/\text{min}$, $t_{1/2}$ for the second step is 2 min. $[B]_{max} = 1.44 \times 2\ \text{min} \times 0.5\ \mu M/\text{min} = 1.44\ \mu M$. At 2, 4, 6, and 8 min, [B] will be 0.7, 1.1, 1.3, and 1.36 μM, respectively.

TWO-STEP REACTIONS WITH BOTH STEPS FIRST ORDER

$$A \xrightarrow{k_1} B \xrightarrow{k_2} C \qquad \text{[Eq. 2-41]}$$

This is a common situation in metabolic assays. It is useful to know the rate of appearance of C as a function of k_1 and k_2. When the first step is much slower than the second, the time lag curve is the same as that given for the previous case (step 1 zero order, step 2 first order). Departure from this simple situation is not serious until the rate of the first step approaches or exceeds the second. Therefore, in most analytical situations, the maximal time lag caused by the second step can be taken as 1.5–2 half-times (for the second step).

For those who may need more exact information, the following more detailed analysis is presented. We will approach this problem by regarding the time course of the first step as primary and asking what *time lag* in the overall reaction is caused by the second step. The kinetic equations are more complicated than when the first step is of zero order, but fortunately approximations and graphical representations will suffice for most purposes.

The appearance of C is the same as the disappearance of A + B. The integrated rate equation is:

$$\frac{[A] + [B]}{[A]_o} = \frac{k_2 e^{-k_1 t} - k_1 e^{-k_2 t}}{k_2 - k_1} \qquad \text{[Eq. 2-42]}$$

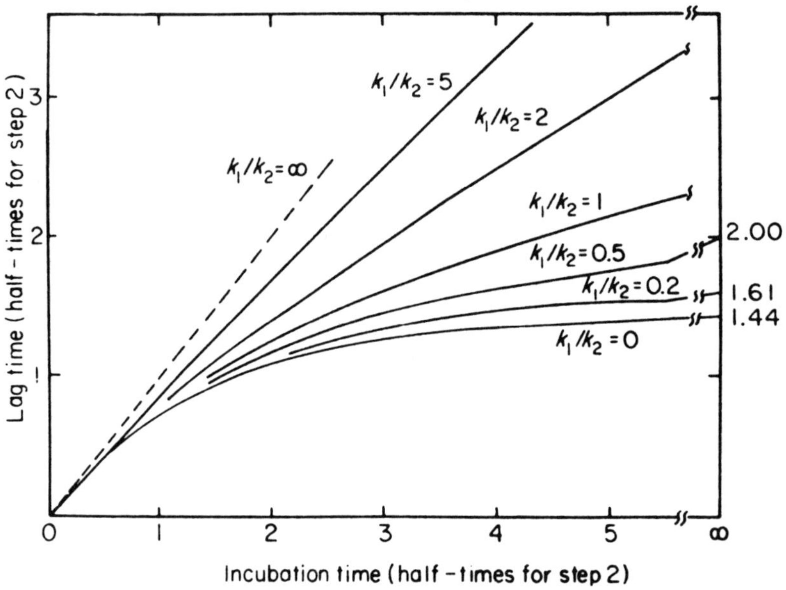

Fig. 7. Lag in a two-step reaction in which both steps are first-order. Time is recorded in multiples of the half-time for the second step. Examples of use: Let $k_1/k_2 = 2$ and the step 2 half-time be 3 min. After incubation for 12 min (four times the half-time for step 2), the lag in appearance of product is 7.5 min (2.5 half-times for step 2).

The equation for the disappearance of A is:

$$[A]/[A]_o = e^{-k_1 t} \qquad \text{[Eq. 2-43]}$$

This is the exponential form of Eq. (2-2).

The time lag in appearance of C can be calculated from the difference between these two equations. For example, let $k_1 = 0.5$/min and $k_2 = 1$/min. It is calculated from Eq. (2-2a) that the half-time for disappearance of A would be 1.39 min. It is calculated from Eq. (42) that the half-time for disappearance of (A + B) is 2.46 min. Therefore, after an elapsed time of 2.46 min, the lag is 2.46 - 1.39 = 1.07 min.

Curves of time lags are given in Fig. 7 for k_2 held constant and k_1 varied from 0.2 to 5 times k_2. When k_1 is smaller than k_2 the lag approaches a finite fraction of the half-time for step 2 (Table 5). If the ratio of k_1 to k_2 is made successively smaller, the lag curve approaches that obtained when the first step is zero order. That is, the maximum lag approaches

Table 5
Lag Time for Two Sequential First-Order Reactions[a]

	Half-time			98% time			
k_1/k_2	A decrease	C increase	Lag at $t_{1/2}$ for C	A decrease	C increase	Lag at 98% time for C	Maximum lag time
0.05	20	21.47	1.47	113	114	1.47	1.48
0.1	10	11.52	1.52	56.5	58	1.52	1.52
0.2	5	6.58	1.58	28.2	29.8	1.59	1.61
0.4	2.5	4.05	1.55	14.1	16.0	1.82	1.84
0.6	1.67	3.20	1.53	9.4	11.5	2.14	2.20
0.8	1.25	2.68	1.43	7.05	9.50	2.45	2.51
0.9	1.11	2.53	1.42	6.25	8.90	2.65	3.70
1.0	1.0	2.42	1.42	5.63	8.40	2.77	*b*
2.0	0.5	1.77	1.27	2.82	6.64	3.82	*b*
5.0	0.2	1.31	1.11	1.13	5.95	4.82	*b*
10.0	0.1	1.15	1.05	0.56	5.80	5.24	*b*

[a]The reaction sequence is $A \xrightarrow{k_1} B \xrightarrow{k_2} C$. The half-time for the second step is held constant at unity. The half-time of the first step is varied, as shown, from 0.1 to 20 times that of the second step. The times are given for 50 and 98% decrease in A, and 50 and 98% appearance of C. The times are recorded as multiples of the half-time of the second step.

[b]These lag times increase without theoretical limit.

1.44 half-times (for step 2), and the lag is 1/2, 3/4, 7/8, and so forth, of the maximum after 1, 2, 3, and so forth, half-times (of step 2).

If the half-time for step 2 is known, the half-time and even the whole reaction curve for step 1 can be calculated with the aid of Table 5 and Fig. 7. As an example, suppose step 2 has a half-time of 2 min and the overall reaction has a half-time of 8 min (four times longer). Table 5 shows that the first step is about 0.4 times as fast as the second step ($k_1/k_2 = 0.4$). If the first step is known to be first order, this would indicate a half-time of 2.5 min. To see if the first step did in fact follow a first-order curve, its time curve could be calculated by deducting the lag times indicated in Fig. 7 along an interpolated curve for $k_1/k_2 = 0.4$. In the present example, readings made at 2, 4, and 6 min (1, 2, and 3 half-times) would be corrected by 0.75, 1.3, and 1.4 half-times, or 1.5, 2.6, and 2.8 min. This procedure would give a close approximation to the time curve for step 1, even if step 1 turned out not to be first order.

Another use for Table 5 and Fig. 7 would be to guide the choice of assay conditions. They allow prediction of amounts of enzymes needed to complete a reaction in a specified time. For example, from Table 5 it is easily seen that for the overall reaction to be 98% complete in about 10 min the half-times for step 1 and step 2 could be (in minutes) 1.7 and 0.17, 1.5 and 0.6, both 1.2, or 0.17 and 1.7. Note that the most efficient use of the enzymes occurs when lhe half-times are about equal and that interchange of the two half-times does not affect the overall time curve.

Other Factors Affecting Enzyme Kinetics

The preceding has been devoted to the relationships of substrate concentrations to reaction velocity. It should be stressed that V_m, Michaelis constants, and equilibrium constants are all subject to the influence of temperature, pH, ionic strength, and the presence of activators and inhibitors. A brief comment on temperature and pH is in order.

Temperature

Ordinarily V_m is increased by increasing temperature until the point of enzyme instability is reached. Temperature coefficients vary greatly from one enzyme to another, and it is sometimes overlooked that enzyme activity does not fall to zero in an ice bath. Usually the V_m at 0° is 3–10% of that at 38° and, in the case of some enzymes, may be as high as 20% or more.

Michaelis constants also would be expected to increase with temperature. Therefore, temperature may have a less favorable effect on rates as low substrate levels (i.e., in the usual situation encountered in analytical work). In some cases, rates may actually decrease with increasing temperature.

pH Effects

The effect of pH on V_m is, of course, extremely variable from enzyme to enzyme. For some enzymes, the pH optimum is extremely sharp, and for others very broad. The pH for maximum velocity does not necessarily coincide with the pH for minimum Michaelis constant. Consequently, the pH optimum for an analytical reaction with low substrate level may be quite different from that for an enzyme assay carried out at near-saturating substrate concentration.

CHAPTER 3

Constriction Pipets

Constriction pipets were originally described by Levy *(1)* working in the laboratory of Linderstrøm-Lang. The large range of sizes in which they can be made (Fig. 1) and the ease of use make them particularly valuable for microchemical analyses. In fact, for all of the analytical applications of the type considered in this book, the most important tool is the constriction pipet. The construction, calibration, and use of this instrument therefore deserve extended presentation. At present, there are commercial sources for standard types and larger sizes of constriction pipets. Smaller sizes and pipets of modified design for special purposes are not readily available.

The Role of Surface Tension in Pipeting

Surface tension represents a force of great importance in handling small liquid volumes. It is usually useful, but is sometimes a nuisance. The purpose of this section is to show how surface tension governs the use of constriction pipets and dictates their construction.

Gravity is the force responsible for emptying large pipets and burets. In contrast, delivery of a small constriction pipet is accomplished almost entirely by air pressure, and the wall is swept clean by the opposing surface tension. A 10-µL pipet can be delivered equally well vertically, horizontally, or even upside down.

If it were not for surface tension, a small pipet would deliver only a small fraction of its contents. Imagine a column of liquid standing in a tube of 1 or 2 mm bore and then being forced to flow out slowly. The fastest flow will tend to occur down the middle of the tube. This, if unopposed, would result in a small channel down the middle, leaving a thick stagnant layer next to the wall. However, as flow begins and the center of the meniscus tries to move faster than the periphery, surface tension exerts a powerful force to maintain the hemispherical shape of the surface (the shape of the smallest surface area). Thus, surface tension tries to force the liquid near the wall to keep up with that in the

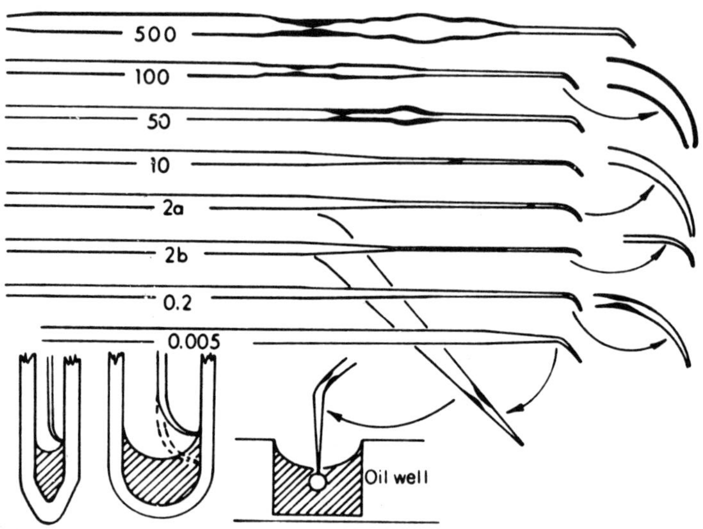

Fig. 1. Design of assorted types of constriction pipets. The volumes are indicated in microliters. The upper five pipets and the 0.2-µL pipet are for general use. The pipet labeled 2b is for use in a slender tube (2–3 mm bore). Note the finer tip and shorter bend as compared to 2a. This relationship between the tip and tube size is shown at the lower left. Note that ideally the pipet tip and meniscus have the same radius of curvature. A 0.005-µL-oil well pipet is illustrated at the bottom. (The pipet with the tip actually in an oil well is of larger size.)

center. (With pipets below 0.001 µL in vol, the effect of surface tension is so great that it may be difficult to apply sufficient pressure by mouth to empty the pipet. In this case, the tremendous sweeping force is readily appreciated.)

For effective removal of liquid near the wall, the flow must be slow enough to give surface tension time to overcome the viscosity of the liquid. A rule of thumb for the flow rate was given by Peters and Van Slyke *(2)* for large pipets and burets. This rule is that with aqueous fluids a linear speed of 1 cm/s will leave behind a sufficiently thin film of liquid (1% of less of the contained vol) to permit good reproducibility.

We have made tests to see whether the rule applies to very small pipets and to measure more exactly the relationship between flow rate and residual fluid. Measurements were made with tubing of five sizes ranging from 0.06 to 1.9 mm in bore. These would encompass the diameters of pipets ranging from 0.003 to 200 µL. Within experimen-

Table 1
Residual Fluid in Pipet as a Function of Delivery Rate

Linear rate, cm/s	Residual fluid, %
0.2	0.4
0.4	0.57
0.8	0.8
1.0	0.9
1.2	1.0
2.0	1.3
4.0	1.8
6.0	2.4

tal limits, the residual fluid for all sizes was found to be related to flow rate (cm/s) by the same simple formula

$$\text{Residual fluid} = 0.9\% \sqrt{\text{flow rate}} \qquad \text{[Eq. 3-1]}$$

Experience with pipets down to 0.000,1 µL indicates that the formula applies even at this level. Table 1 gives some examples calculated from this formula.

It will be seen that for rates near 1 cm/s, a 20% difference in rate makes only a 0.1% difference in the vol delivered. This fact helps explain why it is relatively easy to achieve a high degree of precision with even the smallest constriction pipets.

It may seem surprising that the same rule governs the fraction of fluid left behind over the 10,000,000-fold range of useful constriction pipets. For a 500-µL pipet, 1% of the vol represents a 7-µm film on the wall. For a 0.000,1-µL pipet, 1% of the vol represents only a 0.05-µm film.

Effect of Changes in Surface Tension and Viscosity

From what has been said, it is clear that either a decrease in surface tension or an increase in viscosity of the liquid to be pipeted will increase the amount of residual fluid. The greatest difficulties arise if there is both low surface tension and high viscosity, as is true of certain organic liquids. In this case, special pipets may be needed that provide for very slow delivery.

Surface tension plays other roles in pipeting:

1. The effect of surface tension at the upper constriction makes it easy to fill pipets with an exactly reproducible volume.
2. Surface tension acting at the constricted orifice at the tip of the pipet prevents air from escaping after the fluid volume has been delivered. Sudden escape of air would tend to spatter the sample.
3. Surface tension acting at the tip is also useful when drawing liquid into the pipet. If the tip accidentally gets out of the liquid while suction is being applied, air will not be drawn in unless the suction exceeds a critical value determined by the diameter and surface tension of the liquid.
4. Surface tension is relied on to keep liquid from running up the outside of a pipet during pipeting. When liquid runs out of a pipet tip hanging free in the air, the liquid clings to the tip until the weight of liquid overcomes surface tension and a drop falls. With pipets of <10 µL, the entire sample can collect on the outside without falling. If the shaft is fine enough, the liquid can even ascend up the shaft for a considerable distance. If, however, before beginning to expel the liquid, the pipet tip is touched to a larger surface (inside of a beaker or tube), surface tension will pull the emergent liquid to the larger surface, completely preventing liquid from running up the pipet shaft. The finer the pipet tip, the less the tendency for liquid to cling to the tip instead of the surface touched.

Pipeting Technique

Constriction pipets are capable of precision delivery (±0.3% or better) in sizes ranging from the largest useful volumes (up to 1 mL) down to a vol of 10^{-7} mL or even less. It is difficult to make large errors in using a well-constructed 200-µL constriction pipet. With pipets under 20 µL, however, good technique is necessary to achieve precision. It is believed, therefore, that attention to the "fine points" emphasized in this section may prove rewarding.

The pipets are filled and emptied by mouth using very flexible rubber tubing of appropriate diameter attached to the pipet barrel. If the tubing is too stiff it will require undue force to manipulate the pipet, and this will increase the danger of breakage. A suitable tubing is 3/32-in. bore with 1/32-in. wall. For working with pipets under 20 µL a good light is recommended and, if necessary, eyeglasses to permit sharp vision at a distance of 10–12 in. Eyeglasses with +2 or +3 diopters are very useful, particularly if one is a little presbyopic. When ready to fill, the pipet tip is held against the wall of the vessel, but is dipped only a minimum distance into the fluid. The bend of the tip keeps the shaft away from the vessel wall, preventing fluid from running up between shaft and wall. (If the pipet is accidentally dipped too far into the fluid, it should be withdrawn slowly; this permits surface tension to strip off excess liquid that would otherwise cling to the tip. It is clear that any material collecting on the tip will be rinsed off into the sample during delivery.) Fluid is now sucked in to just above the upper constriction and then driven back until the meniscus almost reaches the narrowest portion of the constriction. Before releasing or even decreasing the pressure, the tip must be withdrawn from the liquid surface; otherwise, excess fluid would be drawn into the pipet by capillarity. With a slender 1-μL pipet, this could result in disastrous errors.

To deliver the sample, the pipet tip is placed against the wall of the receiver. If liquid is already present, the tip is ordinarily touched to the wall through the edge of the meniscus, but is never extended any further than necessary below the surface. The tip bend again functions to keep the shaft away from the wall. Pressure is now applied and increased until the critical pressure is reached at which delivery begins. For reasons already given, the delivery rate should be as nearly as possible the same each time the pipet is used. The easiest way to accomplish this is to maintain throughout delivery the same pressure that is necessary to start delivery. If the pipet is properly made this will provide for a suitable flow rate. This requires a proper balance between the diameters of the upper and lower (tip) constrictions. The critical pressure is determined by the diameter of the upper constriction (and by the surface tension of the liquid). The resistance to delivery is determined by the diameter of the tip and shaft, and by the viscosity of the liquid.

If at the end of delivery a very small, reproducible amount of fluid remains in the tip, it is ordinarily preferable not to try to blow this out. If it must be delivered, in order not to spatter, the tip is withdrawn and touched to the wall above the meniscus before blowing out. In any event, pressure must be maintained until the tip is removed from the meniscus; otherwise, fluid will usually be drawn back into the pipet by capillarity. Even if pressure is maintained, the tip should be lifted from the wall before withdrawal, because if this is not done, liquid will follow the pipet tip up the receiver wall.

Special Considerations with Small Pipets and Small Tubes

When pipeting into or out of a slender tube, there exists considerable danger that fluid will inadvertently run up between the pipet and the wall. This is likely to spoil the analysis. Such accidents are avoided in the first place by using a well-made pipet with a bend at the tip that is equal to the radius of the tube (*see* section entitled Making Constriction Pipets). Pipeting into small tubes is also made easier if the tip, as it touches the inside of the tube, is pointed toward the pipeter. This permits exact positioning of the shaft in the middle of the tube. Otherwise, it would be difficult to judge the position of the shaft in relation to the wall; if it should twist sideways, a run-up is sure to occur. Manipulation of the pipet is facilitated if the hands are turned so that the little fingers touch each other. This will also help prevent breakage of pipets, which, if designed for small bore tubes, are of necessity somewhat fragile. To make a clean withdrawal of a sample from a small bore tube, the pipet tip is inserted into the meniscus, rather than into the body of the fluid, and is then moved down the side of the tube to keep pace with the meniscus. Similarly, if a substantial volume of fluid is added into a slender tube, delivery is started at the bottom or at the surface of any fluid already present, and the pipet is drawn up along the side of the tube, keeping the tip always in the meniscus.

Common Pipeting Errors

1. Using a bad pipet (*see* Making Constriction Pipets);
2. Using stiff rubber tubing;
3. Dipping pipet too deep into liquid of reagent or sample;
4. When filling, not removing tip from meniscus before releasing pressure;
5. During delivery, increasing or decreasing pressure from that used to start delivery; and
6. Blowing air after the sample.

Rinsing Pipets

Since a pipet constructed and used properly leaves a residual fluid vol of 1% or less after delivery, it is usually satisfactory (in fact preferable) not to clean or rinse the pipet between samples in a series. If an error of 2% (from failure to rinse) is acceptable, it is clear that only when the previous sample had been more than three times *stronger* than the sample concerned would an unacceptable error arise. (Because in this case the pipet will be filled with 1% of concentration 3 = 0.03, plus 99% of concentration 1 = 0.99, the sum is concentration 1.02). Rinsing is nevertheless desirable before pipeting blanks, since the carryover from a sample would have a large relative effect, and any error in the blank affects the evaluation of every other sample. When rinsing is believed necessary, the rinsing fluid (ordinarily water) should be delivered slowly to leave a minimum of liquid in the pipet. Cleaning during the analysis, other than by rinsing with water, is not recommended unless the pipet is not delivering properly because of accummulation of protein or lipid.

Making Constriction Pipets

Properties of a Good Pipet

Precision and convenience are enhanced by proper construction. The constriction should be narrow enough and abrupt enough so that the meniscus will stop at an easily reproducible point. The constriction should be wide enough so that it is not easily obstructed and an excessive amount of force is not needed to drive the meniscus past the constriction. The region around the constriction should be such that, once delivery is started, the orifice remains free until delivery is complete. This condition can usually be attained by narrowing the tubing above and below the constriction. This prevents fluid from collecting in the neighborhood of the constriction. If, instead, the constriction is made abruptly in the middle of a wide tube (Fig. 2), fluid will not be swept away cleanly during the delivery, and will drain back into the constriction to give a jerky and less precise delivery.

The tip itself should be related to the size of the pipet. It should taper abruptly to a sufficiently fine point, so that the amount of adherent liquid will be small in comparison to the delivered volume. Most important of all, the diameter of the upper constriction and the size of the tip orifice need to be balanced against each other to provide a suitable rate of delivery, and the upper constriction must be slightly larger than the lower constriction (i.e., less force ought to be needed to start delivery than to blow air out of the tip after delivery).

The shape of the bend at the pipet tip is not critical, except for pipets to be used with small tubes. The ideal bend is an approximation of a quarter circle. In slender tubes, the meniscus is hemispherical. When the quarter circle of the pipet tip crosses a meniscus of the same radius, the shaft and fluid are at right angles to each other. This minimizes any ten-

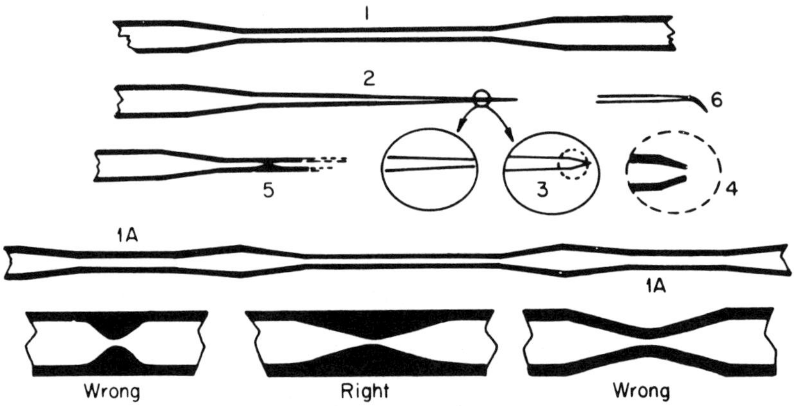

Fig. 2. Steps in making a pipet. Two pipets are made from each piece of tubing. Steps 1–6 are shown for making a pipet of 10 µL or less. Step 1A shows the extra step for large pipets in order to narrow the tubing above and below the future constriction. At the bottom are shown one correct and two faulty constrictions; that on the right is too fragile, that on the left would trap fluid above and below the constriction to give jerky delivery.

dency for fluid to run up the shaft (Fig. 1). The optimal bend at the tip will place the shaft at the center of the tube (Fig. 1), where it will be least likely to trap fluid against the wall. As smaller and smaller tubes are used, this problem of preventing fluid from running up between shaft and wall becomes more and more difficult. A practical limit is reached with tubes of about 1.5-mm bore. With tubes in this range, the danger of a "run-up" is minimized if the pipet shaft is made more slender than usual for the last few millimeters, and if the shaft in turn tapers at the very tip to a somewhat finer point than usual.

Construction

When making constriction pipets for routine use, Pyrex™ tubing is recommended, because it is stronger than soft glass. For even greater durability, quartz tubing may be used, and it is especially desirable for pipets of 3 µL or less.

For most of the glass-blowing described, tinted eyeglasses should be worn. Ordinary sunglasses are adequate when working with Pyrex, but special dark glasses are essential when using quartz. For making and bending the tips, less light is produced, and glasses may be unnecessary even with quartz.

As in all glass-blowing, success is increased by using a flame of correct size and temperature. Two burners are ordinarily required, a multiflame burner or blowpipe torch for large and middle-sized flames, and a special microtorch.

Tubing of 4 or 5 mm od is cut into 20- or 25-cm lengths, each of which will make two pipets. (The 4-mm tubing is suitable for pipets of 150 µL or less, and the 5 mm for 200 to 500 µL pipets.) For pipets of 10 µL and larger, the tubing in the region of each of the future constrictions is thickened (for strength) and narrowed for a length of 10–20 mm to a bore of 0.5–2 mm, depending on the intended volume (Fig. 2). This is accomplished by rotating the tube in a broad oxygen-gas or oxygen-air-gas flame with care not to pull the ends apart. The distance between the two thickened regions is determined by the desired pipet size.

The portion of tubing between the two future constrictions is heated, usually with little or no thickening, and drawn down to a diameter of 1.5–3 mm. The tubing is always removed from the flame to make the pull. The center portion is narrowed further to a diameter of 0.5–0.8 mm, and the tubing finally separated at the center. For pipets in the 1–10 µL range, it is satisfactory to slightly thicken the original tubing in the center and pull it down without any special thickening in the region of the future constrictions.

The actual tip of each pipet is made in a small cool flame, provided by the microtorch. Air rather than oxygen may be preferable for smaller Pyrex pipets to avoid collapsing the wall. A small area of glass is heated, and the glass pulled to give a sharp taper (Fig. 2). This provides a fine tip without undue fragility. With a diamond point, the tip is cut off at a point judged suitable for the volume of the pipet concerned. The upper end of the pipet is fire-polished.

The exact position for the constriction is determined in the case of 1–500 µL pipets, by drawing into the pipet a measured volume of water. A graduated pipet of suitable size (0.1, 0.2, or 1 mL) is partially filled with water and attached almost horizontally to a ring stand (Fig. 3). The tip of the new pipet is touched to the tip of the graduated pipet, and the desired volume of water is drawn into the thickened (constriction) area. A mark is made with a wax pencil at the point indicated, and a temporary label with the approximate volume is attached to the barrel of the pipet.

Fig. 3. Drawing in a measured volume of water to determine the constriction site. The same arrangement is used for calibrating the finished pipet *(see text)*.

The constriction is put in with the same microburner used to pull the tip, using a small hot oxygen flame. It is well to experiment with burner tip sizes, gas, and oxygen supply until the optimal flame for this purpose is obtained. The constriction is placed slightly above the pencil mark to compensate for shrinkage just below thc constriction. The pipet is heated, first in the cooler part of the flame and gradually in the hotter part of the flame as the glass thickens. The pipet is rotated or rocked in the flame to ensure uniform thickening. It is necessary to avoid pushing or pulling in the process; this can be done if the left hand holds the pipet, but the right hand merely cradles the smaller segment. It is easier to make the constriction if one's hands are supported at the proper height above the flame by resting on wooden blocks. This also helps to control lateral motion of the pipet in the flame, which could unduly lengthen the constricted region. When the glass is cool, the constriction is tested by drawing water into the pipet. The constriction can be made smaller by further heating, but cannot be made larger.

The delivery rate is adjusted at this stage by either fire polishing the tip if delivery is too fast or by cutting it off slightly if delivery is too slow. Narrowing the constriction will also accelerate delivery by increasing the force needed to start delivery. Fire polishing of the tip is done at the edge of a very cool flame. If fire polishing is carried too far, the pipet can usually be rescued by pulling out a new tip.

The final step is to bend the tip. A very cool flame is used, and the tip of the pipet gradually brought into the flame while pressure is applied to the tip with a flat piece of metal, such as a flattened nail or, better, a piece of platinum. No more heat then necessary should be applied for the bending in order not to narrow or seal off the tip.

Quartz Pipets

The directions given above are for pipets made from Pyrex. If quartz tubing is used, a hotter flame is required, and the working temperature range is narrower, so that the quartz must be pulled more quickly when it is out of the flame. Because quartz is a better conductor of heat than Pyrex, it is not necessary to rotate the tubing as much for the initial pull or constriction.

Oil Well Pipets

Pipets for use in oil wells are provided with a very slender tip, with a 45° bend 5 or 10 mm from the end, but usually without a bend below this (Fig. 1 and Chapter 12). The slender tip is to minimize the amount or adherent oil and to make it easier to strip off the aqueous volume when the pipet is withdrawn through the oil–air interface. Because of the fragility of the slender tip, quartz is usually preferable to glass.

Smaller Pipets

In making pipets of <1 µL, it is also desirable, as said earlier, to use quartz instead of glass, not only because it is stronger, but because it is actually easier to handle in construction of these smaller pipets. It is also desirable to use magnification. Most suitable is a wide-angle (dissecting) microscope with a long working distance and a zoom lens. For most purposes, a total magnification of from 5 to 15× will be adequate.

The smaller the pipet, the smaller the flame that will be needed. A flame with an exceedingly small hot zone can be achieved by using a torch tip with an orifice of 0.3–0.5 mm and by increasing the ratio of oxygen to gas until it is just below the point of extinguishing the flame. We have found an oxygen–gas mixture to be preferable to an oxygen–hydrogen mixture. With hydrogen, the flame appears to be just as small, but the hot zone is not so well localized. Whatever the flame size, the effective width can be narrowed by working on the edge of the flame.

When making very small pipets, it is desirable to keep the bore as wide as possible by reducing the length as well as the bore. If the bore is too narrow, surface tension makes delivery difficult. This is accentuated by the fact that the smaller the bore of the pipet, the smaller the constriction must be. This results from the need to provide a sharp difference between the pressure needed to move the liquid down to the constriction and that needed to make the constriction "let go." In designing small pipets, a rough guide is to keep the ratio between the bore at its largest point and the length between constrictions in the range from 1:25 to 1:50. This means that the distance between upper constriction and tip for a 0.001-µL pipet may be only 1 or 2 mm.

Because the smallest pipets are very short, it is usually convenient to change the order of construction from that given for larger pipets. The tubing (quartz) is pulled down to a suitable diameter for the constriction, and cut off 8 or 10 mm below the site of the constriction. The constriction is now made with the free end as a handle, which can be supported as necessary by another piece of quartz. If the free end is too short to pull with the fingers, a longer piece of fine quartz is now attached, the flame is enlarged, and the body and tip are now pulled out. In contrast to the situation with larger pipets, the pulling can often best be done directly in the cooler part of the flame.

Calibration of Pipets

Three ways of calibrating constriction pipets are offered gravimetric, colorimetric, and volumetric. For volumes of 20-µL and over, the gravimetric method is the most accurate (0.1–0.3%). The volumetric method is by far the most rapid and is accurate to 0.5–1% down to perhaps 5 µL. The colorimetric method is accurate to 0.5% or better down to 0.002 µL. It is usually desirable to have a selection of larger pipets calibrated accurately by the gravimetric method, whereas the bulk of the pipets in the 5–50 µL range will be sufficiently accurate if calibrated volumetrically. Under 5 µL, the colorimetric method is required.

Although it is desirable to have available accurately calibrated pipets, in most analyses this is not essential, since the pipet volumes will cancel out of the final calculation if standard solutions and samples are handled with the same pipets.

The calibration of pipets is one of the best ways to test pipeting technique. Failure to obtain satisfactory checks means faulty technique or bad pipets. Since the gravimetric method is the most precise, it offers the most critical test, but only for the larger pipets. Also, there is one fault that will not be detected gravimetrically—that of leaving too much fluid on the outside of the pipet—since this would be expected to evaporate before delivery and not appear in the weight. Colorimetric calibration will detect this fault, because the tip is washed off in the diluent. Colorimetric calibration is also more likely to reveal bad technique since it can be used with the smallest pipets, for which pipeting faults can cause very serious errors.

Gravimetric Calibration

Constriction pipets are calibrated by delivery of water into a vessel already containing a little water. The increase in weight of the vessel is measured on an analytical balance. A narrow-mouthed vial without stopper and with a capacity of not more than 20 times the volume of the pipet is recommended. The pipet to be calibrated is delivered three or four times into the vessel, and the increase in weight recorded for each delivery. The pipet should be carefully touched to the wall before starting delivery just as in ordinary use. Care should be taken not to blow air after the sample, since this would drive moist air out of the calibration vessel. The vessel should be handled as little as possible with the fingers. In order to correct for possible evaporation, sham deliveries are made and any weight change in the vessel used as a correction for calibration. The successive weights of the water delivered by the pipet should agree with ±0.1 – ±0.3%. The volume is calculated from the weight and the apparent specific gravity of water in air at the temperature at which the calibration is done. (Amounts of water that appear to weigh 1 mg in air at 21, 25, and 29°C are equivalent to 1.003, 1.004, and 1.005 µL, respectively.)

Colorimetric Calibration

The pipet to be calibrated is used to deliver a volume of concentrated *p*-nitrophenol into a much larger accurately measured volume of diluent. The resulting solution is read at 400 nm in a spectrophotometer, and the reading is compared with readings of dilutions made with previously standardized pipets. (*p*-Nitrophenol is chosen because it is highly colored, but unlike many colored materials is not strongly adsorbed to surfaces. Full color is obtained at any pH above 9, but above pH 11, oxidation becomes noticeable.)

For pipets in the 0.5–50 µL range, the stock solution is 0.5% *p*-nitrophenol in 0.05*M* Na_2CO_3. It is carefully protected from evaporation throughout the period of calibration. The diluent is 25 m*M* $NAHCO_3$-25 m*M* Na_2CO_3 (pH 10). The *p*-nitrophenol is pipeted into an approx 1000-fold larger volume of diluent measured with accurate volumetric pipets. It is recommended that the constriction pipets be calibrated in duplicate or triplicate, and the diluting volumetric pipets be themselves calibrated for this purpose.

Two standard solutions are made with 100 mL of diluent in a volumetric flask. To one is added 100-µL and to the other 120-µL of the *p*-nitrophenol *after* the flasks have been diluted to 100 mL. The volumes of *p*-nitrophenol, although they need not be exactly 100 and 120 µL, must be accurately known and are added with constriction pipets standardized gravimetrically.

The standards are read in a spectrophotometer, with appropriate correction for the readings of the cuvets with carbonate buffer alone. From the corrected readings and the dilution of the standards, a factor is obtained for other pipets. For example, suppose the two standard volumes of *p*-nitrophenol were 100.6 and 120.4 µL of *p*-nitrophenol and the corrected spectrophotometer readings were 0.506 and 0.608.

$$\text{Factor 1} = \frac{100.6\ \mu\text{l}}{100\ \text{ml} \times 0.506} = 1.988\ \mu\text{l/ml}$$

$$\text{Factor 2} = \frac{120.4\ \mu\text{l}}{100\ \text{ml} \times 0.608} = 1.980\ \mu\text{l/ml} \qquad \text{[Eq. 3-2]}$$

The factor to be used would be the average of the two standard calculations, or 1.984. The use of two standards provides a check on the proportionality of the spectrophotometer or the possibility that one of the standardizing pipets is in error.

The use of the factor is illustrated by the following example: A pipet that was previously estimated to be about 3 µL is delivered into exactly 3.01 mL of carbonate buffer. The corrected triplicate readings are 0.515, 0.513, and 0.512. The calculation is:

$$3.01 \text{ ml} \times 0.513 \times 1.984 \text{ µl/ml} = 3.06 \text{ µl} \qquad \text{[Eq. 3-3]}$$

Calibration within 0.3% is easily attainable.

For pipets larger than 50 µL, the same principle is followed, but weaker *p*-nitrophenol (0.05%) is used and the dilutions are approx 1 to 100 instead of 1 to 1000 in order to keep the total volumes within convenient limits.

For pipets smaller than 0.5 µL, the *p*-nitrophenol concentration can be increased to 5%, which is prepared in 0.4*M* KOH (to increase the solubility). The dilution is increased to 1–10,000 for pipets in the 0.05–0.5 µL range (total volume 0.5–5 mL). Still smaller pipets can be calibrated by increasing the dilution to 1–25,000 and, if necessary, using microspectrophotometer cells. Thus 0.002 µL of 5% *p*-nitrophenol diluted with 50 µL of carbonate solution will give an od of nearly 0.3, which permits 0.5% precision.

Although it is beyond the scope of this section, still smaller pipets can be calibrated fluorometrically with strong quinine solutions. Others have found it convenient to calibrate pipets with highly radioactive solutions.

Calibration by Volume

Pipets ranging from 5–200 µL can be rapidly calibrated using a series of graduated pipets. The graduated pipets are selected for uniform bore, determined by withdrawing successive aliquots of water with a constriction pipet previously calibrated by weight or color. If the pipets are not exactly uniform, it can nevertheless serve if it is itself first calibrated. A pipet of 0.1-mL volume with 1000 subdivisions/mL is of suitable size for calibrating constriction pipets of 5–50 µL. A similar 0.2-mL pipet is appropriate for constriction pipets of 30–150 µL. For larger pipets, a 1-mL graduated pipet with 100 divisions/mL can be used.

There is danger of breaking the tip of the constriction pipet on the tip of the calibrated pipet. This can be prevented by superimposing a tip made of several short pieces of polyethylene (PE) tubing of decreasing size.

The recommended calibration procedure is as follows. The graduated pipet is mounted on a ring stand as in Fig. 3 and filled with water. The angle is adjusted so that water will not spontaneously flow out, but will fill the pipet exactly to the tip. The constriction pipet to be calibrated is filled from a small beaker and then delivered back into the beaker at its proper delivery rate. (This creates a space in the pipet exactly equal to the volume that it ordinarily delivers.) The tip of this pipet is now carefully inserted at an angle into the tip of the graduated pipet and filled at a uniform rate. This rate should be such that the meniscus moves down the graduated pipet not faster than 1 cm/s. (*See* section entitled The Role of Surface Tension in Pipeting.) The rate of movement of the meniscus of the graduated pipet should be the same for all pipets calibrated, and should be the same when the graduated pipet is itself calibrated or checked for uniformity.

By observing these precautions and by careful estimation of the meniscus position without parallax, readings can be made reproducibly to ±0.1-µL. With pipets of <20 µL, precision can be increased by repeating the withdrawal two to four times between readings, thus bringing the total volume change up to at least 20 µL. Significant evaporation must either be ruled out or corrected for on the basis of sham pipetings.

Cleaning and Storage of Constriction Pipets

After use, if a pipet is to be employed subsequently with a similar solution, there seems to be little reason for rigorous cleaning; a simple rinse with water should suffice. Otherwise, thorough cleaning is desirable. The following procedure is recommended for routine purposes.

The narrow constrictions and fine tips of the pipets preclude the usual procedure of pipet cleaning. Instead, each pipet is cleaned individually by drawing and expelling a succession of cleaning agents into the pipet. If many pipets are to be cleaned, the process can be facilitated by using the vacuum line in the laboratory. A suction flask is interposed between the pipet and line to trap the cleaning fluids. Pressure tubing is used between the line and flask and pipet. A short section of rubber tubing of the type used for pipetting is joined to the pressure tubing with a section of glass tubing. The end of the pipet is inserted in the tubing, and the pipets filled with suction. The cleaning fluids must fill the pipet both below and above the constriction. In some instances, complete filling is made easier if the negative pressure is reduced by pinching the rubber tubing. The cleaning solutions can be kept in wide-mouth, glass-stoppered, 60-mL bottles. The pipet is first rinsed in distilled water to remove any residual material, and the following solutions drawn through in sequence:

1. 0.1*M* NaOH;
2. Deionized (or glass-distilled) water (bottle No. 1);
3. Fuming HNO_3;
4. Deionized water, at least three rinses (bottle No. 2);
5. Deionized water, at least three rinses (bottle No. 3); and
6. Acetone.

The acetone is used to dry the pipet for storage, but will probably leave material on the inner walls of the pipet. Before use, the pipet should be rinsed with deionized water and, if possible, with the reagent to be used.

Because the constrictions and tips are very small, the orifices may become plugged during use if certain precautions are not taken. Pipets that have been used for protein solutions should be rinsed two or three times with water immediately after use. Any concentrated solution is likely to clog the pipet if left to dry on the walls around the tip and constriction. If the pipet is to be left overnight, the last traces of fluid are removed from the tip by sucking in air after rinsing. The danger of leaving material at the orifice is thus minimized.

Pipets with very small constrictions (those of <1-µL vol or oil well pipets) are only occasionally cleaned by this routine because of the danger of plugging with particles in the cleaning fluids. Instead, they are rinsed in and out successively with the above fluids by suction from the mouth. Each fluid is drawn up a short distance above the constriction. The water rinses are drawn up further than the alkali or acid, to be sure the latter are completely removed.

If a pipet becomes plugged, the first step is to determine whether it is at the tip, the constriction, or both. The pipet is first dipped into water. If the tip is open, water will usually enter to a slight degree because of capillarity. To test the constriction, water is introduced above the constriction with a small plastic wash bottle or another pipet. The fluid is shaken down to the constriction and pressure applied by mouth through rubber tubing. If the constriction is open, the pressure will force some water through. If it is not open, the water may dissolve the material if it is allowed to remain in the pipet a few minutes. It is useful to know what substance is responsible for plugging the pipet. If the material is a protein, it is better to fill the pipet with 0.1*M* NaOH than with acid. In either case, the cautious application of heat may hasten solution of the plug. This can be done by using a match flame or immersing the pipet in hot water (60°C). Once the constriction is open, the pipet can be filled from above by shaking it like a thermometer. If the tip is closed, the same procedure is followed. The fluid is allowed to remain in contact with the plugged orifice and can be very gently heated if neccssary. Once the openings are clear, the pipet can be cleaned in the standard way.

Occasionally, it may be found that the pipets do not deliver properly after cleaning and bubbles form on the walls when the pipet is filled. This usually means that the pipet is not thoroughly clean. If the material last used was protein, the pipet should be filled with 0.1*M* NaOH, and gently heated with a match flame or by immersing it in hot water (60°). It is helpful in many instances to fill the pipet with fuming nitric acid and allow it to remain for several minutes. Such procedures are followed by routine cleaning.

If the pipet is dirty on the outside, it can be immersed as far as necessary in the cleaning solutions. The pipet should not be handled on the outside below the constriction and particularly not near the tip. There is a temptation to warm the pipet by rubbing it with one's fingers and thus hasten drying. Such practices must be scrupulously eschewed to avoid contaminating the outside of the pipet.

The pipets are stored in small cabinets when not in use to protect them from dust and fumes. A convenient cabinet can be made by using small metal four-drawer tool cabinets ($10 \times 10 \times 11\ ^1/_2$ in.) available at most hardware stores. The metal partitions are replaced with two parallel wooden strips, about 2 in. wide, notched to hold the pipets in a horizontal position. During active use, pipets are conveniently kept on the table top in grooved wooden racks (Fig. 4).

Tubes and Racks

Recommendations for tubes used in fluorometric analyses are given in Chapter 1, page 7. Special racks for these tubes are made from Lucite (Plexiglas) to minimize scratching, which can occur when metal

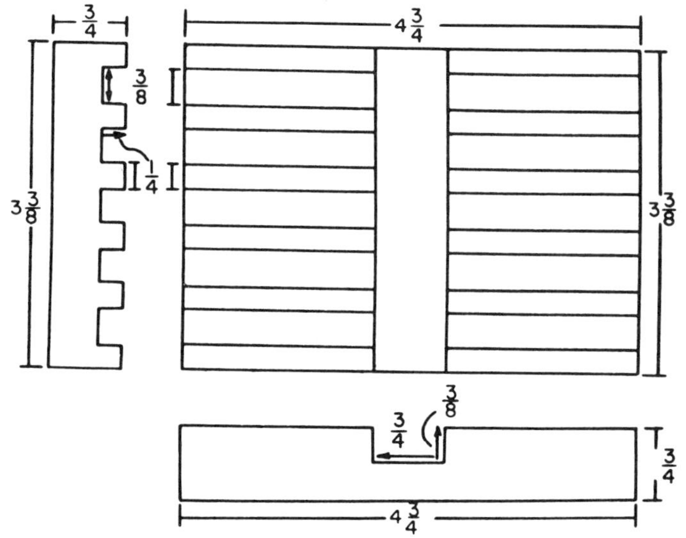

Fig. 4. Wooden rack for holding pipets on the desk top. (Measurements are given in inches.)

racks are used. Racks holding 50 tubes are a convenient size; with 10 holes/row, it is easy to keep track of samples without numbering the tubes. A volume of 1 mL fills these tubes sufficiently for reading in the fluorometer. Rapid easy mixing becomes difficult with volumes much greater than about 1.2 mL. For reactions involving two steps, the first step can often be carried out in the fluorometer tube, if the volume is in the 50–200 μL range. Before reading, 1 mL of reagent or diluent is added.

For reactions carried out in a volume of 25–200 μL, 7 × 70-mm or 6 × 50-mm tubes are suitable. In this case, since the tubes are not used for fluorometric or optical measurement, metal racks can be used and are preferable because of greater heat stability.

Chapter 4

Preparation of Tissues for Analysis

Often the preparative steps in a tissue analysis are the most critical. In the case of a metabolite assay, the most hazardous period is usually between the moment the blood supply is cut off and the moment enzyme action is finally stopped. In the case of an enzyme, with a few important exceptions, there is not likely to be much change for many minutes or even hours. Instead, the biggest problem may be to render the enzyme fully accessible for assay, or to prevent loss during or after homogenization. In a few cases, if the objective is to determine the state of activity as it was in vivo, the problem may be to prevent a specific activity change of the enzyme (e.g., the conversion of phosphorylase *b* to *a*, and so on).

Preparation for Enzyme Assays

It is easier to give specific directions for preparing tissue for metabolite assays than for enzyme assays. Each enzyme is a separate problem, and only trial can determine the conditions necessary for releasing and preserving full activity. In consequence, this section will be mainly limited to some generalities. In specific cases, the literature should be consulted if possible.

The majority of enzymes are not affected by freezing when the tissue is intact. This is even true of many enzymes that are unstable to freezing after isolation. Therefore, suitable homogenates can usually be prepared from either fresh or frozen tissues. If the frozen tissue is to be stored before homogenization, the temperature should be –50° or below, since losses are much more likely to occur at higher temperature (e.g., –20°).

Many soluble enzymes can be prepared in fully active form by merely homogenizing tissues at 0° to 4° in hypotonic buffer. Water alone is sometimes used, but in general, a buffer at the pH of maximum enzyme stability is much safer. This is particularly true of a tissue with an active glycolytic system, which may continue to form acid in the homogenate. The use of hypotonic rather than isotonic buffer favors disruption of cells and organelles. A glass homogenizer, hand or motor driven, is preferable to a mechanical blender, which is likely to denature some of the enzyme. Even though an enzyme may be soluble, it is strongly urged that whenever possible the assay made be on the whole homogenate rather than an extract because a portion, and possibly a variable portion, may be lost owing to incomplete tissue disintegration.

Many insoluble enzymes are also fully active in simple homogenates. If not, they can frequently be solubilized or made accessible to the substrate by sonication, freezing, and thawing, or by the use of a detergent. Unfortunately, each of these could also cause some enzyme destruction.

A number of enzymes require stabilization by additives. The most common additives are chelators, such as EDTA, and sulfhydryl compounds, such as mercaptoethanol or the more stable dithiothreitol (Cleland's reagent). Stability is sometimes increased by addition of the substrate or coenzyme.

Stability is, of course, favored almost always by keeping the temperature close to 0°. If the enzyme withstands freezing and thawing, it can usually be stored in the homogenate without loss for long periods at –50° or below.

Stability can also be affected by dilution. The original homogenate is usually prepared for convenience at a tissue dilution of not more than 10- or 20-fold. Often further dilution is required before assay. Many enzymes become unstable because of surface denauration if the total protein concentration falls much below 0.2 mg/mL (corresponding to 500–1000-fold tissue dilution). Protection can usually be afforded by adding bovine plasma albumin to give this concentration.

Preparation for Metabolite Analyses

Most metabolite analyses are performed on extracts prepared from rapidly frozen tissues. There are two objectives: to freeze the tissue fast enough to prevent significant change in the levels of the metabolites concerned and to extract the frozen tissue without allowing opportunity for enzyme action.

The first objective is difficult to achieve except with very small animals or with anesthetized animals. The reason, as emphasized by Dawson *(1)*, is the slow rate of heat flow through water-filled material.

If the surface is brought to the temperature of liquid N_2 it requires 2–5 s to freeze the tissue at a depth of 1 mm, and 10–20 s at a depth of 10 mm. Wollenberger et al. *(2)* introduced the use of metal tongs, chilled to liquid N_2 temperature with which a small organ or portion of a larger one could be crushed flat with nearly instantaneous freezing.

This method is not suitable for all organs, and satisfactory results may often be obtained by quickly removing a small piece of tissue from an anesthetized animal and immersing it in a very cold liquid. It may even be satisfactory in some cases to freeze in this manner an entire small animal, such as a mouse. Alternatively, the tissue can be exposed under anesthesia and flooded with the cold liquid. Less of the liquid will be needed if it is possible to place a wall of foil around the organ to retain the freezing liquid.

Possibly the earliest valid analyses in brain for such labile metabolites as P-creatine and ATP were made by Kerr and Ghantus *(3)*, who exposed the brain of the dog under anesthesia and poured over it liquid N_2 while maintaining artificial respiration. In this way, the circulation to deeper parts of the brain was maintained until the freezing point arrived.

Liquid N_2 is not as satisfactory for rapid freezing of small samples as an organic liquid chilled with liquid N_2, because the gas evolved from the boiling N_2 forms an insulating layer around the tissue. Isopentane and even liquid propane have been used, but constitute a fire hazard since O_2 will eventually condense in the liquid N_2. For many purposes, Freon-12 (CCl_2F_2), which freezes at –150°, is a satisfactory and readily available substitute. The Freon in a metal container is suspended in liquid N_2 with stirring until it begins to freeze. To provide sufficient heat capacity, it is recommended that at least 25–50 mL of Freon be used/g tissue. When the tissue is immersed, the Freon should be stirred vigorously and, unless the sample is very small, the container lowered into the liquid N_2 *(see also* Chapter 9).

Ferrendelli *(4)* has shown that, with larger structures, liquid N_2 is superior to Freon for freezing at depths > 3 or 4 mm. Apparently the colder temperature of liquid N_2 (–190°) ultimately more than compensates for the brief early delay at the surface owing to the gas evolution.

Once frozen, the tissue can be safely stored for several weeks (and probably several months) at–50° or below. At higher temperatures, changes resulting from enzyme action can occur surprisingly rapidly.

Tests made wilh mouse brain show a 20% loss in ATP in 24 h at –20°, 14 h at –15°, and 20–30 min at –10°.

Glucose-6-P was lost even more rapidly. It was at least 20% gone in an hour at –20°, 30 min at –15°, and 10 min at –10°. Changes were also observed in AMP, fructose-1,6-P_2, and 3-P-glycerate in a few hours at –20°. P-Creatine fell to half in 8 h at –15°. Even glucose had fallen by a third after 6 h at –10°. On the other hand, none of these metabolites changed during a week's storage at –35°. Clearly, in the case of tissues to be analyzed for labile constituents such as these, the temperature should not be raised above –35°, except as necessary for short periods during preparation of extracts.

Preparation of Tissue Extracts

Most metabolites can be satisfactorily assayed in protein-free extracts prepared with $HClO_4$ or trichloracetic acid. $HClO_4$ is usually preferred because most of it can be easily removed by precipitation as the potassium salt. Trichloroacetic acid is also a potent inhibitor of several enzymes used in metabolite assays. (An important example is glucose-6-P dehydrogenase, which is used in many different analytical systems and which is exceedingly sensitive to trichloroacetate inhibition.)

Once a tissue has been frozen it must not be thawed even for a few seconds, if drastic changes in many metabolite levels are to be avoided. If frozen tissue is added to an acid precipitant at 0°, the tissue must in principle thaw, however briefly, before the acid penetrates and stops enzyme action. The following procedure, using $HClO_4$ as the precipitant, avoids this.

Perchloric Acid Extracts

If the tissue needs to be dissected before extraction, this is conveniently done in a room or cryostat maintained at –15° to –20°, since at much lower temperatures, tissues are exceedingly hard and brittle. As the data shown above indicate, the tissue should be held no longer than necessary in this warmer temperature range. Samples are either powdered at liquid N_2 temperatures or cut into pieces no larger than 10 mg. A weighed portion of the powdered or fragmented tissue is placed on top of approx 3 mL/g of 3*M* $HClO_4$ previously frozen and kept on dry ice in a tube of appropriate size. The tubes are transferred to an alcohol bath maintained at –8° to –10°, and the samples stirred or agitated until the acid completely penetrates the powder or tissue fragments (5–15 min). The temperature must not be below –10° for complete extraction of the ice (2*M* $HClO_4$ freezes at about –12°). When it is certain that no frozen tissue remains, 1 mL of H_2O is added for each 0.3 mL of $HClO_4$, and the sample is repeatedly mixed at 4° for 5 or 10 min. If there is a possibility of bone contaminating the sample, 1 m*M* EDTA is included in the water. Otherwise, if even a small amount of bone is present, calcium phosphate will precipitate later upon neutralization and remove part of the ATP, fructose bisphosphate, and probably other phosphorylated compounds.

The samples are centrifuged at 5000*g* for 10 min, and the supernatant fluid removed from the protein precipitate. The removal of all protein is critical, because a number of enzymes survive the brief acid treatment at this low temperature. The presence of enzymes in the neutralized extract may lead to disturbances in the subsequent analyses for metabolites. For example, during the analysis of P-creatine, which requires added ADP, the presence of myokinase, which is very stable toward acid, would lead to erroneously high results. Similarly, at least some of the ATPase can survive the acid treatment and give low ATP results if not completely removed.

As a rule, the $HClO_4$ extracts are partially or completely neutralized before proceeding with analyses. If acid-labile metabolites are to be assayed, this should be done promptly. In the case of P-creatine, for example, the tolerance limit is about an hour in either 2*M* $HClO_4$, at –10° or in 0.6*M* $HClO_4$ at 0°.

Neutralization can be made in several ways, depending on the properties of the particular metabolite. One standard procedure is to neutralize with a 5 or 10% excess of 2*M* $KHCO_3$. (This is about 0.35 vol per vol of acid extract, but the exact amount should be checked with the particular $HClO_4$ and $KHCO_3$ solutions used.) After most of the evolved CO_2 has been dissipated, the $KClO_4$ precipitate is removed at 0–4° by centrifugation. Approximately 0.05*M* $KClO_4$ remains in solution at 0°; the solubility is considerably greater at room temperature. The pH at this point is about 6, but later may fall to 7 or 7.5 as more CO_2 comes off. The majority of metabolites, including ATP, P-creatine, and most members of the citric acid cycle, are stable in such an extract and may be stored indefinitely at –50°. There is always the danger, however, that traces of certain enzymes may remain. Consequently, it is recommended that extracts not be allowed to stand for long periods unfrozen even at 0°. A case in point is ATP, which has on occasion been found erroneously low in certain samples as the result of the action of ATPase that had not been completely removed. A better method of neutralization that avoids CO_2 evolution is to use a mixture containg 2*M* KOH and 0.4*M* imidazole base. This is added in such proportion (0.29 mL/mL of extract) as to neutralize 90% of the $HClO_4$ with KOH and the rest with imidazole, leaving the solution buffered at pH 7. To be sure the $HClO_4$ and KOH are closely matched, check the final pH of a simulated extract made with 0.8 mL of water in place of 1 g of tissue.

The Use of Methanol-HCl in Preparing Extracts

At the risk of being repetitious, one more variant on the preparation of acid extracts is offered. This permits somewhat quicker extraction of frozen tissue at even lower temperature than that above and offers the possibility of measuring glycogen on the same tissue sample used for other metabolite analyses *(5)*. (Glycogen in many tissues appears partly in the acid extract and partly in the acid precipitate.) The weighed frozen tissue (fragment or powder) is homogenized at –20 to –30° with 2 vol of 0.1*M* HCl in methanol in a loose fitting glass homogenizer of suitable size (or with a little more trouble the tissue can be dispersed with a stirring rod in an ordinary test tube). After the solvent has penetrated the tissue completely, the temperature is raised to 0°, 1 mL of 0.02*M* HCl in H_2O is added for each 0.1 mL of methanol, and the homogenization or stirring is continued briefly to ensure thorough mixing.

A portion of the sample can now be set aside for glycogen assay. The enzymatic assay is made on this whole homogenate after heating 10 min at 100° (Chapter 6). Protein is precipitated from the rest of the solution by adding/mL of homogenate 0.125 mL of 10 m*M* EDTA in 3*M* $HClO_4$. After centrifuging, the supernatant extract is neutralized by one of the procedures given above, except that because the total acid is only 60% as great as above the amount of base is reduced in proportion.

If the methanol concentration exceeds 15% after $HClO_4$ addition, protein precipitation will not be complete. Methanol, like a number of polar organic solvents, increases solubility of proteins in acid.

Fluorescence of Tissue Extracts

Extracts prepared as described are somewhat fluorescent and contribute to the total fluorescence blank. Brain extracts, for example, are equivalent to at least 50 μmol of NADH/kg. Extracts of other tissues, such as liver and kidney, are even more fluorescent (the equivalent of 300–500 μmol of NADH). This presents a serious problem in direct measurement of metabolites present at levels of 20 μmol/kg or less, of which there are many. The problem can be reduced by using the strong alkali method for measuring Ihe final pyridine nucleotide, and with enzymatic cycling, it can be practically eliminated. However, there are, as has been discussed, certain advantages of following the reactions directly in the fluorometer. A substantial fraction of the tissue blank is contributed by riboflavin nucleotides FMN and FAD. FAD is only about 10% as fluorescent as FMN at pH 7. It is easily hydrolyzed in acid to FMN. Therefore, acid extracts that are not neutralized promptly will show increased fluorescence.

Imidazole

In the case of analyses conducted near neutrality, the riboflavin component of the blank can be specifically reduced by the use of 50–200 m*M* imidazole buffer. Concentrations of 50, 100, and 200 imidazole reduce the fluorescence of FMN at pH 7 approx 30, 60, and 80%. Imidazole preparations usually contain fluorescent impurities. Sigma Chemical Co. offers a preparation with greatly reduced fluorescence, but even this contributes substantially to the blank reading when the concentration is 200 m*M*. The fluorescence of imidazole solution can be reduced by charcoal treatment. (This capacity of imidazole to quench ribonavin fluorescence is apparently related to the fact that flavin-adenine dinucleotide is much less fluorescent than flavin monophosphate). The imidazole portion of the adenine ring must associate with the isoalloxazine ring to form a nonfluorescent adduct *(6).*

Adsorption of Blank Materials

An alternative and more effective method of reducing the tissue fluorescence blank is to treat the extract with Florisil or charcoal. In either case, certain metabolites will be adsorbed, and the treated extracts are useful only for special analyses.

Florisil

To prepare Florisil for use, it is boiled in 10 vol of 1*M* HCI for 10 min, rinsed acid-free with water, and dried at 50°. The "fines" are removed in the rinsing process. The acid extract before neutralization is passed through a 2-mm internal diameter column containing 50–60 mg of Florisil/mL of extract (approx 7 mg of tissue). The diameter of the column should be such as to provide a column at least 20 mm long. Alternatively, the extraction can be carried out in a test tube with similar proportions of extract and Florisil. The sample is repeatedly mixed for a total of 5 min, and the Florisil allowed to settle or spun down in a centrifuge. Florisil treatment reduces the tissue fluorescence as much as 90% or more, and permits quantitative recovery of dicarboxylic and tricarboxylic acids. Phosphorylated sugars and nucleotides are partially adsorbed.

Charcoal

The $HClO_4$ extract can be treated with 10 mg of dried acid-washed charcoal (Norit)/mL of extract. After adding the charcoal, the sample is brought to pH 6.8–7.0 with the KOH-imidazole-KCl mixture and filtered through acid-washed glass fiber paper. Treatment with charcoal also removes up to 90% of the tissue fluorescence and allows complete recovery of the phosphorylated intermediates, but removes nucleotides.

Measurement of Pyridine Nucleotides in Tissue

The measurement of pyridine nucleotides in tissues *(7)* has presented special difficulties in preparing the material for analysis without loss owing to enzymatic action, and without oxidation of NADH and NADPH. It seems worthwhile to present the preparation in detail in order to illustrate the kinds of unexpected problems that may arise and to emphasize the point made earlier that the preliminary preparation may be more crucial than the assay itself.

From the stability properties of oxidized and reduced pyridine nucleotides (Chapter 1), it would be supposed than an acid extract would be suitable for measuring NAD^+ and $NADP^+$, and that a heated alkaline homogenate or extract would be suitable for measuring NADH and NADPH. Unfortunately, if any blood (hemoglobin) is present during heating in alkali, there may be some oxidation of reduced nucleotides leading to low results. Conversely, during the preparation of acid homogenates, there is a special danger of oxidation, which leads to erroneously high values for the oxidized forms. This second phenomenon is again attributable to hemoglobin, which if present can stoichiometrically oxidize the reduced nucleotides before they are completely destroyed.

These problems are solved by making a single homogenate in cold dilute alkali containing cysteine. One portion is acidified after adding ascorbic acid, which protects NADH and NADPH fom oxidation by acid decomposition products of hemoglobin. This acid portion is then heated at a pH and temperature chosen to give maximal destruction of the very acid-stable enzyme NADase and minimal destruction of NAD^+ and $NADP^+$. Acidification of the alkaline homogenate must be done promptly, before there is time for significant destruction of NAD^+ or $NADP^+$.

Another portion of the aLcaline homogenate is heated to destroy NAD^+ and $NADP^+$. The cysteine present serves to protect NADH and NADPH from oxidation in the process. The preparation of the tissue is facilitated by the fact that the actual assays arc made by enzymatic cycling and therefore the tissue can be highly diluted.

Since the cycling assay (Chapter 5) does not distinguish between oxidized and reduced nucleotides, lolal NAD and total NADP can be measured directly on a third portion of the homogenate, which is not heated and which is analyzed promptly. (It is necessary to measure NADP with NADP Cycle II, because cycle I reacts to a small degree with NAD as well; *see* Chapter 5). Because of the extreme dilution during cycling, NADase need not interfere even though it is not destroyed. Liver is used as the specific example, but the same procedure can be applied to other tissues, providing the final tissue dilution in the cycling reagent is adjusted to give the proper concentration.

Rats are lightly anesthetized with etha, the abdominal cavity opened, and a piece of liver removed and quick-frozen in Freon 12(CCl_2F_2) chilled to its freezing point (–150°) in liquid nitrogen. A 50-mg sample (weighed in a room or cryostate maintained at –20°) is homogenized rapid at 0° in 5 mL of 0.04*M* NaOH containing 0.5 m*M* cysteine (NaOH-cysteine). The homogenate is treated subsequently as outlined in Table 1. When the levels are much lower than in liver, the original homogenates can be made al dilutions of 1:50 or 1:20 instead of 1:100, and in the case of NADH and NADPH, the subsequent dilution in

Table 1
Preparation of Liver for Analysis of Nicotinamide Adenine Dinucleotide and Nicotinamide Adenine Dinucleotide Phosphate[a]

Nucleotide measured	Volume of standard or sample, µL	NaOH-Cysteine[b], mL	Ascorbic Acid, 1.2*M*, µL	0.02 H2SO$_4$ 0.1*M* Na$_2$SO$_4$, mL	Min at at 60°	Aliquot for cycling, µL
Total NADP	200	4	–	–	–	2
NADPH	200	4	–	–	10	2
NADP$^+$	200	–	5	2	30	4
Total NAD	200	4	–	–	–	2
NADH	200	1	–	–	10	4
NAD$^+$	200	–	5	2	30	1

[a]The homogenates are prepared as described in the text. For total NAD (NADP), the samples are kept at 0° and added to the appropriate cycling reagent within 30 min. Standard cycling procedures in 100 µL of the appropriate cycling reagent are used as described in Chapter 5. Standards are provided by adding stock pyridine nucleotide solution to NaOH-cysteine at 0° to give concentrations of 5 and 10 µ*M* for NAD$^+$, 0.5 and 1 µ*M* for NADH, 0.5 and 1 µ*M* for NADP$^+$, and 3 and 6 µ*M* for NADPH. The standards are prepared at the same time as the homogenates and treated as much alike as possible.

[b]Added to the 200-µ*M* sample before heating or taking the aliquot for cycling.

[c]Added to the 200-µ*M* aliquot before heating.

NaOH-cysteine can be reduced as well. However, when hemoglobin concentrations are very high, the dilution at this step should not be reduced too far; otherwise, some oxidation of NADH and NADPH may occur on heating. For example, with whole blood, the dilution at this step must be at least 1:200. For NAD^+ and $NADP^+$ assays, the dilution in H_2SO_4-Na_2SO_4 should probably not be reduced below that shown (eleven-fold), unless the ascorbic acid concentration is increased or hemoglobin is very low. The aliquot used for cycling can be increased to 4 or 5 µL. If it is increased more than this, it is important this be done for standards as well, because the cycling rate may be affected. This is particularly true for $NADP^+$ assays, since sulfate decreases the cycling rate. Obviously, much smaller tissue sample could be used, since only a minute fraction of the sample is taken for cycling.

Part 2

Specific Methods and Procedures

Chapter 5

Enzymatic Cycling

Introduction

Biochemists have an insatiable demand for analytical sensitivity. This is because some of the most interesting biological materials are present in tissues at very low concentrations and some crucial biochemical questions concern amounts of materials in single cells or parts of cells. In many (perhaps most) cases, the needed sensitivity can be provided by methods based on radioactive labeling. There are, nevertheless, other situations in which tracer methods are not sensitive enough, are difficult to apply, or a direct chemical assay would simply take less time. For example, it might be difficult to measure the concentration of glucose-6-P in a microgram of tissue with radioactive methods. With enzymatic cycling, this is no problem.

Enzymatic cycling offers a means for theoretically unlimited amplification of analytical sensitivity for a great many substances of biochemical interest. Accurate measurement in the 10^{-15}–10^{-17} mol range is completely practicable. The most useful cycling methods are those for NAD and NADP, because they can be used to increase the sensitivity of enzymatic methods for a host of other substances, including those presented in this book.

A parallel advantage of the amplification of the analytical signal is that unrelated fluorescence, optical absorption, or turbidity from other tissue components can be completely diluted out.

The cycling principle is simple and may be illustrated by the NADP cycle, which uses glucose 6-phosphate dehydrogenase (G6PDH) and glutamate dehydrogenase (GDH):

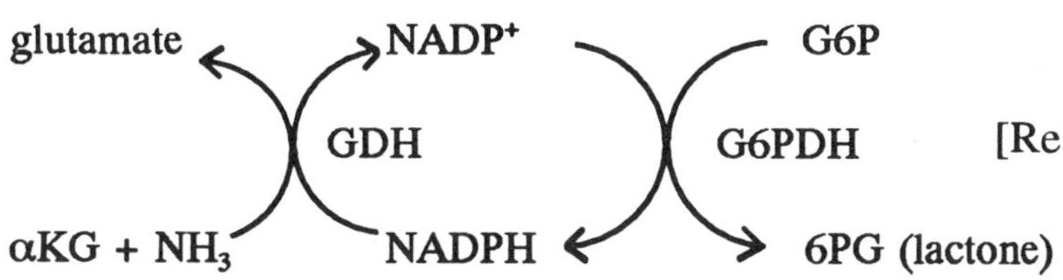

[Re. 5–1]

(the lactone breaks down to 6PG, in part during the cycling step and *in toto* during the heating step.) Abbreviations: αKG, α-ketoglutarate; G6P, glucose 6-phosphate; 6PG, 6-phosphogluconate.

With large amounts of the two enzymes, each molecule of NADP can be alternately oxidized and reduced at least 25,000 times/h. After an hour (or more if needed), the reactions are stopped with heat, after which the 6PG is measured enzymatically (the indicator step):

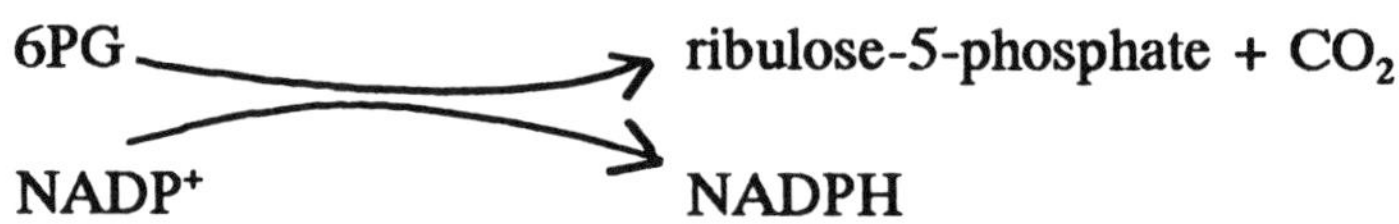

[Re. 5–2]

The NADPH is determined by its absorption or fluorescence. The latter is much to be preferred because of its convenience and greater sensitivity.

With large numbers of samples, the cycling step adds only a minute or two per sample, the reagent and procedure are simple, and precision and linearity are excellent. As long as standards are incubated together with the samples, there is no need to control time or temperature carefully.

To maintain linearity, all that is necessary is to keep the pyridine nucleotide level well below the Michaelis constants of the two enzymes. Cycling rates are adjusted by varying enzyme concentrations. Upper rate limits are determined by the kinetic constants of the enzymes *(see* section entitled Cycling Kinetics, *below).*

As suggested, the real value of the NAD and NADP cycles is to amplify other assays. The sensitivity of measurement of any substance that can be induced to reduce NAD^+ or $NADP^+$, or oxidize NADH or NADPH can be amplified with one of the pyridine nucleotide cycles. Because the cycles do not distinguish between the oxidized and reduced forms of the nucleotides, whichever form was the substrate in the primary assay must be destroyed before cycling the product. This is done with either acid or alkali as described in Chapter 1, and illustrated repeatedly in Chapters 6 and 7.

The enormous sensitivity made possible by enzymatic cycling is not achievable without some increase in attention to analytical details. See the end of this chapter for suggestions on how to obtain very useful sensitivity.

Cycling Kinetics

(It is extremely useful to understand the kinetic basis of enzymatic cycling rates, but it must be emphasized that analytical calculations should be based on actual results obtained with standards carried through the whole procedure and not on predicted results from kinetic constants.) The discussion will be limited to pyridine nucleotide cycles, but the principles are the same for other cycles as well.

For an enzymatic cycle to give results proportional to pyridine nucleotide concentrations, these concentrations must be far below the Michaelis constants for both enzymes. Therefore, the important kinetic parameter is the activity of each enzyme toward low levels of the nucleotide. A measure of this parameter is the apparent first-order rate constant, k, which with "normal" kinetics is equal to V_{max}/K_m. With V_{max} expressed as specific activity (mmol/min/mg, i.e., U/mg), and K_m expressed as micromolar concentration, k becomes the "specific first-order rate constant" (k_{sp}). For example, k_{sp} for G6PDH from *Leuconostoc mesenteroides* is about 75/min. This means that, at a 1 mg/L concentration (1 μg/mL), this enzyme would convert $NADP^+$ to NADPH at an initial rate of 75-fold/min.

With low enzyme levels, the k for each enzyme would be $k_{sp} \times$ mg of enzyme/L. Consider the following cyclic reaction where A and B are the alternating forms of the cycled nucleotide and [A] + [B] = [C], the total nucleotide:

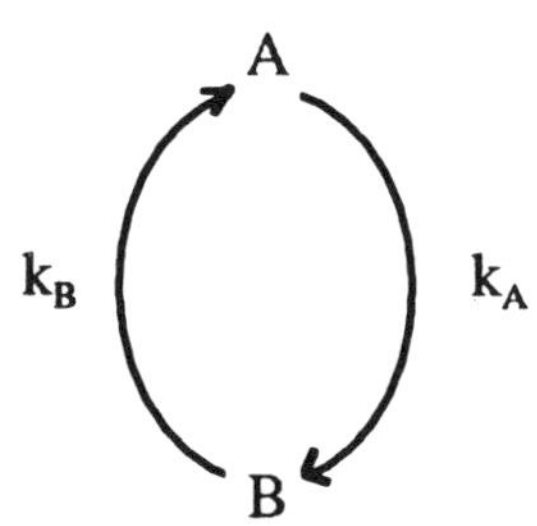

[Re. 5–3]

By definition, during the steady state of the cycling process:

$$k_A[A] = k_B[B] = k_C[C] = k_C([A] + [B])$$

Rearranging: $$k_C = \frac{K_A[A]}{[A] + [B]}$$ [Re. 5–4]

Substituting $[B] = \frac{k_A}{k_B}[A]$: $$k_C = \frac{k_A k_B}{k_A + K_B}$$ [Re. 5–5]

For example, in the NADP cycle, with 5 μg/mL of glucose-6-P dehydrogenase (k_{sp} about 75/min) and 50 μg/mL of glutamate dehydrogenase (k_{sp} about 3/min), $k_A = 5 \times 75 = 375$/min, and $k_B = 50 \times 3 = 150$/min. Therefore, the calculated cycling rate would be $375 \times 150/(375 + 150) = 107$/min.

Observed rates agree well with this formulation at low enzyme levels. However, this formulation applies only to the turnover of free nucleotide. When enzyme concentrations are high enough on a molar basis to be a substantial fraction of their Michaelis constants, they will tie up a significant proportion of the pyridine nucleotide. Consider the case of a cycle in which enzyme A reacts with NAD^+ and enzyme B reacts with NADH. If the molar concentration of active sites of enzyme A approaches its K_m for NAD^+, the total free NAD will fall and the overall cycling rate will be less than calculated above. Cha and Cha *(1)* have analyzed this situation. If enzyme B is held constant, as enzyme A is increased, k_a will reach a maximum (as free NAD^+, but not free NADH, approaches zero), and the cycling rate levels off, but theoretically will not decrease, no matter how high A is raised. Cha and Cha did not consider the more complicated (but usual) situation in which enzyme A binds both forms of the nucleotide. Here, if enzyme B is held constant, as enzyme A is increased, the cycling rate will go through a maximum, and then diminish as enzyme A binds more and more of both NAD^+ and NADH and reduces the amount of free nucleotide to low levels. Thus, a low affinity of each enzyme for its pyridine nucleotide product is desirable for high cycling rates and may be more important than the k_{sp} itself.

NADP Cycle I (*Leuconostoc* G6PDH)

Two cycles are offered for measuring NADP. The first uses G6PDH from *Leuconostoc mesenteroides* as the reducing enzyme. For a number of reasons, this is the best of the two cycles. However, *Leuconostoc* G6PDH also reacts with NAD (at about 10% of the rate with NADP). This is of no disadvantage in the most useful situation (i.e., when the $NADP^+$ or NADPH has been generated in an enzymatic assay for another substance), since only in rare cases will a significant amount of NAD be present. To cover the less common situation when both NAD and NADP may be present, a second cycle is also described, that is completely specific for NADP, because it substitutes yeast G6PDH.

Table 1
Effect of Time, Temperature, and Enzyme Concentrations on Cycling Yield with *Leuconostoc* G6PDH

GDH U/mL	G6PDH U/mL	Temperature, °C	Time, h	Cycles, approx
0.5	0.5	38	1	1000
1.5	1.5	38	1	3000
5	5	38	1	10,000
1.5	1.5	38	3	10,000
5	5	24	1.5	10,000
0.5	0.5	24	2.0	10,000
5	5	15	2.5	10,000
12	25	38	1	30,000
12	25	38	4	100,000
5	5	24	20	100,000
12	25	15	20	200,000

Both cycles use glutamate dehydrogenase as the partner enzyme. The chemical equations for the cycle and for indicator steps were given in the Introduction section.

Cycling Reagent

This consists of imidazole buffer, pH 7.0 (50 m*M* imidazole, 50 m*M* imidazole-HCl), 2 m*M* glucose-6-P, 7.5 m*M* disodium α-ketoglutarate, 0.1 m*M* ADP, 25 m*M* ammonium acetate, and 0.1% bovine serum albumin, plus sufficient *Leuconostoc* glucose-6-P dehydrogenase (G6PDH) and beef liver glutamate dehydrogenase (GDH) to give the required cycling rate (Table 1).

If either free α-ketoglutaric acid or its monosodium salt is substituted for the disodium salt, 15 or 7.5 m*M* NaOH, respectively, should be included so as to maintain the required pH. For the same reason, the sample to be assayed should not contribute more than 10 m*M* strong base to the cycling reagent. If the final measurement is to be made in a spectrophotometer or with small cycling volumes (10 µL or less), the amplification needed to yield enough product to measure accurately might use up most of the glucose 6-P. In this case, the glucose 6-P is increased to 5 m*M*.

All components except the enzymes can be stored as a 10– or 20-fold concentrated stock solution for at least several months at –40° or below and for several weeks at –20°. The most unstable component is α-ketoglutarate. The enzymes are added just before use, and the reagent is then kept on ice to minimize the reagent blank from any trace of NADP that may be present.

If the GDH should contain appreciable NADP, this can be removed with charcoal *(see* NAD Cycle). The *Leuconostoc* G6PDH preparations have not contained appreciable amounts of NADP, but if there should be any present, it could probably be removed the same way. If either enzyme is provided as a suspension in ammonium sulfate, most of this is removed by centrifugation before use; final sulfate concentration should not exceed 5 m*M*.

Indicator Reagent

This consists of imidazole buffer, pH 7.0 (50 m*M* imidazole base; 50 m*M* imidazole-HCl), with 30 m*M* ammonium acetate, 2 m*M* $MgCl_2$, 0.1 m*M* EDTA, 100 μ*M* $NADP^+$, and 0.1 U/mL of Torula yeast 6-P-gluconate dehydrogenase (6PGDH). For the spectrophotometer, $NADP^+$ is increased to 300 μ*M* and the 6PGDH to 0.5 U/mL. It is convenient to store a 20-fold concentrated stock solution at –40° or below with all components present, except the enzyme. This is probably stable indefinitely.

Effect of Time, Temperature, and Enzyme Activity (Table 1)

The cycling yield will vary with each of these three variables, but none of them needs to be exactly controlled, provided only that standards and blanks are included with each set of samples. Thus, high precision can be obtained with cycling overnight at room temperature, even if meanwhile room temperature should vary markedly. (Standards should be made up in a blank solution simulating the samples as closely as possible in composition and volume.)

As illustrated in Table 1, cycling rates are nearly proportional to the concentration of the enzymes if these are varied together and do not exceed about 10 U/mL. The rates at 24, 15, and 4° are about 65, 40, and 20%, respectively, of those at 38°. The rates fall off with time, primarily because of inherent instability of $NADP^+$ and NADPH, particularly the latter. The fall off is roughly 1, 3, 6, and 12%/ h at 4, 15, 24, and 38°, respectively. Therefore, for very long incubations, temperatures well below 38° are necessary (Table 1).

Table 2
Useful NADP Cycling Conditions.
A. Fluorometer
Cycling Step in 0.1 mL, Indicator Step in 1 mL

Total cycles	NADP quantity, pmol	NADP cycling step concentration, n*M*	NADPH indicator step concentration, μ*M*
1000	1–10	10–100	1–10
3000	0.3–3	3–30	0.9–9
10,000	0.1–1	1–10	1–10

Sample Fluorometric Procedure with Cycling in 100 μL (About 10,000 Cycles; 0.1–1 pmol)

Cycling reagent is prepared with 5 U/mL of GDH and 5 U/mL G6PDH added just before use; *see* precautionary note above. Reagent volumes of 100 μL are placed in 10 × 75 m*M* fluorometer tubes in an ice bath. Samples and standards are added in volumes of 20 μL or less. For highest precision, all volumes are kept the same within 2 μL. The rack of tubes is incubated 1 h in a water bath at 38° and then heated 3 min in another bath at 95–100°. After bringing to room temperature, 1 mL of the indicator reagent is added to each tube, and the fluorescence is read when all the 6-P-gluconate has reacted (10–30 min). Table 2 gives examples of $NADP^+$ and NADPH quantities and concentrations at successive steps for three typical cycling protocols.

Danger of Incomplete Heat Inactivation of G6PDH After Cycling

One precaution is essential because of the unusual stability of *Leuconostoc* G6PDH. If the pH of the cycling reagent is higher than 7.2, this enzyme will not be completely inactivated by heating at 95°, and in fact will slowly reactivate further after cooling and during the indicator step *(2)*. Because the indicator reagent contains a large excess of $NADP^+$ and the cycling reagent always has an excess of glucose-6-P, a progressive increase in NADPH at the indicator step will occur. This is easily detected by an upward drift in fluorescence continuing long after the 6PGDH reaction is over. Because of this danger, care should be taken that samples to be analyzed do not contribute enough NaOH or alkaline buffer to shift the pH into the danger zone. If this pH control becomes a problem, the G6PDH can always be completely and permanently killed by adding an excess of NaOH before heating, as is done in any event after cycling small volumes in an oil well *(see* section on Sample Fluorometer Procedure).

Useful Cycling Conditions

The most useful range for cycling in 0.1-mL vol with a 1 mL fluorometric step is shown in Table 2. This is based on the provision of at least a 1-n*M* NADP concentration at the cycling step (i.e., 0.1 pmol) and no more than a 10-μ*M* concentration of NADPH at the indicator step (i.e., the approximate upper limit for a linear relationship between NADPH concentration and fluorometer response) (Chapter 1). At the other extreme, 100 n*M* NADP is approaching the upper limit for cycle linearity between NADP and 6-P-gluconate yield, although up to 200 n*M* is still satisfactory when necessary. Similarly, Table 2 conditions are designed with 1 μ*M* NADPH as the lower limit for measurement at the indicator step, whereas with care, accurate readings can be made down to 0.2 μ*M* (Chapter 1). Table 2 does not specify the time and temperature conditions to achieve the cycles required. Table 1 illustrates the flexibility in this regard. For most purposes, an hour at 38° is most convenient. However, one way to reduce a cycling blank owing to moderate contamination of the enzymes with the NADP is to reduce the enzyme levels and provide a corresponding increase in time (Table 1).

Spectrophotometric Procedures

Almost the same reagents and incubation conditions are satisfactory if a spectrophotometer is used instead of a fluorometer for the indicator step. Because higher final NADPH levels are required for similar precision, more 6-P-gluconate must be produced in the cycle. Therefore, the concentration of G6P in the cycling reagent is raised from 2 to 5 m*M*, $NADP^+$ in the indicator reagent is increased to 300 μ*M*, and the 6PGDH to 0.5 U/mL. The cycling and indicator reaction steps are carried out in 3–mL tubes as described for the fluorometer, and then transferred to the spectrometer cells after time enough for the 6PGDH reaction to be complete. With higher levels of the cycling enzymes, turbidity from heat denaturization may require removal by centrifugation before transfer. Table 3 indicates the most useful range of cycling conditions for spectrophotometer based assays. These are designed for final readings in the 0.09–0.9 OD range.

Table 3
Useful NADP Cycling Conditions.
B. Spectrophotometer
Cycling Step 0.1 mL, Indicator Step in 1 mL

Total cycles	NADP quantity, pmol	NADP cycling step concentration, n*M*	NADPH indicator step concentration, μ*M*
3000	5–20	50–200	15–60
10,000	1.5–15	15–150	15–150
30,000	0.5–5	5–50	15–150
100,000	0.15-1.5	1.5–15	15–150

Extending Cycling Sensitivity by Reducing the Cycling Volume

Table 2 indicates a lower limit for sample size of 0.1 pmol in a cycling volume of 100 μL if 1 n*M* NADP is regarded as a practical lower concentration limit during cycling. To decrease the lower limit below 0.1 pmol and still keep the cycling blank of manageable size, it is necessary to decrease the cycling volume. By working in a Teflon rack under oil (Chapter 12), it is quite feasible to carry out a pyridine nucleotide analytical step, destroy the excess substrate nucleotide with acid or alkali, and amplify the product nucleotide in a cycling volume of 2 μL or less. The following is an example of cycling in 10 μL of reagent.

Sample Fluorometric Procedure with Cycling in 10 μL (100,000 Cycles, 0.01–0.1 pmol of NADP)

The sample itself might be generated directly in the oil well beforehand or added to the oil well from a reaction carried out in another vessel, and is assumed to be no more than 2 μL in volume. With the samples already in the oil wells and the oil rack at room temperature, 10 μL of the cycling reagent are added at timed intervals (e.g., every 10 s), and cycling is stopped in the same sequence and at the same time intervals by adding 2 μL of 0.1*M* NaOH. The rack of samples is then heated 20 min at 80° in an oven, after which a 10-μL aliquot is transferred from each well into 1 mL of indicator reagent in a fluorometer tube.

Any combination of enzyme concentrations, time, and temperature (Table 1) to provide 100,000 cycles may be used to suit analytical convenience. The reaction is stopped with alkali and heat rather than heat alone, because heating to 95° at neutral pH causes coagulation of the enzymes and would make pipeting difficult. In addition, this makes

Table 4
Useful NADP Cycling Conditions.
C. Fluorometer
Cycling Step in 10 µL, Indicator Step in 1 mL

Total cycles	NADP quantity, pmol	NADP cycling step concentration, nM^a	NADPH indicator step concentration, µM
10,000	0.1–1	10–100	0.8–8
30,000	0.04–0.4	4–40	1–10
100,000	0.01–0.1	1–10	0.8–8

[a]The cycling step is carried out in an oil well with an aliquot transferred to a fluorometer tube for the indicator reaction.

Table 5
Useful NADP Cycling Conditions.
D. Fluorometer
Cycling Step in 2 µL, Indicator Step in 1 mL

Total cycles	NADP quantity, pmol	NADP cycling step concentration, nM^a	NADPH indicator step concentration, µM
10,000	0.1–0.4	50–200	0.8–3
30,000	0.03–0.3	15–150	0.8–8
100,000	0.01–0.05	5–25	0.8–4
200,000	0.005–0.03	2–12	0.7–4

[a]The cycling step is carried out in an oil well with an aliquot transferred to a fluorometer tube for the indicator reaction.

the pH less critical since, as stated above, heating in alkali removes any danger of incomplete destruction of G6PDH. Tables 4 and 5 suggest the useful limits of cycling in 10 and 2 µL, respectively.

Double Cycling

Sensitivity can be increased even further by double cycling. For example, after what would have been the indicator (6PGDH) step of the NADP cycle, the excess $NADP^+$ is destroyed with alkali and heat, and the NADPH formed in this 6PGDH step is then recycled. If the first and second amplifications are 1000- and 10,000-fold, respectively, the overall amplification would be 10,000,000. To minimize the blank, the cycling enzyme levels and therefore cycling rates are kept lower at the first step, during which the NADP levels will also be very low, but the enzymes can be increased for the second cycle, during which much higher NADP levels will be present. The following example illustrates this point.

Sample Double-Cycling Protocol (0.2–2 fmol) of NADP, 10,000,000 Cycles

First Cycle

Samples no larger than 0.5 µL in vol are placed in oil wells at room temperature and 2-µL vol of cycling reagent are added at convenient time intervals. After cycling for 2 h at room temperature, the reactions are stopped with 1 µL of 0.15*M* NaOH added at the same intervals, followed by heating the oil well rack for 20 min at 80°. The cycling reagent contains only 0.3 U/mL of each enzyme plus sufficient NaOH or HCl to compensate for any acid or base in the sample that would shift the final cycling pH beyond acceptable limits. The low enzyme levels make it easier to keep the cycling blank in the 0.1 n*M* NADP range and still give about 1000 cycles. Initial NADP at 0.1–1 n*M* concentration (0.2–2 fmol) will yield 6-P-gluconate at 100–1000 n*M* concentration, and about 15–150 n*M* NADPH at the first 6PGDH step.

6PGDH Step 1

Ten microliters of "indicator" reagent are added to each oil well and the reaction then stopped (after allowing generous time for completion) with 1 µL of 1.5*M* NaOH followed by heating the oil well rack for 20 min at 80°. Because the 6-P-gluconate at this step will be much less than 1 µ*M*, the $NADP^+$ in the 6PGDH reagent is reduced to 10 µ*M*. This lessens the danger that a significant amount of $NADP^+$ will escape destruction.

From this point on, the procedure is no different from that of ordinary single cycling. A 10-µL aliquot from each oil well is added to 100 µL of cycling reagent with cycling enzyme levels to provide 10,000 cycles. The subsequent treatment is identical to that described for measuring 0.1–1 pmol of NADP (in this case, NADPH).

An example of the concentrations expected at each step in the above protocol may be useful, if the starting sample contains 1 fmol of NADP. (Account is taken for losses from portions left behind at two of the steps):

1. Original sample (0.1 µL) 10 n*M* NADP;
2. First cycle (2.1 µL) 0.48 n*M* NADP, yields 480 n*M* 6PG (1000-fold);
3. First 6PGDH step (12.3 µL) 81 n*M* NADPH;
4. Second cycle (110 µL) 7.4 n*M* NADPH, yields 74 µ*M* 6PG (10,000-fold);
5. Final indicator step (1.1 mL) 7.3 µ*M*.

Table 6
Effect of Time, Temperature, and Enzyme Concentrations on NADP Cycling Yields with Baker's Yeast G6PDH

GDH, U/mL	G6PDH, U/mL	Temperature, °C	Time, h	Cycles, approx
0.5	0.3	25°	1	500
2	1.5	25°	1	2000
5	3	25°	1	4000
10	6	25°	1	8000
0.5	0.3	38°	1	1000
2	1.5	38°	1	4000
5	3	38°	1	8000
10	6	38°	1	15,000
5	3	25°	4	10,000
2	1.5	38°	4	10,000

NADP Cycle II (Baker's Yeast G6PDH)

In contrast to G6PDH from *Leuconostoc mesenteroides,* Baker's yeast enzyme appears to be completely unreactive with NAD *(3)*. It is therefore substituted when a significant amount of NAD is present. The only disadvantage compared to *Leuconostoc* G6PDH is that the maximum cycling rate is lower (about 15,000/h) and prolonging the time beyond an hour is only moderately useful because of a marked fall off in rate with time, owing to slow destruction of NADPH by the G6PDH.

The cycling reagent is the same as for NADP cycle I, except for change from imidazole to Tris buffer, pH 7.4 (80 m*M* Tris-HCl: 20 m*M* Tris base). This is an improvement *(4)* over pH 8.0 as originally proposed *(3)*. Imidazole buffer causes erratic G6PDH losses. The indicator reagent is the same as for NADP cycle I. Relationships between cycling rates, enzyme levels, and temperature are given in Table 6. The cycling rate at 0°C is about 5% of that at 38°.

NAD Cycle

This cycle devised by Kato *(5)*, is much superior to the original NAD cycle *(3)* given in the first edition of this book. The cycle enzymes are alcohol dehydrogenase (ADH) and malate dehydrogenase (MDH):

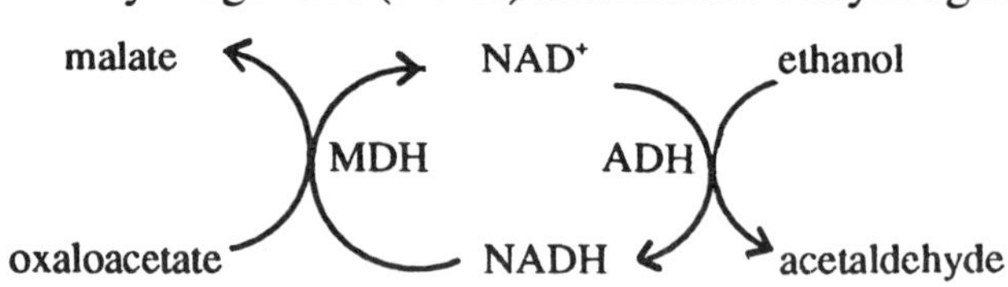

[Re. 5-6]

The indicator step measures the malate generated, after first destroying the oxaloacetate of the cycle reagent with heat. The indicator reaction is pulled to completion with aspartate aminotransferase (AAT):

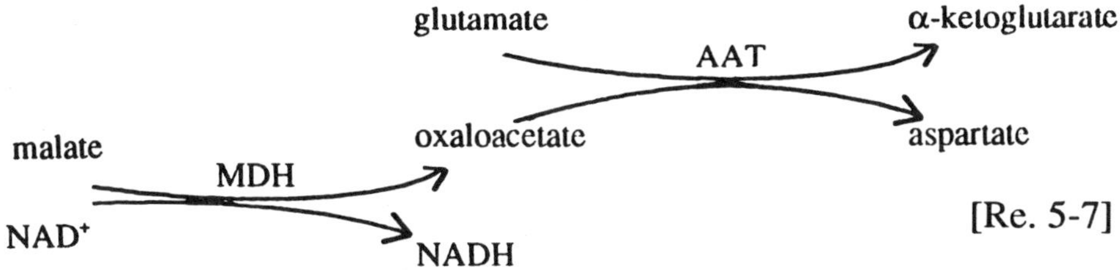

[Re. 5-7]

Cycling Reagent

This consists of 100 m*M* Tris-HCl, pH 8.0 (1:1 base acid), 300 m*M* ethanol, 2 m*M* oxaloacetate, 2 m*M* mercaptoethanol, and 0.02% (w/v) bovine serum albumin, plus sufficient baker's yeast, ADH, and beef heart MDH to give the desired cycling rate. The reagent, complete except for oxaloacetate and enzymes, can be prepared and stored at –70° or below. One molar oxaloacetic acid in 0.5*M* HCl is stored separately at –70°C. Oxaloacetate is unstable at neutral pH at room temperature (loss in the cycling reagent is 4, 2, and 0.4%/ h at 25, 15, and 0.5°, respectively). The enzymes and oxaloacetic acid are added to the reagent within a few minutes of use. $(NH_4)_2SO_4$ present in enzyme suspensions is largely eliminated by centrifugation, and the precipitated enzymes are dissolved in 20 m*M* Tris-HCl (pH 8.0) before addition. Horse liver ADH is not a satisfactory substitute for the yeast enzyme because of its very low activity.

Removal of NAD from Enzymes

Preparations of ADH and MDH have sometimes been found to contain significant amounts of NAD. This can be corrected by charcoal treatment as follows. A 2% suspension of Norite is prepared in 20 m*M* phosphate buffer (pH 7). Into a centrifuge tube is placed 1 mL of this suspension for each 3–5 mg of protein to be treated. The Norite is centrifuged; the supernatant buffer is discarded and replaced by the enzyme solution, which has been diluted to a concentration of 3–5 mg/ mL with the 20 m*M* phosphate buffer. (It will usually be necessary to first remove by centrifugation some of the $[NH_4]_2SO_4$ solution in which the enzyme has originally been suspended.) The charcoal suspension is incubated with occasional mixing for 30 min at 38°, centrifuged, and the enzyme precipitated from the supernatant fluid at 0–4° by addition of solid $(NH_4)_2SO_4$. The final $(NH_4)_2SO_4$ concentration (including any of the salt that may have been carried over from the original suspension) needs to be 2.5–3*M* in the case of ADH and 3.3–3.6*M* in the case of MDH. The precipitated enzyme is resuspended for storage in either 2.5*M* $(NH_4)_2SO_4$ (ADH) or 3–4*M* $(NH_4)_2SO_4$ (MDH). There may be some loss of enzyme during this treatment. In specific cases, charcoal treatment has reduced NAD contamination of ADH from 4 pmol/mg to 1 pmol/mg, and that of MDH from 36 pmol/mg to 1.6 pmol/mg.

Indicator Reagent

This consists of 50 m*M* 2-NH_2-2-methylpropanol-HCl buffer, pH 9.9 (1:1, base:acid), 200 µ*M* NAD^+, 10 m*M* glutamate, 4 U/mL pig or beef heart MDH, and 1 U/mL beef liver aspartate aminotransferase. The reaction time at 25°C is 3–6 min. The fluorescence blank need not exceed the equivalent of 0.6 µ*M* NADH.

Sample Cycling Procedure (0.2–2 pmol of NAD^+ or NADH)

Reagent is prepared with 15 U/mL of ADH and 15 U/mL of MDH (added at the last minute to avoid increase in the cycling blank). The sample (10 µL or less) is added to 100 µL of cycling reagent in a 3 mL fluorometer tube (10 × 75 m*M*) in an ice bath. (Samples, blanks, and standards should have the same vol ± 2 µL.) Because the cycling rate is about a third as fast at 0° as at 25° *(see below)*, all samples should be added within 5 min of each other. Standards containing 0.2–2 pmol (2 µL of 0.1–1 µ*M* NAD) may be used, according to the expected range of samples.

All tubes are incubated together at 25° for 1 h, then heated 3 min at 95–100°. After cooling to room temperature, 1 mL of indicator reagent is added, and readings made after the reaction is complete. The expected cycling rate is about 5000/h. Therefore, the final amount of NADH will be 1–10 nmol giving a concentration of 1–10 µ*M* in the fluorometer. (Concentrations >10 µ*M* exceed the linear fluorescence range for NADH; Chapter 1.)

Modifications of Cycling Procedure

The cycling rate can be adjusted upward or downward according to need, by increase or decrease in the enzyme concentrations (Table 7). Because of the temperature characteristics of ADH activity *(see Effect of pH and Temperature)*, the cycling rate is fastest at 25–30°. The total

Table 7
Effect of Time, Temperature, and Enzyme Concentrations on NAD Cycling Yields

ADH, U/mL	MDH, U/mL	Temperature, °C	Time, h	Cycles, approx
2.5	2.5	25°	1	1000
5	5	25°	1	2000
20	20	25°	1	8000
50	50	25°	1	20,000
100	100	25°	1	40,000
15	15	25°	4	25,000
50	50	25°	4	70,000
100	100	25°	4	150,000
15	15	15°	24	90,000
50	50	15°	24	250,000
50	50	2°	24	150,000

amplification (cycling yield) increases with time, but is not proportional with time for more than 4 h at 25°, partly because of the instability of oxaloacetate. Therefore, for greatest amplification, the cycling temperature is decreased as suggested in Table 7. Note that ratio of MDH to ADH is increased at 4°C, because of the larger temperature coefficient of MDH.

Even with charcoal treatment *(see above)*, there may be some residual NAD in both enzymes. If this presents a problem, it is better to reduce the enzyme levels and increase the incubation time. Table 7 illustrates the flexibility in this regard.

Another way to reduce the cycling blank is to reduce the cycling volume. Cycling in volumes less than 50 µL can be carried out in smaller tubes or oil wells, with transfer of an aliquot into 1 mL of reagent for final reading, as illustrated by the following example.

Measurement of 10–100 fmol of NAD with Cycling Volumes of 5 µL

Reagent

This is the standard reagent described above with 25 U/mL of ADH and 25 U/mL of MDH.

Conduct of the Assay

Samples in volumes of 0.25 µL or less are placed in oil wells under oil, and 5 µL of cycling reagent are added sequentially to each well over a period of 15 min or less. The oil well rack is then incubated in a cold room or ice box at 15° for 24 h to give an estimated 150,000 cycles. The reaction is stopped by adding, in the same sequence, 5 µL of 0.1*M* NaOH. The oil well rack is then heated for 15 min in an oven at 80°C; 8-µL aliquots are transferred into 1-mL vol of indicator reagent for the final reading.

Kinetics of MDH and ADH

The kinetic constants for MDH are much more favorable than for ADH. The K_ms for NADH and oxaloacetate are 5 and 3 µ*M*, respectively, under cycling conditions, whereas in the case of ADH, the K_ms for NAD^+ and ethanol are 150 µ*M* and 30 m*M*, respectively. In addition, the V_{max} of MDH (U/mg) is tenfold greater than that of ADH. As in the NADP cycle, the pyridine nucleotide concentrations will always be far below their respective K_ms and cycling rates will be governed by their respective first-order rate constants, k (i.e., V_{max}/K_m if V_{max} is given as µmol/min/L (i.e., U/L) and K_m as µ*M* (i.e., µmol/L), k will have the dimension min^{-1} (i.e., the turnover/min). In a typical case with 3 U/mL of ADH (3000 U/L) and 3 U/mL of MDH (3000 U/L), the calculated k values would be 3000/150 = 20/min for ADH and 3000/5 = 600/min for MDH. The calculated cycling rate, from $k = (k_1 \times k_2)/(k_1 + k_2)$ (*see* section entitled Cycling Kinetics *above*) would be $(20 \times 600)/(20 + 600) = 19$/min or 1160/h.

Note that equal unit concentrations of the two enzymes are given for this example and for Table 7. This is in spite of the fact that the k value/unit of enzyme activity is 30 times higher for MDH than for ADH. The reason is to take full advantage of the opportunity to make ADH activity the limiting factor.

Effect of pH and Temperature on the NAD Cycle Kinetics

The apparent first-order rate constant for NADH with MDH increases moderately (60%) from pH 7.5 to 9. The rate for NAD^+ with ADH goes through a maximum at pH 8–8.5. Because ADH is the rate-limiting enzyme, and because stability decreases at a more alkaline pH, a pH of 8.0 was selected for the cycling reagent.

At pH 8.0, the first-order rate constant for NADH with MDH increases an average of 7%/degree from 0° to 40°C. In contrast, ADH increases only 2.6%/degree from 0° to 25°C and decreases above 30°C.

From these data, it is clear that the optimal temperature and the overall temperature coefficient will depend on the relative amounts of the two enzymes. With the recommended ratio for the two enzymes, ADH is 95% rate limiting. Therefore, maximal cycling rates are obtained at 25°–30°C, and the rate at 0°C is roughly a third of that at 25°C.

Maximum Cycling Rates

As the above indicates, the cycling rate increases with ADH concentration. This increase is almost linear up to 50 U/mL, but at higher concentrations (tested up to 1000 U/mL), the rate levels off at about 1000/min (60,000/h). Half-maximal values of the k for ADH are obtained with about 100 U/mL, which corresponds to approx 5 μ*M* ADH. This leveling off phenomenon is most easily explained by binding of NADH to the enzyme, which would lower the amount of free nucleotide available for cycling. In keeping with this explanation is the observation that very high levels of ADH decrease the apparent first-order rate constant for MDH. For example, the rate constant for 40 U/mL of malate dehydrogenase was 3500/min with 140 U/mL of ADH, but only 1000/min with 600 U/mL of ADH.

Achieving High Useful Sensitivity (the Blank Problem)

Although there is no theoretical limit to the amplification by enzymatic cycling, there are practical limits in regard to measuring NADP and NAD *per se*, and, more important, in regard to application of a cycle to measure other things. These limits are usually a function of the blank.

Blanks from the Pyridine Nucleotide Itself

Often the most significant component of the overall cycling blank is the pyridine nucleotide itself (present as a contaminant). This is usually contributed by the enzymes, but may easily result from adventitious pyridine nucleotide contamination of other components of the cycling reagent. This may easily happen because of the extremely low concentrations that can be measured. Most of the pyridine nucleotide can be removed from the enzymes with charcoal. However, some usually remains, and there is, therefore, finally a trade off between increasing amplification with more enzyme and keeping the blank low. In our experience, when using high enzyme levels, it is difficult to keep the cycling blank for either NAD or NADP much below 0.5 n*M* concentration. A sample that adds another 0.5 n*M* might be taken as the practical lower limit *(see below)*. Obviously, whatever the concentration limit, the smaller the cycling volume, the smaller the amount of NAD or NADP that can be measured: A blank concentration of 0.5 n*M* corresponds to 0.5 pmol in 1 mL, 25 fmol in 50 µL (a convenient vol for final fluorometric reading in 1 mL), and 1 fmol in 2 µL.

As the volume and, therefore, absolute amount of pyridine nucleotide is decreased, the amplification must be increased to obtain a satisfactory final reading at the indicator step. With a 0.1-mL cycling vol, there is no increase in useful sensitivity in cycling more than 20,000-fold. With a 10 µL-vol, 200,000 cycles would be required at the above arbitrary lower useful sensitivity limit.

With double cycling, still smaller volumes for the first cycling step can be used to advantage. Suppose the problem is to measure 2 $\times 10^{-17}$ mol (0.02 fmol) of NADP. Here there is great gain in keeping the cycling blank to a minimum at the first cycling step. As indicated above, this is easier when low enzyme levels are used. If the enzymes are reduced to give a cycling rate of only 1000/h, the NADP contamination can probably be reduced with care to 0.1 n*M* (equivalent to 0.02 fmol in 0.2 µL), whereas with a 0.5 n*M* blank, the volume would have to be reduced to about 0.05 µL for a successful assay. As illustrated above, after the 1000-fold amplification, the second cycle can be conducted in 5 or 10 µL (NADP concentration 2–4 n*M*) with no particular blank problem, even at very high cycling rates.

Locating a Cycling Blank

If contamination of the cycling reagent with pyridine nucleotide (PN) proves to be the problem, and one or both enzymes are suspected, the following protocol will locate the offender(s). Prepare three reagents, (a), (b), and (c), in small amounts from a single stock with enzymes A and B in the proportions (relative to their usual proportions) of (a) 3:1, (b) 1:3, and (c) 1:1. With each reagent run three tubes:

1. A reagent blank incubated 60 min = R_1;
2. A standard incubated 60 min that will give a reasonable signal even with Reagent (c) = R_2; and
3. A reagent blank that is stopped in the usual manner, but without incubation = R_3.

The true PN equivalence of the cycling blank, for each of the three reagents will be given by readings $(R_1–R_3)/(R_2–R_1)$ multiplied by the nucleotide content of the standard equals the PN content of each reagent. Subtracting the PN in Reagent (c) from the PN in Reagents A and B gives the PN in 2× amounts of enzyme A and B, respectively. This procedure permits distinction between changes in blank readings owing to changes in cycling rates from changes owing to adding more or less of the nucleotide concerned.

Blanks Not Owing to the Nucleotide Cycle

When a pyridine nucleotide cycle is to be used to amplify the sensitivity of other assays, the blank problem may be more difficult. This is not because of contributions of fluorescence or absorption from the original sample. (As already pointed out, one great advantage of high amplification is that such contributions are reduced to insignificance relative to the final signal.) What has to be avoided is anything in the precycling reagents that would result in an NAD or NADP residue in the blank. The possible causes are too numerous to mention, but in general, it is difficult to reduce this blank to <0.01–1 μ*M* (calculated at the precycling enzymatic step), and for an assay of such ubiquitous substances as lactate or P_i, a 1-μ*M* blank would be a success. In any event, the problem is always reduced (as in the case of the cycling blank) by reduction in volume. The fact that the cycling blank can usually be kept much lower than the precycling reaction blank means that the sample can be diluted substantially after the precycling step without adding significantly to the overall blank.

ATP-ADP Cycle (6)

This is a cycle for measuring the sum of ATP plus ADP. By adding a preliminary step, it can also include AMP to give an amplification assay for total adenylate.

The cycle enzymes are hexokinase (HK) and pyruvate kinase (PK):

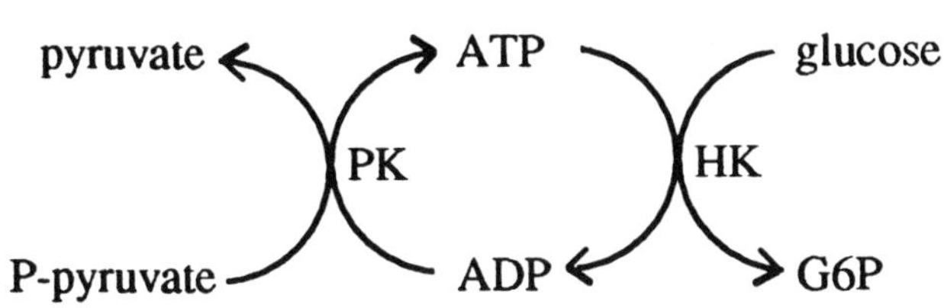

[Re. 5-8]

The indicator reaction is:

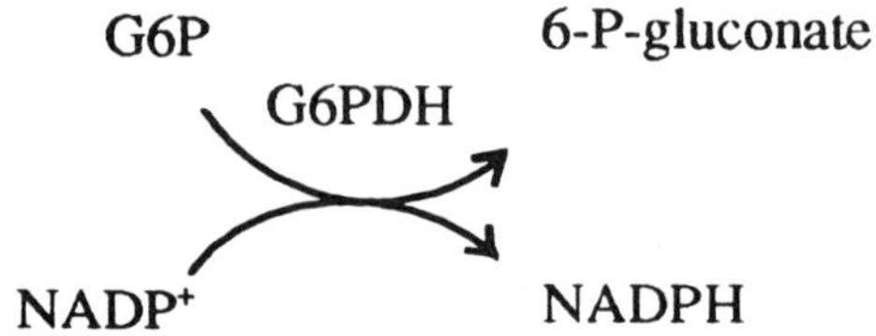

[Re. 5-9]

To measure total adenylate, hexokinase is replaced by adenylokinase (AK) and incubated until the following reaction goes to completion:

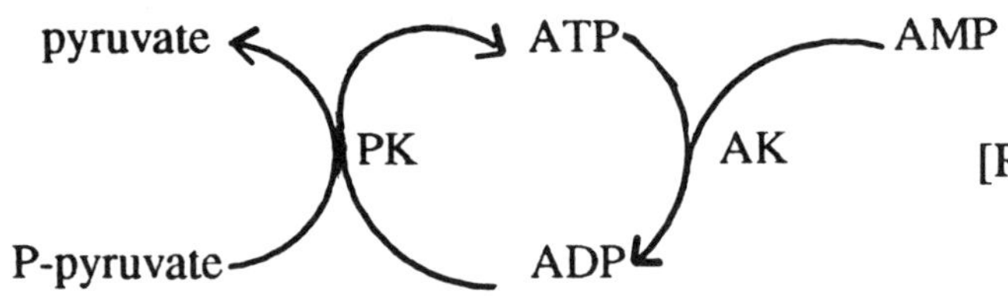

[Re. 5-10]

Hexokinase is then added to start the cycling reaction.

Cycling Reagent

This consists of imidazole-acetate buffer, pH 7.4 (15 m*M* imidazole-acetate, 35 m*M* free base), P-pyruvate, 200 μ*M*; glucose, 1 m*M*; $MgCl_2$, 2 m*M*; potassium acetate, 50 m*M*; and bovine plasma albumin, 0.02%; plus rabbit muscle pyruvate kinase and yeast hexokinase, according to the amplification needed *(see below)*.

Comment

Chloride is kept low because it lowers the V_{max} and raises the ATP K_m for hexokinase and to a lesser degree for pyruvate kinase as well *(see also below)*. The reagent can be prepared in bulk without the enzymes and stored at –50° or lower.

Indicator Reagent

This consists of Tris-HCl, pH 8.1 (25 m*M* Tris base, 25 m*M* Tris-HCl), $NADP^+$, 100 μ*M*, yeast or *Leuconostoc mesenteroides*, glucose-6-P dehydrogenase, 0.05 U/mL, and EDTA, 2 m*M*. The enzyme should be added shortly before use and not frozen in the reagent.

AMP to ATP Converting Reagent

This is the same as the cycling reagent except for the inclusion of 0.02 μ*M* ATP to start the reaction, and substitution of 25 U/mL of rabbit muscle adenylate kinase in place of hexokinase. The amount of pyru-

Table 8
Effect of Time, Temperature, and Enzyme Concentrations on Cycling Yield with the ATP-ADP Cycle

Hexokinase, U/mL	Pyruvate kinase, U/mL	Temp, °C	Time, h	Cycles, approx
2	2	25°	1	500
4	4	25°	1	1000
10	10	25°	1	2000
25	25	25°	1	5000
5	5	25°	4	5000
10	10	25°	4	10,000
1	1	15°	24	3000
3	3	15°	24	10,000
10	10	15°	24	20,000

vate kinase is much greater than necessary for this conversion, and the glucose is not needed, but neither do they cause harm, and having them present prepares for the cycling step when they both will be needed. It is, of course, necessary in the final calculation to correct for the ATP that was added to start the AMP conversion.

Sample ATP-ADP Cycling Procedure (1–10 pmol)

Step 1. Samples, standards, and blanks in volumes of 20 µL or less (but all the same within 2 µL) are added at regular intervals to 100 µL of cycling reagent in 3-mL fluorometer tubes at room temperature. The reagent with 4 U/mL of each enzyme will provide about 1000 cycles in 1 h. One of the enzymes should be withheld until just before use to minimize any blank reaction.

Step 2. Starting 1 h after the first sample was added, 1 mL of indicator reagent is added serially to each tube at the same time intervals. The EDTA in the indicator tube immediately stops the cycling reaction.

Step 3. Fluorescence readings are made when the indicator reaction is complete (20 min or less).

Amplification up to 20,000-fold can be obtained by increased enzyme levels and prolonging cycling time (Table 8). For incubations longer than 4 or 5 h, the temperature is lowered to stabilize the enzymes. This extends the useful sensitivity into the 0.05–0.5 pmol range. There is little advantage in increasing the cycling temperature above 25°.

For still smaller samples, the best solution is to carry out the ATP-ADP cycling and indicator steps in the same or smaller volumes and then amplify the resulting NADPH with the NADP cycle. An example is presented to illustrate the possibilities.

Sample Total Adenylate Assay with the ATP-ADP Cycle (2–10 pmol)

The assay is the same as that described for ATP-ADP, except for a preliminary step to convert AMP to ATP as shown above.

Step 1. This is identical to Step 1 of the Sample ATP-Cycling Procedure above, except for substitution of the AMP to ATP converting reagent for the ATP-ADP cycling reagent, but the timing (about 1 h) is not critical.

Step 1A. Ten microliters of hexokinase (200 U/mL) in 50 m*M* imidazole-HCl buffer, pH 7.4, are added at limited intervals to each tube at room temperature.

Step 2. Starting 1 h after hexokinase was added to the first sample, 1 mL of indicator reagent is added serially to each tube at the same time intervals.

Step 3. Fluorescence readings are made when the indicator reaction is complete (20 min or less).

Comment

The AMP to ATP converting reagent contains 2 pmol of ATP (100 µL of 20 n*M* ATP), which sets a practical lower limit to the amount of total adenylate that can be measured without losing highest precision. For kinetic reasons, the ATP level in the converting reagent cannot be reduced very much below 20 n*M* without a great delay in AMP conversion, unless the sample itself contributes at least this much ATP. (For more details, *see* section entitled ATP-ADP Cycle Kinetics, *below*.) This also means that to measure total adenylate at much lower levels than 2 pmol, the AMP to ATP conversion step volume has to be reduced at least in proportion.

ATP-ADP Cycle with NADP Cycle Supplement (1–10 fmol)

ATP-ADP Cycling Reagent

The above cycling reagent with enzyme levels to provide about 1000 cycles in 1 h at 25°.

Step 2 Reagent

This consists of the above indicator reagent with $NADP^+$, 20 µ*M* and MgCl, 5 m*M*.

Procedure

Step 1. Add samples, standards, and blanks in volumes of 0.2 µL or less to 5 µL of the complete cycling reagent under oil in an oil well rack.

Step 2. After 60 min at 25°, add 5 µL of the ATP-ADP cycle indicator reagent.

Step 3. After 30 min, add 5 µL of 0.2*M* NaOH, and heat the oil well rack 20 min in an oven at 80°.

Step 4. Transfer 10-µL aliquots to fluorometer tubes containing 100 µL of NADP cycling reagent with enzyme levels to give about 1500 cycles in 60 min at 25° (Chapter 5).

Step 5. Add 1 mL NADP cycle indicator reagent (Chapter 5), and read when reaction is complete.

Note that the overall useful amplification is about a million-fold. Because the ATP-ADP cycle blank can easily be kept down to 0.1 n*M* or less, it would not be difficult to measure 10–100 attomol by using volumes of 1 µL for Step 1 and increasing amplification at Step 1 to 10,000-fold and at Step 4 to 15,000-fold to give overall useful amplification of 100,000,000.

ATP-ADP Cycle Kinetics

The K_m for ATP in the cycling reagent is about 50 µ*M*, and that for ADP is about 100 µ*M*. The recommended P-pyruvate and glucose levels are nearly saturating; therefore, the respective calculated first-order rate constants with 1 U/mL (1000 U/L) would be 1000/50 = 20/min (1200/h) for ATP and 1000/100 =10/min (600/h) for ADP. Substitution of Cl^- for acetate as the anion of the buffer lowers the V_{max} about 20% for both hexokinase and pyruvate kinase, and raises the K_m for ATP 40%, resulting in a 50% decrease in the ATP first-order rate constant and a 10% decrease for that of ADP.

GTP-GDP Cycle (1,7)

This is a cycle for measuring the sum of GTP plus GDP. The cycle enzymes are succinate thiokinase (STK) and pyruvate kinase (PK):

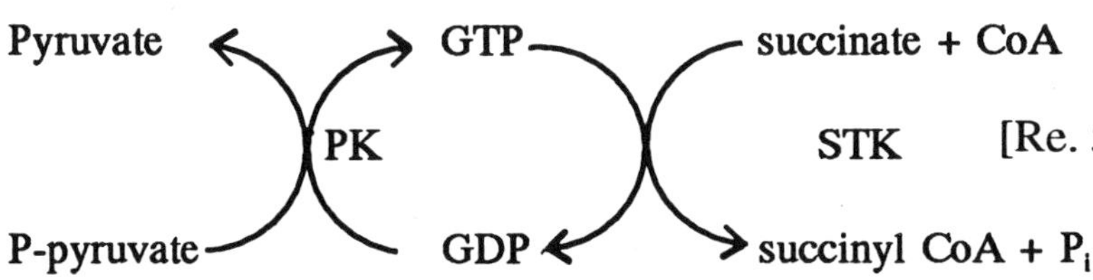

[Re. 5-11]

Table 9
Effect of Time, Temperature, and Enzyme Concentrations on Cycling Yield with the GTP-GDP Cycle

Succinate thiokinase, U/mL	Pyruvate kinase, U/mL	Temp, °C	Time, h	Cycles, approx
0.5	40	25°	1	2000
0.5	40	25°	2	4000
0.5	40	25°	4	8000
0.12	10	38°	1	1250
0.25	20	38°	1	2500
0.5	40	38°	1	4000
0.5	40	38°	2	8000
0.5	40	38°	4	15,000
1.0	80	38°	1	6000

The indicator reaction is:

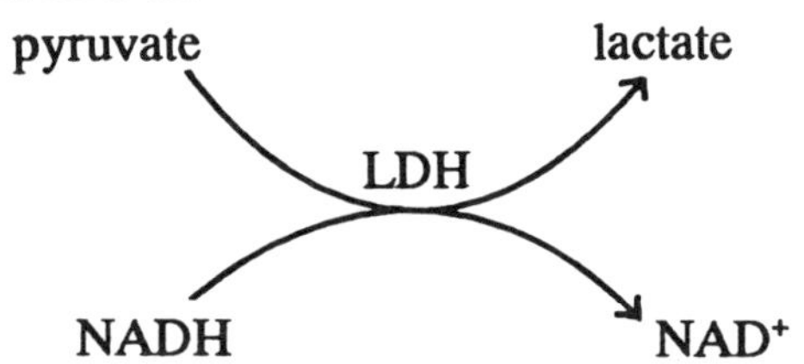

[Re. 5-12]

The NAD^+ is measured by heating in strong NaOH after destroying excess NADH with HCl. A simpler, but more restricted protocol is also offered for the indicator reaction.

The cycle and indicator reactions can be combined into a single step, but this results in a high cycling blank, which is much reduced by separating the two reactions. The cycling step is stopped, and the PK and STK destroyed by bringing the pH to 10.5 and heating to 95°C. (Heating at a less alkaline pH would partially hydrolyze the P-pyruvate; heating at a more alkaline pH would begin to destroy NAD^+.)

Cycling Reagent

This consists of Tris-acetate buffer, pH 8.5 (15 m*M* Tris-acetate, 35 m*M* Tris base); P-pyruvate, 200 µ*M*; disodium succinate, 1 m*M*; $MgCl_2$, 2 m*M*; potassium acetate, 80 m*M*; coenzyme A, 0.2 m*M*; and dithiothreitol, 0.5 m*M*; bovine plasma albumin, 0.02%; plus rabbit muscle pyruvate kinase and pig heart succinate thiokinase according to the amplification needed *(see Table 9)*.

Indicator Reagent

This consists of imidazole-HCl buffer, 1*M* imidazole-HCl, 0.2*M* imidazole base; NADH, 2 m*M*; and beef heart lactate dehydrogenase, 4 U/mL (20 μg/mL).

Sample GTP-GDP Cycling Procedure (0.05–5 pmol)

Step 1. Samples, standards, and blanks, in volumes of 10 μL or less (but all the same within 2 μL) are added at regular intervals to 100 μL of cycling reagent in 10 × 75 m*M* fluorometer tubes at room temperature (22–25°). The reagent with 40 U of pyruvate kinase and 0.5 U of succinate thiokinase will provide about 2000 cycles in 1 h. One of the enzymes should be withheld until just before use to minimize any blank reaction.

Step 2. Starting 1 h after the first sample was added, 5 μL of 1.5*M* 2-amino-2-methyl-1-propanol free base, is added serially to each tube at the same time intervals (to stop the reaction), and the rack of tubes is heated 3 min at 95–100° to destroy the enzymes.

Step 3. Ten microliters of indicator reagent are added.

Step 4. After 10 min, 10 μL of 2*M* HCl are added.

Step 5. After at least 10 min, 1 mL of 6*M* NaOH containing 10 m*M* imidazole is added followed by heating 10 min at 60°. Cool exactly to room temperature and read the fluorescence.

The simpler protocol indicator reaction is as follows:

Steps 1 and 2. The same as above.

Step 3. Add 1 mL of 10 μ*M* NADH in 50 m*M* imidazole-HCl buffer, pH 6.8 (30 m*M* imidazole-HCl, 20 m*M* imidazole base), and read the fluorescence.

Step 4. Add 0.2 U (1 μg) of lactate dehydrogenase and read again after 10 min.

Comment

With this simpler direct protocol, care must be taken that the final pyruvate concentration is not >10 μ*M* in the indicator reagent (or <1 μ*M*). This could easily happen if the sample contains more than expected GTP + GDP, or the cycling rate is greater than planned for.

Table 9 provides examples of cycling conditions to provide up to 15,000-fold amplification. Note that cycling rates are approximately doubled by raising the temperature from 25 to 38°, and that amplification is almost proportional to time for at least 4 h.

An amplification of 15,000-fold moves the useful assay range down to 10–500 fmol. However, because the cycling reagent blank may be 1–2 n*M*, to take full advantage of this sensitivity, it would be necessary to reduce the cycling vol to 20 µL (for example) in a 7 × 100 m*M* serological tube: adding 1 µL of 1.5*M* 2-amino-2-methyl-1-propanol for Step 2; 2 µL of indicator reagent for Step 3; 2 µL of 2*M* HCl for Step 4; and then transferring a 20-µL aliquot to 1 mL of 6*M* NaOH in a fluorometer tube for Step 5.

Chapter 6

A Collection of Metabolite Assays

This chapter contains protocols for measuring 44 different metabolites. In each case, a spectrophotometric assay and at least one fluorometric assay have been presented. In many cases, additional fluorometric procedures are given for increasing sensitivity to permit measurement of as little as a few femtomoles of material. The reader should have no difficulty designing similar adaptations of any of the methods to achieve whatever sensitivity is required. Chapter 8 may be useful for this purpose.

The spectrophotometric assays, unless noted otherwise, are designed for measuring 20–150-μ*M* concentrations of the metabolite. The direct fluorometric methods are planned for 1-mL vol of reagent. Unless otherwise noted, enzyme activities under reagent headings given as U/mL (IU/mL) are at 25°.

Most of the methods are adaptations of spectrophotometric methods from many authors. Although we will not try to trace the history of each one, the following is an attempt to record the first edition of most of the methods. ADP, Kornberg and Pricer *(1)*; alanine, Pfleiderer et al. *(2)*; aspartate, Pfleiderer et al. *(3)*; ATP, Kornberg *(4)*; citrate, Moellering and Gruber *(5)*; dihydroxyacetone-P, Thorn et al. *(6)*; fructose and fructose-6-P, Slein *(7)*; fructose-1,6-P_2 via glycero-P-dehydrogenase, Slater *(8)*; fructose-1,6-P_2 via glyceraldehyde-P-dehydrogenase, Vishniac and Ochoa *(9)*; glucose, Slein et al. *(10)*; glucose-6-P, Slein *(7)*; glucose-1,6-P_2, Paladini et al. *(11)*; glutamate, Albers et al. *(12)*; glycero-P, Bublitz and Kennedy *(13)*; isocitrate, Ochoa *(14)*; lactate (method I), Noll *(15)*; P-pyruvate, Kornberg and Pricer *(1)*; pyruvate, Kubowitz and Ott *(16)*; ATP via P-fructokinase as in total nucleotide triphosphate method, Slater *(8)*; UDP-glucose, Strominger et al. *(17)*.

ADP and AMP

$$\text{AMP} + \text{ATP} \xrightarrow{\text{myokinase}} 2\ \text{ADP} \qquad \text{[Re. 6-1]}$$

$$\text{ADP} + \text{P-pyruvate} \xrightarrow[\text{kinase}]{\text{pyruvate}} \text{ATP} + \text{pyruvate}$$

[Re. 6-2]

$$\text{Pyruvate} + \text{NADH} + \text{H}^+ \xrightarrow[\text{dehydrogenase}]{\text{lactate}} \text{lactate} + \text{NAD}^+$$

Spectrophotometer, 15–120 nmol/mL

Reagent: Imidazole-HCl buffer, pH 7.0 (30 m*M* imidazole base, 20 m*M* imidazole-HCl); $MgCl_2$, 2 m*M*; KCl, 75 m*M*; ATP, 100 µ*M* (needed only for AMP assay); NADH, 50–150 µ*M* (at least a 20% excess); P-pyruvate, 300 µ*M*; myokinase (rabbit muscle), 0.36 U/mL (1 µg/mL); lactate dehydrogenase (beef heart), 0.4 U/mL (2 µg/mL); and pyruvate kinase 0.3 U/mL (2 µg/mL).

Reaction times: Pyruvate, 1 min or less; ADP, 2–4 min; and AMP, 3–6 min.

Conduct of the Assay

Unless pyruvate is to be determined in the sample, lactate dehydrogenase can be incorporated in the reagent. After adding the sample, time is allowed for pyruvate to be reduced, and a reading is made. ADP and AMP are then measured in sequence by the decrease in reading upon addition of first pyruvate kinase and then myokinase (2 mol of NAD^+ are produced/mol of AMP).

Comment: ADP Method

Because P-pyruvate may contain traces of pyruvate, its concentration is kept low, particularly when measuring low levels of ADP. Samples that contain more than 1% pyruvate should not be used. If a satisfactory sample is not available, pyruvate can be removed at neutral pH by adding a slight molar excess of H_2O_2 over the pyruvate and heating for a few minutes at 100°.

The method for ADP is not completely specific, since other nucleotide diphosphates react with P-pyruvate, although at slower rates. According to Strominger *(18)*, the relative rates with ADP, GDP, IDP, UDP, and CDP are 100, 19, 12, 3, and 2, respectively, at substrate levels of 0.04 m*M*.

Comment: AMP Method

Most NADH preparations contain substantial amounts of 5'AMP, sufficient in some cases to give a very high blank and use up a major fraction of the NADH in the assay. A satisfactory lot can probably be selected. If not, it is possible to remove the AMP with phosphatase in the following manner.

To a 5-m*M* NADH solution at pH 9.0, add 3 U/mL (70 μg/mL) of alkaline phosphatase (Sigma, St. Louis, MO, type III, bacterial) and incubate 20 min at 38°. Add NaOH to give a 20-m*M* excess (pH must be at least 12); heat 2 min at 100° to destroy the phosphatase. The NADH can be stored at –50° at this pH. If the ATP contains disturbing amounts of either ADP or AMP, the amount of ATP can be lowered with compensatory increase in myokinase. ATP actually increases during the overall reaction sequence; therefore, the intial level can be less than that of the AMP to be measured.

Fluorometer, Direct Assay, 0.1–8 nmol/mL of ADP, 0.1–4 nmol/mL of AMP

Reagent: The reagent is the same as in Spectrophotometer, except as follows: ATP, 5 μ*M*; NADH, 0.2–10 μ*M* (20–75% excess), P-pyruvate, 20 μ*M* (10 μ*M* with <1 μ*M* substrate); pyruvate kinase, 0.75 U/mL.

Reaction times: Pyruvate, 2 min or less; ADP, 3–6 min; and AMP, 3–6 min.

Conduct of the Assay

See Spectrophotometer *above*. Ordinarily in living systems, ADP is present at substantially higher levels than AMP. This tends to decrease the precision of AMP measurement. It is important to be sure that the reaction with ADP is complete before adding myokinase. If other dinucleotides are present in considerable amounts, there may be a drift as they slowly react. In this case, it may be wise to increase the amount of pyruvate kinase. If necessary, two parallel samples can be run; both receive pyruvate kinase until ADP has reacted, after which only one receives myokinase. The difference should provide a more accurate measure of AMP.

Kinetics

Pyruvate kinase has a V_{max} of about 150 μmol/mg/min with ADP as substrate. The K_m for ADP is 200 μ*M*, and that of P-pyruvate is 30 μ*M*. The K_m for GDP is 280 μ*M*; the V_{max} is 28 μmol/mg/min.

Myokinase is a very active enzyme. The V_{max} of commercial preparations tested has been up to 600 µmol/mg/min. Noda and Kuby *(19)* reported a V_{max} of three times this. Under analytical conditions, the K_m for AMP is about 50 µ*M*, and that of ATP close to 100 µ*M*. ATP samples almost always contain significant amounts of ADP. Therefore, ATP levels are kept relatively low. With 5 µ*M* ATP and 1 µg/mL of myokinase, observed halftimes for AMP are approx 1 min (the other two enzymes in excess). Myokinase dilutions keep better in the presence of 2 m*M* dithiothreitol.

Adenylic Acid (5'AMP) Method II

$$(Glc)_n + P_i \xrightarrow{\text{phosphorylase } \underline{a}} \text{glucose-1-P} + (glc)_{n-1}$$

$$\text{Glucose-1-P} \xrightarrow{\text{P-glucomutase}} \text{glucose-6-P} \qquad \text{[Re. 6-3]}$$

$$\text{Glucose-6-P} + NADP^+ \xrightarrow[\text{dehydrogenase}]{\text{glucose-6-P}} \text{6-P-gluconolactone} + NADPH + H^+$$

This method for the measurement of 5'AMP is based on the observation that, when glycogen and P_i are present at low levels, the activity of muscle phosphorylase *a* (EC 2.4.1.1.) becomes a sensitive and specific indicator of 5'AMP concentration *(20,21)*. The NADPH generated is measured by its absorption or fluorescence, or can be cycled for greater sensitivity.

Spectrophotometer, 50–500 pmol/mL

Reagent: Imidazole acetate buffer, pH 7.0 (30 m*M* imidazole base, 20 m*M* imidazole acetate); NaSCN, 10 m*M*; β-mercaptoethanol, 3.5 m*M*; magnesium acetate, 2 m*M*; bovine plasma albumin, 0.02%; glucose-1,6-P_2, 0-5 µ*M*; glycogen, 1 m*M* as glucosyl units; and if tissue extracts are present, 5 µ*M* Ap5A, the adenylate kinase inhibitor (*see* final Comment). Just before use are added P-glucomutase, 0.4 U/mL (2 µg/mL); glucose-6-P dehydrogenase (baker's yeast, 0.15 U/mL (1 µg/mL) and phosphorylase a (rabbit muscle) 0.01–0.02 U/mL (0.5–1.0 µg/mL).

P_i solution: Na_2HPO_4, 30 m*M*.

Indicator reagent: 2-Amino-2-methyl-1,3-propanediol (base, 1*M*; HCl, 0.05*M*) and $NADP^+$ 3 m*M*.

Conduct of the Assay

The reaction is conducted in 1-mL vol in 10×75 mm tubes. Samples, blanks, and a series of standards (in steps of two) are added in volumes up to 10 µL. To all tubes are added 10 µL of the P_i solution to start the reaction.

The tubes are incubated for 1.5–2 h at room temperature, and the reaction stopped by the addition of 50 µL of indicator reagent. The final NADPH concentration should be in the range of 25–100 µ*M*.

Fluorometer Direct Assay, 50–500 pmol

Reagent: The same as Spectrophotometer, except NaSCN, 20 m*M*; and phosphorylase *a*, 0.005–0.01 U/mL (0.25–0.50 µg/mL).

P_i solution: The same as for Spectrophotometer.

Indicator reagent: The same as for Spectrophotometer, except 50 µ*M* $NADP^+$.

Conduct of the Assay

To 1 mL of reagent in 10×75 mm tubes are added blanks, standards, and samples. Standards and samples should be in the range of 0.02–0.05 µ*M* in steps of two, all of which are carefully matched in terms of salt concentration. The reaction is started by adding 10 µL of the P_i solution at intervals of 15–20 s, and subsequent readings are spaced accordingly. The tubes are read at 20, 40, and 60 min.

Fluorometer Indirect Analysis, 4–40 pmol

Reagent: The reagent is the same as Spectrophotometer, with phosphorylase *a*, 0.02 U/mL (1 µg/mL).

Indicator reagent: 1-Amino-2-methyl-1,3-propanediol buffer, pH 9.2 (35 m*M* propanediol base, 15 m*M* propanediol HCl) and NADP, 50 µ*M*.

Procedure

Step 1: Samples and standards are added in volumes of 10 µL or less to 100 µL of reagent in 10 × 75 mm tubes in a rack in an ice bath. To all tubes is added 1 µL of 30 m*M* Na_2HPO_4, and the tubes incubated 1 or 2 h at 25°.

Step 2: The rack is returned to an ice bath, and 1 mL of the indicator reagent is added. The tubes are incubated at room temperature and then read in the fluorometer. The reaction for glucose-1-P and glucose-6-P require 30 min or more because of the dilution of the auxiliary enzymes.

Cycling Procedure, 1–10 fmol

Reagent: The reagent is like Spectrophotometer, except the buffer is imidazole-HCl, 100 m*M*, pH 7.1 (67 m*M* imidazole base, 33 m*M* imidazole

HCl) to compensate for the acid in the first droplet; $NADP^+$, 75 μ*M*; bovine plasma albumin, 0.1%; and the remaining components are double-strength.

Reaction vessels: Oil wells.

Procedure

Step 1: Standards are 10 n*L* of 5'AMP (0.05–0.4 μ*M*) in 0.02*M* HCl, frozen-dried tissue samples are added to 10 n*L* of the acid. Heat at 80° for 20 min.

Step 2: To all samples are added 40 n*L* of reagent; followed by 25 n*L* of 1 m*M* Na_2HPO_4 at 10-s intervals. Incubate 60 min after the first addition.

Step 3: Add 3 μL of 0.02*M* NaOH to all samples at 10-s intervals. Heat in oven at 80° for 30 min.

Step 4: Transfer 2 μL to fluorometer tubes containing 100 μL of $NADP^+$ cycling reagent with enzyme levels to give about 15,000 cycles in 60 min at 38° (Chapter 5). Heat stop by 2 min in a 95–100° water bath.

Step 5: Add 1 mL of indicator reagent (Chapter 5), and read when reaction is over.

Comment

The phosphorylase reaction is sensitive to ion concentrations, and, therefore, the volume and constituents of the extracting fluid, or blank should be identical for samples and standards. Because of the residual activity of phosphorylase *a* in the absence of added 5'AMP, the range of 5'AMP concentrations that can be added is limited. However, by the addition of 10 or 20 m*M* ClO_4^- or SCN^- (chaotropic ions), the activity is reduced at all 5'AMP concentrations, and the activity without 5'AMP is reduced three or four times more than with high 5'AMP. Consequently, the analytical range can be extended three- or fourfold above the range in the absence of chaotropic ions. The addition of chaotropic ions does not, however, remove the sensitivity to other ions.

Commercial preparations of $NADP^+$ were found to be contaminated with 5'AMP. It is for this reason that the protocols are outlined to introduce the $NADP^+$ in the indicator step, rather than include it in the initial reagents.

It was found that, if NAD^+ were present in the tissue, pretreatment of frozen-dried sections with NaOH resulted in the formation of 5'AMP and ADP. With standard NAD^+, approx 25% was converted to 5'AMP after 30 min at 38°. After the acid treatment, there was no detectable breakdown of ADP to 5'AMP, and that of ATP to 5'AMP was <0.4%.

NAD^+ is of course stable to acid (Chapter 1). If tissue samples are being analyzed, there is always danger that active myokinase will still be present because of its remarkable stability to acid treatment. It is therefore recommended that the potent myokinase inhibitor P^1, P^5 di(adenosine-5')pentaphosphate (Ap5A) be added at 5-μ*M* concentration to prevent formation of AMP from ADP *(22)*.

Under these analytical conditions, phosphorylase activity is maximal at about pH 7. It is reduced to 16% at pH 8.1 (50 m*M* Tris buffer) and 0.3% at pH 9.1. The strong inhibition at pH 9 was used to stop the reaction of phosphorylase without stopping the other enzyme steps.

With P_i, glycogen and 5'AMP present at low concentrations, phosphorylase activity is maximal at 20–25°, so that temperature control is not critical in this range. The velocity falls rapidly below 15°. At 0° with 0.01–0.4 μ*M* 5'AMP, the rate is only 0.4% of that at 25°.

The yield of NADPH from 5'AMP depends on enzyme concentration and time of incubation. With standard reagent, the rate is reasonably constant until 60 μ*M* NADPH is formed. With 0.03 μ*M* 5'AMP and blank equivalent to the same concentration, a yield of 1000-fold can be attained. Standard deviations of 2–3% are to be expected, providing the salt concentrations are carefully controlled. No other nucleotides were found to react in this system. ATP, ADP, CMP, UMP, TMP, IMP, and 3'5'-cycle AMP were tested.

Alanine

$$\text{Alanine} + \alpha\text{-ketoglutarate} \xrightarrow[\text{transaminase}]{\text{alanine}} \text{pyruvate} + \text{glutamate} \qquad \text{[Re. 6-4]}$$

$$\text{Pyruvate} + \text{NADH} + \text{H}^+ \xrightarrow[\text{dehydrogenase}]{\text{lactate}} \text{lactate} + \text{NAD}^+$$

Spectrophotometer, 15–120 nmol/mL

Reagent: Tris-HCl buffer, pH 8.1 (25 m*M* Tris base, 25 m*M* Tris-HCl); NADH, 50–150 μ*M* (at least a 20% excess); α-ketoglutarate, 200 μ*M*; alanine transaminase (pig heart), 8 U/mL (100 μg/mL); and lactate dehydrogenase (beef heart), 0.4 U/mL (2 μg/mL).

Reaction time: Twenty minutes.

Conduct of the Assay

For standardization purposes, both enzymes can be incorporated in the reagent. For samples that may contribute appreciably to the absorption at 340 nm or that contain pyruvate, it is preferable to add the transaminase last.

Fluorometer Direct Analysis, 0.1–8 nmol/mL

Reagent: Tris buffer as for the spectrophotometer, NADH, 0.2–10 μM (20–70% excess over the expected alanine); α-ketoglutarate, 50 μM; alanine transaminase, 8 U/mL; lactate dehydrogenase, 0.2 U/mL.
Reaction time: Twenty minutes.

Conduct of the Assay

Many kinds of biological material contain pyruvate at levels that are significant relative to alanine. Therefore, the sample is added to the reagent containing lactate dehydrogenase only. After allowing a few minutes for the pyruvate to react, a reading is made, and the reaction is started with transaminase. If it is wished also to evaluate pyruvate, lactate dehydrogenase can be omitted from the reagent and added to the fluorometer tube after the sample.

Comment

The large amount of transaminase used adds appreciable fluorescence to the samples. Therefore, with highest sensitivities, the blanks become especially important. Lactate dehydrogenase enhances the fluorescence of NADH. Because the effect may not be linear, excessive amounts of this enzyme should be avoided, and the linearity checked.

Kinetics

The Michaelis constant for alanine is very large, on the order of 30 mM; that for α-ketoglutarate is very low, <20 μM with alanine levels in the analytical range. The high Michaelis constant for alanine requires that an unusually large amount of enzyme be used. The V_{max} is about 80 µmol/mg/min. From the formula $k = V_{max}/K_m$, it can be calculated that k = µg/mL × 80/30,000/min or about 0.25/min with 8 U/mL (100 µg/mL).

Aspartate

[Re. 6-5]

$$\text{Aspartate} + \alpha\text{-ketoglutarate} \xrightarrow[\text{transaminase}]{\text{aspartate}} \text{oxaloacetate} + \text{glutamate}$$

$$\text{Oxaloacetate} + \text{NADH} + \text{H}^{+} \xrightarrow[\text{dehydrogenase}]{\text{malate}} \text{malate} + \text{NAD}^{+} \qquad \text{[Re. 6-6]}$$

Levels of aspartate in biological materials are ordinarily several orders of magnitude higher than those of oxalacetate. Consequently, no appreciable blank from oxaloacetate should be encountered. The instability of oxaloacetate is no problem in the coupled assay.

Spectrophotometer, 15–120 nmol/mL

Reagent: Imidazole-HCl buffer, pH 7 (30 m*M* imidazole, 20 m*M* imidazole-HCl); NADH, 50–150 μ*M* (at least a 20% excess); α-ketoglutarate, 200 μ*M*; aspartate transaminase (pig heart), 2 U/mL (10 μg/mL); and malate dehydrogenase (pig heart), 1.4 U/mL (2 μg/mL).
Reaction time: Ten minutes.

Conduct of the Assay

For standardizing aspartate solutions, both enzymes can be incorporated in the reagent and sample added last, after an initial reading. For samples that may contribute appreciable absorption at 340 nm, it is preferable to add the transaminase last.

Fluorometer Direct Assay, 0.1–8 nmol/mL

Reagent: Imidazole buffer as for the spectrophotometer, NADH, 0.2–10 μ*M* (20–75% excess over expected aspartate); α-ketoglutarate, 50 μ*M*; aspartate transaminase, 2 U/mL; and malate dehydrogenase, 0.35 U/mL *(see* Comment*)*.
Reaction time: Ten minutes.

Conduct of the Assay

Ordinarily, malate dehydrogenase is incorporated in the reagent; the first reading is made after sample addition, and the second reading after transaminase addition plus time for the reaction.

Malate dehydrogenase enhances the fluorescence of NADH. Because the effect is nonlinear under most circumstances, the amount of this enzyme used should be kept low, and the linearity of the assay checked.

Kinetics

The Michaelis constant for aspartate is large and increases with high levels of α-ketoglutarate. It ranges from about 1 m*M* with 50 μ*M* α-ketoglutarate or less to 4 m*M* with 1 m*M* α-ketoglutarate or more.

Therefore, the reaction is first-order at all aspartate levels in the analytical range. On the other hand, the Michaelis constant for α-ketoglutarate is very small, <1 μ*M* at low levels of aspartate, and approaches 75 μ*M* with very high (20 m*M*) aspartate levels. Because of this situation, the α-ketoglutarate requirement is not more than that which will provide a safe excess over the aspartate to be measured. High levels actually inhibit.

The V_{max} of the transaminase at 25° is about 180 μmol/mg/min. However, because of the peculiarity of the kinetics, in analytical situations with modest α-ketoglutarate concentrations, rates are those expected for a V_{max} of about 60 μmol/mg/min. Thus, the apparent first-order rate constant with 10 μg of enzyme/mL would be 10 × 60/1000 = 0.6 min (from $k = V_{max}/K_m$).

Cycling Assay, 5–25 pmol

Reagent: Imidazole buffer, pH 7.0 (120 m*M* imidazole base, 80 m*M* imidazole-HCl); bovine plasma albumin, 0.04%; α-ketoglutarate, 50 μ*M*; ascorbic acid, freshly prepared, 4 m*M*; NADH, 50 μ*M*; beef heart malate dehydrogenase, 0.35 U/mL (0.5 μg/mL); and pig heart aspartate transaminase, 2 U/mL (10 μg/mL).

Reaction vessels: Oil wells.

Procedure

Step 1: NaOH, 1 μL of 0.02*M* for samples and blanks. Standards of 5–25 μ*M* aspartate are made up in NaOH. The samples are introduced through the oil into the droplet, and the entire rack heated at 80° for 20 min.

Step 2: To each sample is added 1 μL of reagent; 30–40 min at room temperature.

Step 3: To each sample is added 1 μL of 0.25*M* HCl; at least 10 min at room temperature.

Step 4: Transfer 2-μL aliquots to fluorometer tubes containing 100 μL of NAD cycling reagent with enzyme levels to give about 500 cycles in 60 min at 25° (Chapter 5). Heat stop by 2 min in a 95–100° water bath.

Step 5: Add 1 mL of indicator reagent (Chapter 5), and read when reaction is complete.

Cycling Assay, 0.1–0.5 pmol

Reagent: Same as for Cycling Assay, 2–25 pmol.

Procedure

Step 1: NaOH, 25 nL of 0.02*M* for samples and standards (4–20 μ*M*); heat at 80° for 20 min.

Step 2: To each sample, add 25 nL of reagent; 30–40 min at room temperature.
Step 3: Add 100 nL of 0.1*M* HCl.
Step 4: Add to each oil well 5 µL of NAD cycling reagent with enzyme levels to give about 20,000 cycles in 2 h at 25° (Chapter 5). Stop with 2 µL of 0.5*M* NaOH and 20 min in 80° oven.
Step 5: Transfer 5 µL from Step 4 to 1 mL of indicator reagent (Chapter 5), and read when reaction is complete.

ATP and P-Creatine

$$\text{P-Creatine} + \text{ADP} \xrightarrow[\text{kinase}]{\text{creatine}} \text{creatine} + \text{ATP}$$

$$\text{ATP} + \text{glucose} \xrightarrow{\text{hexokinase}} \text{ADP} + \text{glucose-6-P} \qquad \text{[Re. 6-7]}$$

$$\text{Glucose-6-P} + \text{NADP}^+ \xrightarrow[\text{dehydrogenase}]{\text{glucose-6-P}} \text{6-P-gluconolactone} + \text{NADPH}$$

A number of tissues contain ATP and P-creatine at comparable concentrations. In this case, it is convenient to measure them in the same sample. With direct analyses, this is done by making successive readings after each enzyme addition. With indirect analyses, the sample is divided after ATP has reacted. NADPH is measured in one portion as a measure of ATP. In the other portion, NADPH is destroyed, after which all three enzymes are added to give NADPH equal to P-creatine. Alternatively, creatine kinase can be added to the second sample without prior acidification to give NADPH equal to the sum of P-creatine and ATP, from which P-creatine can be calculated by difference. Glucose-6-P can also be measured on the same sample, but usually the level is much lower than that of ATP; consequently, a larger sample will be required.

Spectrophotometer, 15–200 nmol/mL

Reagent: Tris-HCl, pH 8.1 (25 m*M* Tris base, 25 m*M* Tris HCl); $MgCl_2$, 1 m*M*; dithiothreitol, 0.5 m*M*; $NADP^+$, 500 µ*M*; glucose, 1 m*M*; ADP, 500 µ*M* (for P-creatine assay only); P^1,P^5–di(adenosine-5')pentaphosphate, 10 µ*M*; creatine kinase (rabbit muscle), 0.9 U/mL (50 µg/mL), measured in the direction of P-creatine formation; yeast hexokinase, 0.28 U/mL (2 µg/mL); and baker's yeast glucose-6-P dehydrogenase, 0.06 U/mL (0.15 µg/mL).

Reaction time: Three to 5 min for glucose-6-P, 4–6 min for ATP, and 10 min for P-creatine.

Conduct of the Assay

Initial readings are made after adding samples with subsequent readings at adequate time intervals after adding sequentially glucose-6-P dehydrogenase, hexokinase, and creatine kinase. One of the greatest dangers in analyzing biological material is that myokinase, which is acid-stable, is not completely removed in preparing the samples for analysis. This would give erroneously high results. To avoid this potential error, it is recommended that the specific myokinase inhibitor, P^1,P^5-di(adenosine-5') pentaphosphate "AP5A" *(22)* be included in the reagent. The addition of this inhibitor has created no analytical problems, and it is specified for all the analyses where myokinase contamination might be a problem.

Comment

In addition to the danger of myokinase contamination, two other dangers have been encountered in ATP analysis of tissues. The first is failure to remove ATPase completely. This can cause loss of ATP from tissue extracts prior to analysis rather than during the assay itself. The second danger, described in Chapter 4, is that ATP may be lost in preparing $HClO_4$ tissue extracts if there is a substantial amount of Ca^{2+} present. Calcium phosphate or carbonate precipitates, formed on neutralization, can absorb ATP. This can be prevented with EDTA.

ADP almost invariably contains some ATP. A sample should be chosen, if possible, that contains not more than 0.5% ATP. (For removal of ATP from ADP, *see* comment after Fluorometer Direct Assay).

Fluorometer Direct Assay, 0.1–10 nmol/mL

Reagent: This is the same as in Spectrophotometer, except as follows: $NADP^+$, 50 µ*M*; glucose, 100 µ*M*; ADP (for P-creatine assay only), 100 µ*M* (30 µ*M* for <1 µ*M* P-creatine); hexokinase, 0.14 U/mL; and glucose-6-P dehydrogenase 0.03 U/mL of crystalline enzyme.

Reaction times: Two to 3 min for glucose-6-P, 4–8 min for ATP, and 15–20 min for P-creatine with 100 µ*M* ADP, and 25–40 min with 30 µ*M* ADP.

Conduct of the Assay

See the Spectrophotometer protocol.

Comment

Glucose and glucose-6-P dehydrogenase are reduced to decrease the progressive blank from the glucose dehydrogenase side activity (Chapter 6). The ADP is decreased for the lowest P-creatine levels to decrease the blank from ATP contamination. If necessary, the ATP can be removed from the ADP as follows: To 1 mL of reagent in a fluorometer tube containing 0.1*M* Tris-HCl, pH 8.0,100 μ*M* glucose, 1 m*M* Mg^{2+}, 30 μ*M* $NADP^+$, and 1 m*M* ADP, add 0.14 U/mL of glucose-6-P dehydrogenase and 0.28 U/mL of hexokinase. The ATP present will be converted to ADP, and NADPH will be formed in equivalent amount. After NADPH formation ceases, it is destroyed by adding 20 μL of 5*M* HCl and then neutralized after a few minutes with an equivalent amount of NaOH. The enzymes will also be destroyed, and the remaining components will not interfere with subsequent analyses. Stronger solutions of ADP can be treated in a similar fashion with NADPH formation followed in a spectrophotometer. If the ADP is 10 m*M*, the glucose should be increased to 1 m*M* and the $NADP^+$ to 0.3 m*M*. If the reaction is too slow, the enzymes can be increased.

Cycling Assay, 1–10 pmol

(This and the other cycling methods for ATP and P-creatine are described for the analysis of frozen-dried sections. *See* Comment below in regard to other applications.)

ATP reagent: Tris-HCl buffer, pH 7.5 (10 m*M* Tris base, 40 m*M* Tris-HCl); $MgCl_2$, 1 m*M* dithiothreitol, 0.5 m*M*; glucose, 100 μ*M*; AP5A, 10 μ*M*; $NADP^+$, 20 μ*M*; bovine plasma albumin 0.02%; hexokinase, 0.28 U/mL; and glucose-6-P dehydrogenase, 0.1 U/mL.

P-creatine reagent: Tris-HCl buffer, pH 8.6 (150 m*M* Tris base, 50 m*M* Tris-HCl; ADP, 80 μ*M*; bovine plasma albumin, 0.02%; Ap5A, 10 μ*M*; hexokinase, 0.6 U/mL; glucose-6-P dehydrogenase, 0.25 U/mL; and creatine kinase, 1.8 U/mL.

Reaction vessels: Oil wells.

Procedure

Step 1: NaOH, 1 μL of 0.1*M* for samples and blanks, and 1 μL of 3–10 μ*M* ATP of P-creatine in 0.1*M* NaOH for standards. Heat 10 min at 60°.

Step 2: Add 5 μL of ATP reagent; 20 min at room temperature. Transfer 3 μL to another tube for P-creatine analysis.

Step 3: To the portion for P-creatine, add 1 µL of 0.125*M* HCl (to destroy NADPH); 10 min at room temperature. Add 3 µL of P-creatine reagent; 30 min at room temperature.

Step 4: To both the ATP and P-creatine samples, add 3 µL of 0.2*M* NaOH; 15 min at 60°.

Step 5: Transfer 5-µL aliquots to fluorometer tubes containing 100 µL of NADP cycling reagent with enzyme levels to give about 3000 cycles in 60 min at 38° (Chapter 5). Heat stop by 2 min in a 95–100° water bath.

Step 6: Add 1 mL of indicator reagent (Chapter 5), and read when reaction is complete.

Comment

The bovine plasma albumin is added to protect the enzymes in the small volume. The buffers used for the ATP and P-creatine are designed so that after addition to the sample the resulting pH will be close to 8. If the original sample is not treated with alkali (or with less alkali as in Cycling Assay *below)*, a buffer at pH 8 would be used for the ATP reagent as in Spectrophotometer and Fluorometer Direct Assay. The reaction times should be checked directly in the fluorometer using all reagents in 200-fold larger volumes.

Cycling Assay, 0.2–1 pmol

ATP reagent: This is the same as in Cycling Assay, 1–10 pmol, but the buffer is 50 m*M* Tris-HCl, pH 8.1, as in Spectrophotometer and Fluorometer Direct Assay.

P-creatine reagent: Tris-HCl buffer, pH 8.1 (100 m*M* Tris base, 100 m*M* Tris HCl); ADP, 25 µ*M*; $NADP^+$, 20 µ*M*; glucose, 100 µ*M*; bovine plasma albumin, 0.02%, AP5A, 10 µ*M*; hexokinase, 0.28 U/mL; glucose-6-P dehydrogenase, 0.14 U/mL; and creatine kinase, 2.24 U/mL.

Reaction vessels: Oil wells.

Procedure

Step 1: NaOH, 0.06 µL of 0.1*M*, for samples and blanks, and 0.06 µL of 5–20 µ*M* ATP or P-creatine in 0.1*M* NaOH for standards; 20 min at 80°.

Step 2: ATP reagent; 0.6 µL, 20 min at room temperature.

Step 3: Transfer 0.3 µL into 5 µL of 0.5*M* NaOH in another oil well rack; heat 20 min at 80°. (These are the ATP samples.)

Step 4: To the remainder of the sample in the original rack, add 0.1 µL of 0.2*M* HCl; 10 min at room temperature.

Step 5: Add 0.6 µL of P-creatine reagent; 45 min at room temperature.

Step 6: Add 5 µL of 0.05*M* NaOH to P-creatine samples; 20 min at 80°.

Step 7: Transfer 4-µL aliquots to fluorometer tubes containing 50 µL of NADP cycling reagent with enzyme levels to give about 25,000 cycles in 2 h at 38° (Chapter 5). Heat stop by 2 min in a 95–100° water bath.

Step 8: Add 1 mL of indicator reagent (Chapter 5), and read when reaction is complete.

Comment

Because of the larger relative vol of P-creatine reagent than in Cycling Assay, 1–10 pmol, $NADP^+$ and glucose are added to the reagent. This is preferable to increasing the concentrations in the ATP reagent.

Cycling, 20–100 fmol

ATP reagent: This the same as in Cycling Assay, 1–10 pmol, except as follows Tris buffer, pH 7.4 (10 m*M* Tris base, 50 m*M* Tris-HCl); bovine plasma albumin, 0.03%.

P-creatine reagent: This is the same as in Cycling Assay, 1–10 pmol, except as follows: ADP, 60 µ*M*; bovine plasma albumin, 0.06%; and glucose-6-P dehydrogenase, 0.12 U/mL (0.3 µg/mL).

Reaction vessels: Oil wells.

Procedure

Step 1: NaOH, 0.025 µL of 0.1*M* for samples and blanks, and 0.025 µL of 14 µ*M* ATP and P-creatine for standards; 20 min at 80°.

Step 2: Add 0.08 µL of ATP reagent; 15 min at room temperature.

Step 3: Transfer 0.05 µL into 0.2 µL of 0.1*M* NaOH in another oil well rack; 20 min at 80°. (This is the portion for ATP.)

Step 4: Add 0.005 µL of 0.5*M* HCl to the remaining part of the sample for P-creatine; 10 min at room temperature.

Step 5: Add 0.025 µL of P-creatine reagent to this portion; 45 min at room temperature.

Step 6: Add 0.2 µL of 0.1*M* NaOH; 20 min at 80°.

Step 7: To all oil wells, add 5 µL of NADP cycling reagent with enzyme levels to give about 100,000 cycles overnight at 15° (Chapter 5). Stop with 2 µL of 0.5*M* NaOH plus 20 min in 80° oven.

Step 8: Transfer 5-µL aliquots into 1 mL of indicator reagent (Chapter 5), and read when reaction is complete.

Kinetics

The V_{max} of hexokinase is about 140 µmol/mg/min. The K_m for ATP with 1 m*M* glucose is 30 µ*M* and that for glucose with 300 µ*M* ATP is 160 µ*M*. From this, it may be calculated with 10 µ*M* ATP or less and saturating glucose levels that the half-time with 1 μg/mL of enzyme should be 0.7 × 30/140 = 0.15 min. With 100 μ*M* glucose, the half-time would be 0.4 min (0.7[30/140] [260/100]). Under assay conditions, the rates are about half this fast.

The kinetic situation is much less favorable for P-creatine. The V_{max} for creatine kinase of a commercial preparation under assay conditions was about 40 μmoles/mg/min with a very large K_m for P-creatine, 2.2 m*M*. With P-creatine in the analytical range, the rates are even lower than this would indicate. With a saturating level of ADP, the half-time for P-creatine with 1 μg/mL of creatine kinase would be about 60 min. The K_m for ADP is approx 150 μ*M*, which constitutes no problem in assays for higher P-creatine levels, but is a problem at the lowest levels because of the possible contamination of ADP with ATP.

Adenylates and P-Creatine by the Luciferin–Luciferase Method

$$\text{LuciferinH}_2 + \text{ATP} + \text{Enzyme (E)} \rightarrow \text{(E)LuciferinH}_2\text{AMP} + \text{PP}_i$$

$$\text{(E)LuciferinH}_2\cdot\text{AMP} + \text{O}_2 \rightarrow \text{Luciferin} + \text{H}_2\text{O} + \text{luminescence} \quad \text{[Re. 6-8]}$$

The commercial availability of purified luciferase and synthetic luciferin, together with the inherent sensitivity of the luminescent technique, prompted the present modifications of existing procedures *(23)*. The emitted light can be measured either using the phototube of the Farrand fluorometer or a luminescence biometer. The reaction is specific for ATP, ADP, and AMP. P-creatine can be converted stoichiometrically to ATP and measured in a series of steps.

ATP and P-Creatine 0.1–80 pmol

Reagent: Imidazole-HCl, pH 7.0 (30 m*M* imidazole base, 20 m*M* imidazole-HCl); $MgCl_2$, 1 m*M*; ADP, 60 μ*M*; bovine serum albumin, 0.02%; Ap5A, 10 n*M*; and creatine kinase, 0.05 U/mL (3 μg/mL) (when P-creatine is to be measured).

Luciferin–luciferase reagent: Glycylglycine buffer, pH 8.1 (25 m*M* glycylglycine base, 25 m*M* glycylglycine-HCl); EGTA, 2 m*M*; $MgCl_2$, 2 m*M*; dithiothreitol, 2 m*M*; bovine serum albumin, 0.04%; Ap5A, 10 n*M*; luciferin, 18 μ*M*; and luciferase, 0.05 U/mL (5 μg/mL).

Procedure

Step 1: Samples and standards in volumes up to 10 µL are added to 100 µL of the imidazole reagent without creatine kinase.

Step 2: A 10-µL aliquot is transferred to 200 µL of the luciferin–luciferase reagent in 5 × 60 mm glass tubes. The samples are added, mixed, and read at exactly timed intervals.

Step 3: To the remaining sample (100 µL) are added 3 µL of a 100 µg/mL solution of creatine kinase, and this is incubated 30–45 min at room temperature.

Step 4: A 10-µL aliquot is transferred to 200 µL of luciferin–luciferase reagent in 5 × 60 mm glass tubes. The samples are mixed and read as in Step 2.

Comment

The concentrations and amounts of samples and standards can be adjusted, the sample diluted in 1 mL of the luciferin–luciferase reagent, and the measurement made in the Farrand fluorometer as described above.

Ap5A is routinely included in the reagent to inhibit the action of myokinase *(22)*. In the presence of ADP, or ATP and 5'AMP, the myokinase will catalyze the reaction:

$$2 \text{ ADP} \rightarrow \text{ATP} + 5'\text{AMP} \qquad \text{[Re. 6-9]}$$

The myokinase may persist as a contaminant in tissue extracts, or in frozen-dried tissue sections, as well as exist as a contaminant of the added enzymes. The possibility of troublesome myokinase activity is critical when substrate levels of ADP are added for the conversion of P-creatine to creatine and ATP in the presence of creatine kinase. Both creatine kinase and luciferase from commercial sources were found to have substantial myokinase activity present. Ap5A is added in the P-creatine conversion step to protect against myokinase contamination in the creatine kinase. Therefore, some Ap5A will be transferred to the detection step. With ADP at 3 µ*M* in the luciferin–luciferase reagent, 10 n*M* Ap5A was sufficient to inhibit myokinase activity. Optionally, Ap5A may be included in the luciferin–luciferase reagent. The addition of Ap5A caused no detectable blanks or analytical problems.

ADP is frequently contaminated with 0.5–1% ATP on a molar basis, that may cause troublesome blanks. The ATP can be converted to ADP enzymatically as described under the ATP-ADP section above.

Total Adenylates: ATP, ADP, and 5'AMP

Reagent: Imidazole-HCl, pH 7.0 (30 m*M* imidazole base, 20 m*M* imidazole-HCl); $MgCl_2$, 2 m*M*; KCl, 75 m*M*; P-enol pyruvate, 3 m*M*; and Ap5A, 10 n*M*. For ATP and ADP, add pyruvate kinase, 20 U/mL (100 µg/mL) and for the sum of ATP, ADP, and 5'AMP, add the pyruvate kinase and myokinase, 8 U/mL (25 µg/mL).

Procedure

Step 1: Blanks, standards of ATP, ADP, and 5'AMP (0.2–10 µ*M*), and samples in volumes up to 10 µL are added to 100 µL of reagent and incubated for 60 min.

Step 2: The luciferin–luciferase reagent is added in a vol of 1 mL, the tubes mixed, and read at timed intervals.

Citrate

$$\text{Citrate} \xrightarrow[\text{lyase}]{\text{citrate}} \text{oxaloacetate + acetate}$$

[Re. 6-10]

$$\text{Oxaloacetate + NADH + H}^+ \xrightarrow[\text{dehydrogenase}]{\text{malate}} \text{malate + NAD}^+$$

A method based on this principle was originally described by Moellering and Gruber *(5)*. Citrate lyase has the peculiarity of being inactivated by its own substrate. In the absence of Mg^{2+} or Zn^{2+}, 30 µ*M* citrate will inactivate the enzyme completely in 5 min. The divalent cations protect the enzyme, but at high levels will inhibit by chelating the substrate. The reagent composition is a compromise between these extremes. At the low levels of citrate in the fluorometric assays, the inactivation problem is much less serious than it is at levels required in the spectrophotometer.

Spectrophotometer, 15–120 nmol/mL

Reagent: Tris-HCl buffer, pH 7.6 (25 m*M* in the free base, 75 m*M* in the hydrochloride); NADH, 50–150 µ*M* (at least a 20% excess); $ZnCl_2$, 40 µ*M*; citrate lyase, 0.12 U/mL (15 µg/mL) *(see* Comment *below)*; and malate dehydrogenase (pig heart), 0.3 U/mL (0.4 µg/mL).

Reaction time: Ten minutes.

Conduct of the Assay

The malate dehydrogenase can be incorporated in the reagent, but because of the danger of inactivation of the citrate lyase, it is preferable to add the lyase last after the sample has been added and a preliminary reading made. If the reaction is not complete within 20 min, it may stop because of enzyme inactivation by unreacted citrate. This would only be expected to occur if the enzyme stock solution had deteriorated. The weaker the enzyme, the longer it is exposed to citrate and, therefore, the greater the inactivation in the test system.

Comment

Citrate lyase, as a stock 0.5% solution, keeps for several weeks at 4° if dissolved in imidazole or triethanolamine buffer, pH 7–7.4, containing 300 μM $ZnCl_2$. When necessary, it is diluted further in either of these solutions. The enzyme is usually supplied as a mixture with a protective salt (e.g., $MgCl_2$).

In addition to citrate, oxaloacetate and malate can also inactivate citrate lyase. However, malate, the final product, is not noticeably inhibitory at levels encountered during analyses.

Fluorometer Direct Assay, 0.1–8 nmol/mL

Reagent: The same buffer is used as in Spectrophotometer, except that at lowest citrate levels the concentration may be cut in half to reduce the blank. NADH 0.2–10 μM (20–75% excess over expected oxaloacetate); $ZnCl_2$, 40 μM; citrate lyase, 0.016 U/mL (2 µg/mL); and malate dehydrogenase, 0.3 U/mL (0.4 µg/mL).

Reaction time: Five to 15 min.

Conduct of the Assay

The citrate lyase is added last after an initial reading. At citrate levels under 5 μM, there is little danger of inactivation of the enzyme, unless it is very weak and, therefore, exposed to substrate for a prolonged period.

Kinetics

Study of the kinetics of citrate lyase is complicated by the required metal, which protects the enzyme against its substrate and at the same time chelates the citrate, thereby reducing its free concentration. Thus, increasing Zn^{2+} concentration from 40 to 400 μM prolongs activity in in the presence of 100 μM citrate, but reduces the initial velocity 50%. With 1 μM citrate the same increase in Zn^{2+} concentration reduces initial velocity 80%. Under prescribed assay conditions, the apparent Michaelis constant for citrate is about 15 μM. The metal concentration requirement is very low, at least at low citrate levels. With 1 μM citrate, the rate was essentially the same with 6 and 40 μM $ZnCl_2$.

Creatine

$$\text{Creatine + ATP} \xrightarrow[\text{kinase}]{\text{creatine}} \text{P-creatine + ADP}$$

$$\text{ADP + P-pyruvate} \xrightarrow[\text{kinase}]{\text{pyruvate}} \text{ATP + pyruvate} \qquad \text{[Re. 6-11]}$$

$$\text{Pyruvate + NADH + H}^+ \xrightarrow[\text{dehydrogenase}]{\text{lactate}} \text{lactate + NAD}^+$$

Because of the high Michaelis constants for both substrates, and low inherent activity in the assay direction, a large amount of creatine kinase and relatively high levels of ATP are required. This introduces problems that will be discussed below.

Spectrophotometer, 15–120 nmol/mL

Reagent: Imidazole-HCl, pH 7.5 (40 mM imidazole, 10 mM imidazole-HCl); $MgCl_2$, 5 mM; KCl, 30 mM; NADH 50–150 μM (at least a 20% excess); P-pyruvate, 200 μM; ATP, 1 mM; creatine kinase (rabbit muscle), 3.6 U/mL (200 µg/mL); pyruvate kinase (rabbit muscle), 0.75 U/mL (5 µg/mL); and lactate dehydrogenase (beef heart), 0.2 U/mL (1 µg/mL).
Reaction time: Ten to 20 min.

Conduct of the Assay

Lactate dehydrogenase and pyruvate kinase are ordinarily incorporated in the reagent. It is usually preferable to add the creatine kinase after the sample has been introduced and an initial reading made. In any event, the kinase should not be added to the reagent long before use because of the slow splitting of ATP by present creatine kinase preparations.

Comment

ATP preparations inevitably contain some ADP. This will use up some of the NADH in the reagent. If the contamination is not excessive, it can be ignored, or more NADH added to compensate. However, if desirable, most of the ADP can be removed *(see below,* under *Comment).* Contamination of P-pyruvate with free pyruvate is ordinarily no problem, because only a modest excess is required.

Fluorometer Direct Assay, 1–8 nmol/mL

Reagent: The reagent has the same composition as in Spectrophotometer, except for the following: NADH, 1–10 µ*M* (20–75% excess); P-pyruvate, 25 µ*M*; and ATP, 200 µ*M*.
Reaction time: Fifteen to 30 min.

Conduct of the Assay

The creatine kinase is added last. Because of the large amount used and the liberation of ADP from ATP, reagent blanks become very important, especially at lowest levels. The timing of samples and blanks must be made the same.

Comment

The most serious problem is the slow ATP splitting by ATPase in current creatine kinase preparations. This may be a side reaction of creatine kinase itself. It appears to have about the same Michaelis constant for ATP. In any event, several lots have all given about the same rate of ATP cleavage (i.e., 1 part in 100,000 parts/min with the amount of enzyme in the protocol [200 µg/mL]). With 200 µ*M* ATP, this yields 0.02 µ*M* ADP/min or 0.3 µ*M* in a 15-min incubation (30% of the reading for a 1 µ*M* sample, but only 4% of an 8-µ*M* sample). This limits the useful sensitivity and means that assays should be carefully timed, with final readings taken as soon as the reaction is complete. There is no gain in reducing the ATP concentration, since this prolongs the incubation time without decreasing the proportionate blank.

A larger relative ATP excess is prescribed for the fluorometer in order to keep the assay time and enzyme requirement in bounds. This makes ADP contamination more serious. The best ATP preparations stored at neutral pH may be satisfactory, except for the lowest concentrations. It is, however, possible to reduce the ADP almost to zero as follows. A 100-m*M* ATP solution is diluted with 10 vol of reagent having the composition of that given in Spectrophotometer above, except for omission of creatine kinase. Additional NADH is added if necessary until a check in the spectrophotometer shows that NADH consumption has ceased. The excess NADH is then destroyed by adding 0.02 vol of 5*M* HCl and allowing this to stand 10 min (not more) at room temperature. If the presence of NAD^+ is not objectionable, the solution is simply neutralized with 0.02 vol of 5*M* NaOH. Otherwise, 0.04 vol of 5*M* NaOH is added, and the mixture is heated 10 min at 60°, and then finally neutralized with 0.02 vol of 5*M* HCl.

Kinetics

The V_{max} of creatine kinase in this direction is only 18 µmol/mg/min. The Michaelis constant is at least 15 m*M* for creatine and 0.5–1 m*M* for ATP. Increase in one substrate decreases the apparent Michaelis constant for the other. With 200 µ*M* ATP, the half-time at 25° is 2–3 min with 200 µg of creatine kinase/mL.

2-Deoxyglucose and 2-Deoxyglucose-6-P

$$\text{2-Deoxyglucose + ATP} \xrightarrow{\text{hexokinase}} \text{2-deoxyglucose-6-P + ADP}$$

[Re. 6-12]

$$\text{2-Deoxyglucose-6-P + NADP}^+ \xrightarrow[\text{dehydrogenase}]{\text{glucose-6-P}} \text{2-deoxy-6-P-gluconate + NADPH + H}^+$$

The methods are based on the fact that glucose-6-phosphate dehydrogenase (G6PDH) can oxidize 2-deoxyglucose-6-P (DG6P) as well as glucose-6-P (G6P), but that it takes 1000-fold more enzyme. Thus, G6P is eliminated with a low level of enzyme, and DG6P is then measured by the NADPH formed with a much greater amount of enzyme. Glucose and 2-deoxyglucose (DG) are similarly distinguished after conversion to G6P and DG6P, respectively, with hexokinase, which reacts with DG almost as fast as with glucose.

The necessary use of very high levels of G6PDH to oxidize DG6P creates the danger of serious errors resulting from side reactions from trace impurities in enzyme preparations. In addition, if tissues are to be analyzed, special precautions are needed to prepare the samples for analysis without artifact. For example, even a few seconds of ischemia or anoxia cause serious conversion of brain DG to DG6P. Rather than make the presentation here too long, the analyst is referred to a paper that deals with both of these problems in detail, and provides a range of procedures for samples containing from 0.1 pmol to 2 nmol of DG or DG6P *(24)*. Here we will limit the assays to spectrophotometric procedures for DG6P and for DG, and a fluorometric procedure for each compound in the 1–10 nmol range.

Commercially available G6PDH, whether from baker's yeast or from *Leuconostoc mesenteroides,* is contaminated with variable low levels of 6-P-gluconate dehydrogenase (6PGDH). However, because DG assays are usually made after removal of glucose and G6P by conversion to 6PG, contamination with 6PGDH can produce serious positive errors in DG values if relatively high levels of glucose are present (as will almost always be the case when measuring tissue DG levels). Fortunately, most of the 6PGDH in *Leuconostoc* G6PDH can be easily removed as follows: The lyophilized enzyme is dissolved, at a concentration of about 5 mg/mL, in 50 m*M* 2-methyl-1-propanol-HCl buffer, pH 9.9 (base:acid, 1:1). It is incubated 40 min at 38° and then brought to about pH 8 with 50 µL/mL of 1*M* Tris-HCl (no Tris base).

The enzyme can subsequently be stored at 4° for indefinite periods, apparently without loss. The enzyme presently available from Sigma Chemical Co. (St. Louis, MO) ("Type XXIV") can be treated in this way without reduction of G6PDH activity. However, for reasons we have not been able to determine, the same treatment of lyophilized G6PDH from another company caused loss of about half of the G6PDH activity, but nevertheless destroyed almost all of the 6PGDH activity. Clearly activity and near freedom from 6PGDH should be checked before use.

Spectrophotometer, 15–100 nmol/mL of DG6P and/or DG

The spectrophotometer assays are mainly designed for standardization purposes, or for samples that contain relatively small amounts of G6P or glucose.

Reagent: Tris-HCl buffer, pH 8.1 (10 m*M* Tris base, 10 m*M* Tris-HCl); $MgCl_2$, 3 m*M*; ATP, 0.5 m*M*; $NADP^+$, 0.5 m*M*; G6PDH (*Leuconostoc mesenteroides*), variable amounts; and hexokinase (baker's yeast) 0.6 U/mL.

DG6P Assay

Step 1: If G6P is possibly present, a relatively small amount of G6PDH (0.02 U/mL) is first added, and a reading made 10 min later.

Step 2: A large amount of G6PDH (25 U) is then added, and a final reading made after another 20–30 min.

DG Assay

Step 1: A low level of G6PDH (0.02 U/mL) is added, and a reading taken 10 min later (in case G6P might be present).

Step 2: The hexokinase (0.6 U) is added, and a reading made after 30 min. This increment is a measure of any glucose that may be present.

Step 3: A large amount of G6PDH is added (25 U), and a third reading made after 30–40 min. This increment measures DG, but would also include any preformed DG6P.

Fluorometer Direct Assay (G6P, DG6P, DG, and Glucose) 2–10 nmol Total

Reagent: The same as for Spectrophotometer, except NADP, 100 μ*M*.

Procedure

Step 1: After an initial reading, 0.02 U of G6PDH is added with a second reading after 10 min. The increment of NADPH is a measure of *G6P*.

Step 2: Add 25 U of G6PDH, and read after 40 min. This increment is a measure of *DG6P*.

Step 3: (Destruction of the high level of G6PDH): Heat 20 min at 80°. Then return exactly to room temperature and read.

Step 4: Add 0.6 U hexokinase and 0.02 U of G6PDH in 10 μL of the reagent buffer, and read after 10 min. This increment is a measure of glucose.

Step 5: Add 10 μL of 4*M* HCl followed after 5 min or more by 10 μL of 4*M* NaOH. This destroys the NADPH from the previous steps.

Step 6: Read, then add 25 U of G6PDH, and read after 40 min. This increment is a measure of *DG*.

Dihydroxyacetone Phosphate Method I (with Glycero-P-Dehydrogenase)

$$\text{Dihydroxyacetone-P} + \text{NADH} + \text{H}^+ \xrightarrow[\text{dehydrogenase}]{\text{glycero-P-}} \text{glycero-P} + \text{NAD}^+ \quad \text{[Re. 6-13]}$$

This is the simpler of the two methods presented, but the alternate glyceraldehyde-P-dehydrogenase method has advantages for *direct* measurement of very low levels of dihydroxyacetone-P, particularly in the presence of fluorescent tissue extracts. For *indirect* measurements of very small amounts of the metabolite, the glycero-P-dehydrogenase method has some disadvantages. *(see* Comment after the indirect procedures.)

Spectrophotometer, 15–120 nmol/mL

Reagent: Imidazole buffer pH 7.0 (30 m*M* imidazole, 20 m*M* imidazole HCl); NADH, 50–150 μ*M* (20–50% excess); and glycero-P-dehydrogenase (rabbit muscle), 0.08 U/mL (2 μg/mL).
Reaction time: Two to 4 min.

Conduct of the Assay

The enzyme is added separately after the sample has been added and an initial reading has been made.

Fluorometer Direct Assay, 1–8 nmol/mL

Reagent: Same as for Spectrophotometer, except NADH, 0.2–10 μ*M* (20–75% excess); and glycero-P-dehydrogenase, 0.02 U/mL (0.5 μg/mL, 1 μg/mL with NADH below 1 μ*M*).
Reaction time: Two to 8 min, depending on final NADH concentration *(see* Kinetics *below).*

Conduct of the Assay

Ordinarily, the reaction is started by addition of the enzyme. Glycero-P-dehydrogenase strongly enhances the fluorescence of NADH. Therefore, in the fluorometric assays, the amount of enzyme is kept minimal. Correction for enhancement, if necessary, is made on the basis of the change in reading of blanks. (With the amount of enzyme indicated in Fluorometer Direct Assay, 1 μg/mL, enhancement is not more than 3% with 1 μ*M* NADH.)

Comment

If the imidazole concentration is increased to 0.2*M* to quench flavin fluorescence of biological extracts, the rate of reaction will be decreased by a factor of 3 or 4. To compensate, the enzyme concentration is increased accordingly, or the reaction time is prolonged.

Because of the very favorable kinetic situation *(see page 137, Kinetics),* the reaction will follow nearly first-order kinetics, except with very low NADH concentrations. Nevertheless, with lowest initial NADH levels (0.2–0.5 μ*M*) the rate will fall off badly if more than half of the NADH is used up, and care must be taken to increase the enzyme or the time of reaction.

Indirect Assay, 10–100 pmol

Reagent: The buffer is the same as for Spectrophotometer. NADH, 2 μ*M*; bovine plasma albumin, 0.01%; and glycero-P-dehydrogenase, 0.04 U/mL (1 μg/mL).

The reagent should be prepared within an hour of use and kept in ice. The NADH is as free as possible of NAD^+ (*see* Chapter 1).

Procedure

Step 1: Reagent volumes of 100 μL in fluorometer tubes at 0°. Samples and standards are added in vol of 20 μL or less. Incubate 15 min at 20–25° (water bath); return to ice bath.

Step 2: NADH destroyed with 20 μL of 2*M* HCl (room temperature 5–10 min).

Step 3: NaOH, 1 mL of 6*M* with 10 m*M* imidazole (rapid mixing); 10 min at 60°.

Step 4: Fluorescence reading with tubes exactly at room temperature. (In regard to tissue analyses, *see* Comment in next section.)

Cycling Assay, 0.5–10 pmol

Specific reagent: The buffer is the same as for the spectrophotometer, 1 μ*M* NADH for 0.5–2 pmol, 5 μ*M* for 10 pmol; bovine plasma albumin, 0.02%; ascorbic acid, 2 m*M*; and glycero-P-dehydrogenase, 0.04 U/mL. The ascorbic acid is added before the NADH from a stock 9% (500 m*M*) solution that is stored at –20°. (*See also* directions for Indirect Assay).

Reaction vessels: Oil wells.

Procedure

Step 1: The sample in 1 μL or less is added to 5 μL of reagent in an oil well and incubated 20 min at room temperature. Standards consist of 1 μL of 1–10 μ*M* dihydoxyacetone-P, depending on range of samples.

Step 2: Add 2 μL of 0.5*M* HCl, and allow to stand at least 30 min.

Step 3: Into fluorometer tubes containing 100 μL of NAD cycling reagent with enzymes to give about 8000 cycles/h at 25° (Chapter 5), transfer either 5-μL aliquots for 0.5–2 pmol samples and their standards and blanks, or 1-μL aliquots for 2–10 pmol samples and the corresponding samples and blanks. Heat stop by 2 min in a 95–100° water bath.

Step 4: Add 1 mL of indicator reagent (Chapter 5), and read when reaction is complete.

Comment

If the original samples are not all 1 μL, a correction is made in the calculation for the differences in volume at the end of Step 2.

Cycling Assay, 0.1–0.5 pmol

Specific reagent: The same as in Cycling Assay, 0.5–10 pmol, except that all components are double-strength; NADH is 4 μ*M*.
Reaction vessels: Oil wells.

Procedure

Step 1: To 0.25-μL vol of samples, blanks, and standards (0.5–2 μ*M*) in oil wells are added 0.25-μL vol of reagent. Incubate 20 min at room temperature.
Step 2: Add 0.2 μL of 0.5*M* HCl, and allow to stand at least 20 min.
Step 3: Add to each oil well 10 μL of NAD cycling reagent with enzyme levels to give about 20,000 cycles in 2 h at 25° (Chapter 5). Stop with 2 μL of 0.5*M* NaOH, and heat 20 min in 80° oven.
Step 4: Transfer 10 μL to 1 mL of indicator reagent (Chapter 5), and read when reaction is complete.

Comment

In the indirect assays of Indirect Assay and Cycling Assay, 0.5–10 pmol, the final analysis is for NAD^+. In general, dihydroxyacetone-P levels in tissues and microorganisms are much lower than NAD^+ levels. Although NAD^+ can be destroyed with alkali, dihydroxyacetone-P will be destroyed even more rapidly. Although one can imagine enzymatic ways of destroying NAD^+ alone (reduction to NADH, cleavage by NADase or pyrophosphatase) for analysis of very small tissue samples, a more satisfactory solution would appear to be to use the alternative assay with glyceraldehyde-P dehydrogenase. In this case, the final analysis is for NADH, which is easily destroyed in the tissue samples with acid without harming dihydroxyacetone-P.

Kinetics

The kinetic values are as follows: V_{max} 40 U/mg; K_{DHAP}, 6 μ*M*; and K_{NADH}, 0.3 μ*M*, in 20 m*M* imidazole buffer, ph 7. The V_{max} is that found with current commercial preparations. The pure enzyme may be more active. Baranowski *(25)* reported a value of 260 U/mL.

At equilibrium, at pH 7:

$$\frac{[GOP]\ [NAD^+]}{[DHAP]\ [NADH]} = 12{,}500 \qquad \text{[Eq. 6-1]}$$

Thus, the low Michaelis constants for both substrates and the equilibrium constant are all very favorable for reduction of dihydroxyacetone-P. Even a slight excess of NADH is sufficient to drive the reaction to completion.

From the kinetic constants, it may be calculated that the reaction being 98% complete in 10 min would require only 0.1 µg/mL of glycerol-P dehydrogenase with 10 µ*M* substrate or less, and only 0.3 µg/mL with 100 µ*M* substrate (Chapter 2, Fig. 4). Both sulfate and phosphate are inhibitory.

Fructose

$$\text{Fructose} + \text{ATP} \xrightarrow{\text{hexokinase}} \text{fructose-6-P} + \text{ADP}$$

$$\text{Fructose-6-P} \xrightarrow[\text{isomerase}]{\text{P-gluco-}} \text{glucose-6-P} \qquad \text{[Re. 6-14]}$$

$$\text{Glucose-6-P} + \text{NADP}^+ \xrightarrow[\text{dehydrogenase}]{\text{glucose-6-P}} \text{6-P-gluconolactone} + \text{NADPH} + \text{H}^+$$

Both fructose and glucose are substrates for yeast hexokinase, but for fructose analysis, more enzyme is required because the Michaelis constant is higher than for glucose.

Spectrophotometer, 15–200 nmol/mL

Reagent: Tris-HCl buffer, pH 8.1 (25 m*M* Tris base, 25 m*M* Tris-HCl); $MgCl_2$, 2 m*M*; ATP, 500 µ*M*; $NADP^+$, 500 µ*M*; hexokinase (yeast), 0.7 U/mL (5 µg/mL); P-glucoisomerase (yeast), 0.35 U/mL (1 µg/mL); and glucose-6-P dehydrogenase (yeast) 0.12 U/mL (equivalent to 0.3 µg/mL of crystalline enzyme).
Reaction time: Ten minutes.

Conduct of the Assay

If glucose is known to be absent and the sample does not contribute appreciably to the light absorption, all enzymes can be incorporated in the reagent, and the sample added last after an initial reading. If glucose is also present, and its value required, only glucose-6-P dehydrogenase is added to the reagent. After adding the sample and making a reading, hexokinase is added and a second reading made after all the glucose has reacted (5 min or less). P-Glucoisomerase is then added to complete the reaction (5 min or less, since fructose has been largely converted to fructose-6-P in the interim).

Comment

There is some danger of contamination of hexokinase or glucose-6-P dehydrogenase with P-glucoisomerase. Even the buffer can become contaminated with this enzyme through the growth of microorganisms. This can be detected by prolonged incubation of glucose with the isomerase omitted. Glucose contamination can also readily occur, but will not distort the fructose value if the assay is made in two steps.

Fluorometer, 0.1–10 nmol/mL

Reagent: The buffer, $MgCl_2$, hexokinase, and P-glucoisomerase are the same as in Spectrophotometer. ATP, 200 μM; $NADP^+$, 50 μM; dithiothreitol, 200 μM; and glucose-6-P dehydrogenase, 0.06 U/mL.
Reaction time: Ten minutes.

Conduct of the Assay

Because of the danger of at least trace glucose contamination, it is almost always preferable to incubate the sample for 5 or 10 min with glucose-6-P dehydrogenase and hexokinase, and take a reading before adding P-glucoisomerase (*See* Spectrophotometer *above*).

Comment

Dithiothreitol is added as a precaution in case of contamination with glutathione and glutathione reductase (*see* the glucose-6-P method, page 162). The danger of contamination with P-glucoisomerase has been mentioned above.

Kinetics

The V_{max} for fructose (220 µmol/mg/min) is almost twice as great as that for glucose, but the K_m is four to six times greater (700 μM). Therefore, three times as much enzyme (or time) is required for fructose as for glucose to react completely. Because the K_m is large, the same amount of hexokinase is appropriate in the spectrophotometer and in the fluorometer.

Fructose-6-Phosphate (Method I)

$$\text{Fructose-6-P} \xrightarrow[\text{isomerase}]{\text{P-gluco}} \text{glucose-6-P} \qquad \text{[Re. 6-15]}$$

$$\text{Glucose-6-P} + \text{NADP}^+ \xrightarrow[\text{dehydrogenase}]{\text{glucose-6-P}} \text{6-P-gluconolactone} + \text{NADPH} + \text{H}^+ \qquad \text{[Re. 6-16]}$$

In direct assays, fructose-6-P can be measured by first allowing the glucose-6-P reaction to proceed to completion and then adding isomerase to give extra NADPH equivalent to the fructose-6-P present. However, the equilibrium position of the isomerase reaction favors glucose-6-P by a ratio of about 3:1. Consequently, fructose-6-P levels in living systems are almost invariably much lower than those of glucose-6-P. This increases the potential error in fructose-6-P measurement. An alternate procedure is to destroy with acid the NADPH produced from glucose-6-P and then measure fructose-6-P alone. To illustrate both procedures, the simpler method is given for the spectrophotometer (since it will usually be used only for standardization purposes), and the two-step procedure is given for the fluorometer.

Spectrophotometer, 15–200 nmol/mL

Reagent: Tris buffer, pH 8.1 (25 m*M* Tris base, 25 m*M* Tris-HCl); NADP$^+$, 500 μ*M*; P-glucoisomerase (yeast), 0.35 U/mL (1 μg/mL); and glucose-6-P dehydrogenase (yeast) 0.12 U/mL (equivalent to 0.3 μg/mL of crystalline enzyme).

Reaction time: Ten minutes.

Conduct of the Assay

The sample is added first and a reading made. Glucose-6-P dehydrogenase is added, and a second reading is made after glucose-6-P has reacted completely (5 min or less). Isomerase is added last. Because, as stated above, the second increment is less than the first, any drift in reading will cause substantial error. It is therefore, desirable to carry through an extra sample that receives only the glucose-6-P dehydrogenase. Any change in reading of this sample during the period of isomerase action can be used as a correction.

Fluorometer, 0.1–8 nmol/mL

Reagent: The buffer is the same as in Spectrophotometer. NADP, 50 μ*M*; dithiothreitol, 200 μ*M*; and glucose-6-P dehydrogenase, 0.06 U/mL.
Reaction time: Five to 10 min.

Conduct of the Assay

Step 1: If a value for glucose-6-P is required, readings are made before and after addition of glucose-6-P dehydrogenase.
Step 2: NADPH is destroyed by adding 10 μL of 5*M* HCl (0.05*M* final concentration). After standing 10 min, 10 μL of 5*M* NaOH are added.
Step 3: Glucose-6-P dehydrogenase is added to give the same concentration as in the original reagent. Ordinarily, the fluorometer sensitivity would be increased two- or threefold at this point. After reading, P-glucoisomerase is added to give a concentration of 0.18 U/mL (0.5 μg/mL). A final reading is made when the reaction is over (6–10 min).

Comment

As discussed more fully for glucose-6-P measurement, dithiothreitol is added to guard against the danger of NADPH reoxidation by oxidized glutathione and contaminating glutathione reductase. This danger is lessened by the three-step procedure described. Another danger that must be guarded against is that of P-glucoisomerase contamination during the first step. P-Glucoisomerase is of high activity in most biological materials. It is essential to test a fructose-6-P standard with the first step reagent. Fructose-6-P can be measured successfully without the acidification step if special care is taken in making the reading and in checking for drift after glucose-6-P has reacted, as described for the spectrophotometric procedure.

Kinetics

The V_{max} of yeast P-glucoisomerase is high, about 350 μmol/mg/min, and the Michaelis constant is of the order of 50 μ*M*. Therefore, the enzyme requirement is low, and only slightly more enzyme is required for spectrophotometer levels than for those in the fluorometer. If glucose-6-P is removed as fast as it is formed, the calculated half-time for 1 μg/mL of isomerase (V_{max} 350 μmol/L/min) would be 0.1 min (0.7 × 50/350). 6-P Gluconate is a potent inhibitor that increases the enzyme requirement.

Fructose-6-P, Method II

$$\text{Fructose-6-P} + \text{ATP} \xrightarrow{\text{fructose-6-P kinase}} \text{fructose-1,6-}P_2 + \text{ADP}$$

$$\text{Fructose-1,6-}P_2 \xrightarrow{\text{aldolase}} \text{dihydroxyacetone-P} + \text{glyceraldehyde-3-P}$$

$$\text{Dihydroxyacetone-P} \xrightarrow{\text{triose-P isomerase}} \text{glyceraldehyde-3-P} \qquad \text{[Re. 6-17]}$$

$$2\ \text{Glyceraldehyde-3-P} + 2\ NAD^+\ \text{(arsenate)} \xrightarrow[\text{dehydrogenase}]{\text{glyceraldehyde-3-P}} 2\ \text{3-P-glycerate} + 2\ \text{NADH} + 2\ H^+$$

This method involves four auxiliary enzymes rather than only two, as described for the preceding protocol. Advantages of this method are that the yield of the final product, NADH, is twice the concentration of the original substrate, and the glucose-6-P, which is always higher than fructose-6-P, need not be removed. However, this method includes fructose-1,6-P_2 and the triose phosphates. In many tissues these metabolites are present at levels much lower than those of fructose-6-P, and the correction is usually <10%. If the tissue values of fructose-1,6-P_2 and the triose phosphates are near those of fructose-6-P, simultaneous analyses will enable determination of the exact levels of the metabolites.

Spectrophotometer, 8–100 nmol/mL

Reagent: Imidazole buffer, pH 7.0 (30 m*M* imidazole base, 20 m*M* imidazole HCl); NAD^+, 1 m*M*; Na_2HAsO_4, 1 m*M*; EDTA, 1 m*M*; β-mercaptoethanol, 2 m*M*; $MgCl_2$, 2m*M*; ATP, 1 m*M*; aldolase (rabbit muscle), 0.09 U/mL (10 μg/mL); triose-P isomerase (rabbit muscle), 2.4 U/mL (1 μg/mL); glyceraldehyde-P dehydrogenase (rabbit muscle), 1.8 U/mL (measured in the direction of glyceraldehyde-P formation); and fructose-6-P kinase (rabbit muscle) 0.6 U/mL (1 μg/mL). The dehydrogenase stock suspension is centrifuged to remove most of the $(NH_4)_2SO_4$ and dissolved in buffer.

Reaction time: Ten min for fructose-6-P.

Conduct of the Assay

For standardization purposes, the substrate can be added last to reagent containing all the enzymes. Otherwise, in direct assays, the sample is added first, and the enzymes are added in succession in reverse order with sufficient time after each addition to allow completion of the step.

Fluorometer, Direct Assay, 0.1–5 nmol/mL

Reagent: The reagent is the same as for Spectrophotometer, except for NAD^+, 400 µ*M* (100 µ*M* with 1 µ*M* fructose-6-P or less); aldolase, 0.05 U/mL; glyceraldehyde-P dehydrogenase, 0.5 U/mL; and fructose-6-P kinase, 0.3 U/mL.

Reaction time: Ten minutes for fructose-6-P.

Conduct of the Assay

The sample is added first, followed by glyceradehyde-P dehydrogenase, triose-P isomerase, aldolase, and fructose-6-P kinase, with incubation times determined by appropriate standards. The aldolase available was found to be contaminated with glucose-6-P isomerase; the aldolase is thus kept to a minimum, and the reaction time after the addition of this enzyme kept to 10 min. Glucose-6-P can also be tested at the expected level in the sample, and corrections made if there is a contribution to the blank.

Cycling Assay, 10–40 fmol

Fructose-6-P Reagent: Imidazole, pH 7.5 (85 m*M* imidazole base, 15 m*M* imidazole HCl); NAD^+, 50 µ*M*; Na_2AsO_4, 2 m*M*; EDTA, 1 m*M*; β-mercaptoethanol, 4 m*M*; $MgCl_2$, 4 m*M*; ATP, 100 µ*M*; bovine serum albumin, 0.04%; aldolase, 0.5 U/mL; glyceraldehyde-P dehydrogenase, 0.9 U/mL; triose-P isomerase, 4.8 U/mL; and fructose-6-P kinase, 0.06 U/mL.

Reaction vessels: Oil wells.

Procedure

Step 1: HCl, 0.1 µL of 20 m*M* for samples and blanks, and 0.1 µL of 0.4 µ*M* fructose-6-P in 20 m*M* HCl for standards. Glucose-6-P standards of 2 µ*M* should be included as a check on P-glucoisomerase contamination. Heat 20 min at 80°.

Step 2: Fructose-6-P reagent; 0.1 µL; 10 min at room temperature.

Step 3: To the samples, add 0.1 µL of 0.15*M* NaOH; heat 20 min at 80°.

Step 4: Add to each oil well 2 µL of NAD cycling reagent with enzyme levels to give about 100,000 cycles overnight at 25° (Chapter 5). Stop with 0.5 µL of *1M* NaOH, and heat 20 min.

Step 5: Transfer 2 µL to 1 mL of indicator reagent (Chapter 5), and read when reaction is complete.

Fructose-6-P and Mannose-6-P

$$\text{Mannose-6-P} \xrightarrow{\text{P-mannose isomerase}} \text{fructose-6-P}$$

$$\text{Fructose-6-P} \xrightarrow{\text{P-glucoisomerase}} \text{glucose-6-P} \qquad \text{[Re. 6-18]}$$

$$\text{Glucose-6-P} + \text{NADP}^+ \xrightarrow{\text{glucose-6-P dehydrogenase}} \text{6-P-gluconolactone} + \text{NADPH} + \text{H}^+$$

In direct assays, fructose-6-P and mannose-6-P can be measured after allowing the glucose-6-P reaction to go to completion. The subsequent addition of P-glucoisomerase, followed by P-mannose isomerase, enables the sequential measurement of fructose-6-P and mannose-6-P. However, the equilibrium position of P-glucoisomerase favors glucose-6-P by a ratio of about 3:1, and the equilibrium of P-mannoseisomerase results in a ratio of 1. The potential error for fructose-6-P and mannose-6-P is thereby very large. An alternate procedure is to destroy the NADPH produced from glucose-6-P, and measure fructose-6-P and then mannose-6-P. To illustrate both procedures, the simpler method is given for the spectrophotometer, since it will be used primarily for standardization purposes,. The alternative method for fructose-6-P and mannose-6-P is given for the fluorometric direct assay and a cycling procedure.

Spectrophotometer, 15–200 nmol/mL

Reagent: Tris buffer, pH 8.1 (25 m*M* Tris base, 25 m*M* Tris-HCl); NADP$^+$, 500 µ*M*; P-glucoisomerase (yeast), 0.35 U/mL (1 µg/mL); glucose-6-P dehydrogenase (yeast) 0.12 U/mL (equivalent to 0.3 µg/mL of crystalline enzyme); and P-mannoseisomerase (yeast), 0.6 U/mL.

Reaction times: One to 3 min for glucose-6-P; 5–10 min for fructose-6-P; 30 min for mannose-6-P.

Conduct of the Assay

The sample is added first and a reading made. Glucose-6-P dehydrogenase is added and a second reading is made after glucose-6-P has reacted completely. P-glucoisomerase is then added. Because, as stated above, the second increment is less than the first, any drift in reading will cause substantial error. It is therefore desirable to carry through an extra sample that receives only the glucose-6-P dehydrogenase. Any change in reading of this sample during the period of P-glucoisomerase action can be used as a correction. Finally, the P-mannose isomerase is added. The same precautions recommended for the preceding step apply here; an extra sample with glucose-6-P dehydrogenase and P-glucoisomerase can be used to correct for any drift.

Fluorometer Direct Assay, 0.1–10 nmol/mL

Reagent: The buffer is the same as in Spectrophotometer. $NADP^+$, 50 µ*M*, and glucose-6-P dehydrogenase, 0.06 U/mL.

Reaction times: One to 3 min for glucose-6-P; 5–10 min for fructose-6-P; 30 min for mannose-6-P.

Conduct of the Assay

Step 1: If a value for glucose-6-P is required, readings are made before and after addition of glucose-6-P dehydrogenase.

Step 2: NADPH is destroyed by adding 10 µL of 5*M* HCl (0.05*M* final concentration). Mix, and after standing 10 min, 10 µL of 5*M* NaOH are added and mixed.

Step 3: Glucose-6-P dehydrogenase is added to give the same concentration as in the original reagent. Ordinarily, the fluorometer sensitivity would be increased two- or threefold at this point. After reading, P-glucoisomerase is added to give a concentration of 0.2 U/mL. A reading is made when the reaction is over.

Step 4: P-mannoseisomerase is added to give a concentration of 0.6 U/mL. A final reading is made when the reaction is over (30 min).

Comment

This fourstep procedure enables the serial measurement of glucose-6-P, fructose-6-P, and mannose-6-P. A danger that must be guarded against is that of P-glucoisomerase contamination during the first step. P-glucoisomerase is of high activity in most biological materials. It is essential to test a fructose-6-P standard with the first-step reagent.

Cycling Assay, 0.1–1 pmol

Glucose-6-P reagent: Tris buffer, pH 8.1 (50 m*M* Tris base, 50 m*M* Tris HCl); $NADP^+$, 30 µ*M*; glucose-6-P dehydrogenase, 0.12 U/mL; and bovine serum albumin, 0.04%.

Fructose-6-P reagent: Tris-HCl buffer, pH 8.5 (75 m*M* Tris base, 25 m*M* Tris-HCl); $NADP^+$, 30 μ*M*; glucose-6-P dehydrogenase, 0.12 U/mL; P-glucoisomerase, 3.5 U/mL; and bovine serum albumin, 0.04%.

Mannose-6-P reagent: This is the same as for fructose-6-P, with the addition of 1.2 U/mL of phosphomannose isomerase.

Procedure

Step 1: HCl, 1 μL of 0.02*M* for frozen-dried samples and blanks, or 1 μL of perchloric acid tissue extract, 1 μL of 1–5 μ*M* glucose-6-P, 1 μL of 0.5–3 μ*M* fructose-6-P, or 1 μL of 0.5–3 μ*M* mannose-6-P; standards all in the same medium as the samples.

Step 2: Glucose-6-P reagent; 1 μL; 20 min at room temperature.

Step 3: HCl, 2 μL of 0.2*M*; 10 min at room temperature.

Step 4: Remove 2 μL to oil wells in another rack for fructose-6-P measurement. Add 2 μL of fructose-6-P reagent; 20 min at room temperature.

Step 5: Add 5 μL of 0.3*M* NaOH; 20 min at 80°.

Step 6: Transfer 5-μL aliquots to fluorometer tubes containing 100 μL of NADP cycling reagent with enzyme levels to give about 10,000 cycles in 60 min at 38° (Chapter 5).

Step 7: Add 1 mL of indicator reagent (Chapter 5), and read when reaction is complete.

Step 8: To the remaining sample at Step 3, add 2 μL of mannose-6-P reagent; 45 min at room temperature.

Step 9: Add 5 μL 0.3*M* NaOH; heat 20 min at 80°.

Steps 10 and 11: These are identical to Steps 6–7 for fructose-6-P.

Kinetics

The V_{max} of yeast P-glucoisomerase is high, about 350 μmol/mg/min, and the Michaelis constant is of the order of 50 μ*M*. Therefore, the enzyme requirement is low and only slightly more enzyme is required for spectrophotometer levels than for those in the fluorometer. If glucose-6-P is removed as fast as it is formed, the calculated half-time for 1 μg/mL of P-glucoisomerase (V_{max} 350 μmol/L/min) would be 0.1 min ($0.7 \times 50/350$). 6-P-gluconate is a potent inhibitor, which increases the enzyme requirement.

The V_{max} of P-mannose isomerase is 60 μmol/mg/min, and the Michaelis constant for mannose-6-P is near 1.5 m*M*. If both the glucose-6-P and frucose-6-P were removed instantaneously, with 0.6 U/mL the half-time would be 1.75 min, with a calculated completion time of 10–15 min; in actual practice, it is advisable to allow 30 min.

Fructose-1,6-Bisphosphate: Method 1

$$\text{Fructose-1,6-P}_2 \xrightarrow{\text{aldolase}} \text{dihydroxyacetone-P} + \text{glyceraldehyde-P}$$

$$\text{Glyceeraldehyde-P} \underset{\text{isomerase}}{\overset{\text{triose}}{\longrightarrow}} \text{dihydroxyacetone-P} \qquad \text{[Re. 6-19]}$$

$$\text{2 Dihydroxyacetone-P} + \text{2NADH} + 2\ H^+ \underset{\text{dehydrogenase}}{\overset{\text{glycero-P}}{\longrightarrow}} \text{2 glycero-P} + 2\ NAD^+$$

Fructose bisphosphate can be measured with the aid of either glycero-P-dehydrogenase or glyceraldehyde-P dehydrogenase. Each has advantages in certain situations.

Spectrophotometer, 8–60 nmol/mL

Reagent: Imidazole, pH 7 (30 m*M* imidazole, 20 m*M* imidazole-HCl); NADH, 50–150 µ*M* (20–50% excess); aldolase (rabbit muscle), 0.02 U/mL (2 µg/mL); triose-P isomerase (rabbit muscle), 1.2 U/mL (0.5 µg/mL); and glycero-P-dehydrogenase (rabbit muscle), 0.16 U/mL (4 µg/mL).
Reaction time: Two to 4 min.

Conduct of the Assay

The enzymes can be added stepwise to measure successively dihydroxyacetone-P, glyceraldehyde-P, and fructose bisphosphate. If fructose bisphosphate alone is to be measured, glycero-P-dehydrogenase and triose-P isomerase can be incorporated into the reagent.

Fluorometer, Direct Assay, 0.05–4 nmol/mL

Reagent: The buffer and triose-P isomerase are the same as for the spectrophotometer. NADH, 0.2–10 µ*M* (20–70% excess); aldolase, 0.004 U/mL (0.4 µg/mL); and glycero-P-dehydrogenase, 0.04 U/mL (1 µg/mL).
Reaction time: Two to 5 min.

Conduct of the Assay

As for Spectrophotometer.

Comment

Because glycero-P-dehydrogenase enhances the fluorescence of NADH, the concentration is kept relatively low. At the recommended concentration of 1 μg/mL, enhancement is about 3% with 1 μ*M* NADH and less than this at higher NADH concentrations.

Fluorometer, Indirect Assay

It is simple enough to design indirect assays for fructose bisphosphate with almost any desired sensitivity. However, with the use of glycero-P-dehydrogenase for the final step, NAD^+ is the product ultimately measured. Most living cells contain much more NAD^+ than fructose bisphosphate. Therefore, to measure smaller amounts of fructose bisphosphate, the alternative procedure with glyceraldehyde-P dehydrogenase as the auxiliary reaction is recommended.

When the presence of NAD^+ is not a problem, sensitive indirect procedures can follow the plan given for dihydroxyacetone-P (Method I), using for the first steps a reagent of the composition given in Fluorometer, Direct Assay, but with added ascorbic acid (2 m*M*) and bovine plasma albumin (0.02%) as needed.

Kinetics

In imidazole buffers at pH 7, the V_{max} for aldolase is relatively low, about 10 μmol/mg/min (10 U/mg), but this is compensated for by an unusually small K_m, 0.6 μ*M*. Therefore, very little enzyme is needed for assay of low levels of fructose bisphosphate, but the amount must be increased at higher levels.

The V_{max} for triose isomerase is very large, 2400 μmol/mg/min, but this is offset somewhat by the large K_m, about 50 μ*M*. The kinetics of glycero-P-dehydrogenase are given in the first method for dihydroxyacetone-P.

Fructose-1,6-Bisphosphate (Method II) Dihydroxyacetonephosphate (Method II), and Glyceraldehydephosphate

$$\text{Fructose-1,6-P}_2 \xrightarrow{\text{aldolase}} \text{dihydroxyacetone-P + glyceraldehyde-P}$$

$$\text{Dihydroxyacetone-P} \xrightarrow[\text{isomerase}]{\text{triose-P}} \text{glyceraldehyde-P} \qquad \text{[Re. 6-20]}$$

$$\text{Glyceraldehyde-P + NAD}^+ \text{ + arsenate} \xrightarrow[\text{GAP dehydrogenase}]{} \text{3-P-glycerate + NADH + H}^+$$

Compared to the alternative methods for these compounds (with glycero-P-dehydrogenase), these methods have the advantage that readings increase with increasing amounts of substance. Each substrate can be measured in succession, if desired, although this may be impractical when the concentrations are widely different. Of the three, glyceraldehyde-P is the most difficult to measure in most biological materials, because the levels are exceedingly low. (The equilibrium ratio between the two triose phosphates is abut 25:1 in favor of dihydroxyacetone-P.) For standardization purposes, the glycerol-P dehydrogenase methods are usually preferable, since the reaction is much faster.

Spectrophotometer, 8–100 nmol/mL

Reagent: Imidazole buffer, pH 7.5 (40 m*M* imidazole, 10 m*M* imidazole-HCl); NAD^+, 1 m*M*; Na_2HAsO_4, 1 m*M*; EDTA, 1 m*M*; mercaptoethanol, 2 m*M*; aldolase (rabbit muscle), 0.09 U/mL (10 µg/mL); triose-P-isomerase (rabbit muscle), 2.4 U/mL (1 µg/mL); and glyceraldehyde-P dehydrogense (rabbit muscle), 1.8 U/mL (50 µg/mL measured in the direction of glyceraldehyde-P formation).The dehydrogenase stock suspension is centrifuged to remove most of the $(NH_4)_2SO_4$ and dissolved in buffer.

Reaction time: Five to 10 min for glyceraldehyde-P, 20–25 min for dihydroxyacetone-P, and only a little longer for fructose diphosphate.

Conduct of the Assay

For standardization purposes, the substrate concerned can be added last to the reagent containing the necessary enzymes. Otherwise, the sample is added first, and the enzymes are added in succession with sufficient time after each addition to complete that step. One mol of fructose bisphosphate, of course, yields 2 mol of NADH.

Comment

Glyceraldehyde-P readily forms Schiff bases, which may dissociate slowly. Consequently, standard solultions may react in a biphasic manner with a rapid initial phase followed by a slower phase. In analyzing glyceraldehyde-P solutions, it is necessary to be aware of the slightest contamination with triose-P isomerase. The presence of a trace of this enzyme would cause part of the substrate to be converted to dihydroxyacetone-P. Once formed, dihydroxyacetone-P would not be converted back to glyceraldehyde-P fast enough to be measured unless a very much larger amount of triose-P isomerase were to be added. The safe procedure is, in fact, to add triose-P isomerase, at the level shown above, after the dehydrogenase reaction has stopped, to see if there is a second increment.

The $(NH_4)_2SO_4$ is removed from the dehydrogenase, because sulfate is inhibitory to both triose-P isomerase and glyceraldehyde-P dehydrogenase. It should not exceed 2 m*M* during the reaction.

Fluorometer Direct Analysis, 0.1–10 nmol (as Triose-P)

Reagent: The reagent is that given in Spectrophotometer, with the following changes: NAD^+, 400 µ*M* (100 µ*M* with 1 µ*M* substrate or less), aldolase, 0.02 U/mL (2 µg/mL); and glyceraldehyde-P dehydrogenase, 0.7 U/mL (20 µg/mL).

Reaction times: Two minutes for glyceraldehyde-P, 10–15 min for dihydroxyacetone-P, and 15–20 min for fructose bisphosphate.

Conduct of the Assay

The sample is added first, followed by glyceraldehyde-P dehydrogenase, triose isomerase, and aldolase in succession, with delay between additions as indicated by the reaction times of appropriate standards. If fructose bisphosphate is present in a substantially higher amount than the triose phosphates, it may be desirable to use a more sensitive setting for the first two steps and then to reduce the sensitivity for the final aldolase step.

Cycling Assay, 2–20 pmol (As Triose-P)

The method given is for the sum of fructose bisphosphate and triose phosphates. For measuring these metabolites individually, *see* Comment *below*.

Specific reagent: The reagent is the same as in Spectrophotometer, except as follows: NAD^+, 30 µ*M*; amd bovine plasma albumin, 0.02%.

Reaction vessels: Oil wells.

Procedure for Freeze-Dried Tissue Samples

Step 1: For samples and blanks, 1 µL of 2–20 µ*M* fructose bisphosphate in 0.02*M* HCl for standards; heat 10 min at 60° (20 min at 80° in the case of oil wells).

Step 2: Specific reagent; 5 µL; 40 min at room temperature.

Step 3: NaOH, 5 µL of 0.1*M*; 15 min at 60° (20 min at 80° in the case of oil wells).

Step 4: Transfer 5-µL aliquots to fluorometer tubes containing 100 µL of NAD cycling reagent with enzyme levels to give about 1000 cycles in 60 min at 25° (Chapter 5). Heat stop by 2 min in a 95–100° water bath.
Step 5: Add 1 mL of indicator reagent (Chapter 5), and read when reaction is complete.

Comment

Glyceraldehyde-P, or the sum of both triose phosphates, can be obtained by using glyceraldehyde-P dehydrogenase alone or in combination with triose-P isomerase. Fructose bisphosphate can be measured alone if the samples are made alkaline after the acid treatment to destroy the triose phosphates. For example, after the acid treatment in Step 1, an equal vol (1 µL) of 0.04*M* NaOH is added, and the samples are heated 5 min at 80°.

Because glyceraldehyde-P levels are exceedingly low in living cells, the indirect procedure with glyceraldehyde-P dehydrogenase and amplification by cycling is the method of choice. However, great care must be taken if true values are to be obtained. The dehydrogenase is not very specific, and the possibility that other compounds are contributing to the result should be examined critically.

Cycling Assay, 0.1–1 pmol (as Triose-P)

Specific reagent: The same as in Cycling Assay, 2–20 pmol, but all components double-strength.
Reaction vessels: Oil wells.

Procedure for Freeze-Dried Tissue Samples

Step 1: HCl, 0.05 µL of 0.02*M*, for samples and blanks; and 0.05 µL of 2–10 µ*M* fructose-1,6-bisphosphate for standards; heat 20 min at 80°.
Step 2: Specific reagent; 0.05 µL; 40 min at room temperature.
Step 3: NaOH, 0.5 µL of 0.1*M*; 20 min at 80°.
Step 4: Add to each oil well 10 µL of NAD cycling reagent with enzyme levels to give about 10,000 cycles in 60 min at 25° (Chapter 5). Stop with 2 µL of 0.5*M* NaOH, and heat 20 min in an 80° oven.
Step 5: Transfer 10 µL to 1 mL of indicator reagent (Chapter 5), and read when reaction is complete.

Kinetics

The kinetic constants for glyceraldehyde-P dehydrogenase are favorable. The V_{max} is 10 µmol mg^{-1} min^{-1}, and the apparent K_m is about 10 µ*M* under analytical conditions. Thus, with a concentration of 1 µg/mL

(10 µmol/L/min), the half-time for low levels of substrate is about 0.7 min. Probably because of partial dissociation, some loss of activity occurs on dilution. This is lessened, but not overcome by the presence of NAD^+ *(26)*. The K_m for NAD^+ is about 30 µ*M* under analytical conditions. Sulfate is inhibitory; a 5-m*M* concentration inhibits the rate 65% (tested with 2 µ*M* glyceraldehyde-P).

With dihydroxyacetone-P analysis, the problem is not triose-P isomerase. This is a very active enzyme, 2400 U/mg measured in the fast direction (dihydroxyacetone-P formation). It is perhaps a third as fast in the reverse direction, and the K_m is of the order of 250 µ*M*. The problem is that, because of the unfavorable equilibrium constant, no matter how much isomerase is added, there can never be more than the 4% of the triose-P present as glyceraldehyde-P. Therefore, the conversion of dihydroxyacetone-P to 3-P glycerate must be at least 25 times slower than for glyceraldehyde-P, and a high level of dehydrogenase is needed. Triose-P isomerase is inhibited 50% by 5 m*M* P_i and 65% by 5 m*M* sulfate (tested with 2 µ*M* dihydroxyacetone-P).

The kinetics of aldolase have been given in Method I for fructose-1,6-bisphosphate. The equilibrium of the aldolase step is concentration-dependent and, at analytical levels, is in favor of the triose phosphates. Hence, with moderate amounts of aldolase, the reaction is practically as fast as with dihydroxyacetone-P. Thus, the rate-limiting step for all three substrates turns out to be glyceraldehyde-P dehydrogenase.

Fumarate

$$\text{Fumarate} \xrightarrow{\text{fumarase}} \text{malate}$$

$$\text{Malate} + NAD^+ \xrightarrow[\text{dehydrogenase}]{\text{malate}} \text{oxaloacetate} + NADH + H^+ \qquad \text{[Re. 6-21]}$$

$$\text{Oxaloacetate} + \text{glutamate} \xrightarrow[\text{transaminase}]{\text{aspartate}} \text{aspartate} + \alpha\text{-ketoglutarate}$$

Spectrophotometer, 15–200 nmol/mL

Reagent: Except for the addition of fumarase, the reagent is the same as that given for the spectrophotometer in Method I for malate. Fumarase (pig heart), 18 U/mL (50 µg/mL, measured from malate to fumarate). If the enzyme is supplied as suspension in $(NH_4)_2SO_4$, it is centrifuged and dissolved in neutral buffer to remove most of the NH_4^+.

Reaction time: Eight to 15 min.

Conduct of the Assay

For standardization purposes, the fumarate can be added last after a preliminary reading. For other purposes, the reagent is prepared with the enzymes omitted. The sample is added; after a preliminary reading, malate dehydrogenase and the transaminase are added. After allowing time for any malate present to react (5 or 10 min), a second reading is made, and fumarase is then added to measure fumarate itself. Because the enzymes may contribute small increments to the reading, blanks are needed. All of the enzymes slowly lose activity at the alkaline pH; therefore, they should not be left in the reagent for more than 20–30 min before starting the reaction.

Comment

Fumarate levels in biological materials are usually much lower than those of malate. At equilibrium in animal tissues, the ratio would be about 1:4. Therefore, to measure accurately the increment in absorption owing to fumarate, it is necessary to be sure the malate reaction is complete before adding fumarase. The best procedure may be to prepare extra samples that are treated identically, except that fumarase is omitted. In order not to decrease the pH, the $(NH_4)_2SO_4$ contributed by the enzyme preparations should not exceed 10 m*M* *(see* Malate; Method I). *(See also* Comment for the fluorometric procedure.)

Fluorometer Direct Assay, 0.1–10 nmol/mL

Reagent: Except for fumarase addition (8 U/mL), the reagent is the same as given for the fluorometer in Method I for malate. Most of the $(NH_4)_2SO_4$ is removed as for the spectrophotometer.

Reaction time: Ten to 20 min.

Conduct of the Assay

See spectrophotometric method. If fumarate is very low compared to malate, precision can be increased by destroying the NADH from the malate as in the next protocol.

Fluorometer, Two-Step Assay

Reagent: The first-step reagent is the same as in Fluorometer Direct Assay, except that fumarase is omitted and no more than 3 m*M* glutamate is used.

Conduct of the Assay

Step 1: The malate reaction is carried out as in the one-step procedure and readings are made if the value of malate is required.

Step 2: First 15 µL of 5*M* HCl are added. After 10 min, 15 µL of 5*M* NaOH are added.

Step 3: Malate dehydrogenase and aspartate transaminase are added first to make sure that no malate remains. A reading is made after 5 or 10 min, after which fumarase is added to complete the reaction.

Comment

The HCl and NaOH should be exactly equivalent. To be sure of this, they are titrated against each other and the concentrations adjusted if necessary. The same micropipet can be used for acid and base additions. To destroy quickly the NADH from the malate, sufficient HCl must be added to reach pH 2 or less. Glutamate and sulfate increase the HCl required. The adequacy can be checked by following the rate of decrease in fluorescence of one sample upon acidification.

Kinetics

In order to favor the malate dehydrogenase step, the analytical reaction is conducted at pH 9.9, which is far from the pH optimum for fumarase (about 6.5; *27*). Under analytical conditions with a commercial preparation, we observed the V_{max} to be 12 µmol/mg/min and the K_m 1250 µ*M*. Thus, with 1 µg/mL of enzyme, the half-time is about 70 min (0.7 × 1250/12). For these reasons, a relatively high level of enzyme is required. The kinetics for malate dehydrogenase and transaminase have been described in Malate: Method I.

γ-Aminobutyric Acid

$$\gamma\text{-aminobutyric acid} + \alpha\text{-ketoglutarate} \xrightarrow[\text{transaminase}]{\text{GABA}} \text{succinate semialdehyde} + \text{glutamate}$$

[Re. 6-22]

$$\text{Succinate semialdehyde} + \text{NADP}^+ \xrightarrow[\text{dehydrogenase}]{\text{succinate semialdehyde}} \text{succinate} + \text{NADPH} + \text{H}^+$$

γ-Aminobutyric acid (GABA) is measured by the sequential reaction of two enzymes, GABA transaminase and succinate semialdehyde dehydrogenase (SSADH). At present, these enzymes are available commercially only as a mixture ("Gabase"), in which the activity of the second enzyme greatly exceeds that of the transaminase.

Spectrophotometer, 15–200 nmol/mL

Reagent: Sodium pyrophosphate buffer, pH 8.6 (70 m*M* $Na_4P_2O_7$, 30 m*M* $Na_3HP_2O_7$); α-ketoglutarate, 5 m*M*; $NADP^+$, 500 μ*M*; dithiothreitol, 0.1 m*M*; EGTA, 2 m*M*; and Gabase, 0.08 U/mL (130 μg/mL of current preparations).
Reaction time: Thirty minutes.

Fluorometer Direct Assay, 3–10 nmol/mL

Reagent: The reagent is the same as in Spectrophotometer, except that the $NADP^+$ is 50 μ*M*.
Reaction time: Thirty minutes.

Conduct of the Assay

The blank owing to the enzyme is the equivalent of 0.1 nmol of NADPH/μg of Gabase, so the sensitivity of the direct fluorometric assay is less than that of other enzymatic fluorometric assays. The reducing agent (dithiothreitol) contributes to the blank, but this is minimized by the addition of EGTA.

Fluorometer, Two-Step Assay, 1–10 nmol/mL

Reagent: Same as Fluorometer Direct Assay, except the concentrations of all the components are doubled.
Reaction tubes: 10 × 75 mm glass tubes.

Procedure for Tissue Extracts

Step 1: Blanks, standards (1–10 nmol), or samples in 0.02*M* HCl are added to the fluorometer tubes in a vol of 25–100 μL.
Step 2: Reagent is added in a volume equal to the sample volume; 30 min at room temperature.
Step 3: One milliliter of 0.1*M* sodium pyrophosphate buffer, pH 8.6, is added, and the tubes read in the fluorometer.

Fluorometer Strong Alkali Assay, 10–100 pmol

Reagent: The same as in Fluorometer, two-step procedure.

Procedure

Steps 1 and 2: The same as in Fluorometer, two-step procedure.
Step 3: Add 100 μL of phosphate buffer, pH 12.3 (125 m*M* Naa_3PO_4, 125 m*M* K_2HPO_4); 15 min at 60°.

Step 4: Add 1 mL 6*M* NaOH, containing 0.03% H_2O_2 and 10 m*M* imidazole; 15 min at 60°. Cool exactly to room temperature before reading.

Cycling Assay, 1–10 pmol

Reagent: Same as in Spectrophotometer, except $NADP^+$ is reduced to 10 μ*M*.
Reaction vessels: Oil wells.

Procedure for Freeze-Dried Tissue Samples

Step 1: HCl, 1 μL of 0.02*M* for samples and blanks; and 1 μL of 1–10 μ*M* GABA in 0.02*M* HCl for standards; 20 min at 60°.
Step 2: GABA reagent; 5 μL; 20 min at room temperature.
Step 3: NaOH, 2 μL 0.5*M*, 30 min at 60°.
Step 4: Transfer 5-μL aliquots to fluorometer tubes containing 100 μL of NADP cycling reagent with enzyme levels to give about 1500 cycles in 60 min at 38° (Chapter 5). Heat stop by 2 min in a 95–100° water bath.
Step 5: Add 1 mL of indicator reagent (Chapter 5), and read when reaction is complete.

Kinetics

The determination of the kinetics of the reaction is complicated by the fact that the two enzymes, GABA-transaminase and SSADH, are presently available only as a mixture. The first-order rate constant for GABA-transaminase increases with increasing pH from 8 to 10, whereas that of SSADH decreases. In the commercial preparations tested, the activity of the SSADH was an order of magnitude greater than the transaminase. Although at a more alkaline pH (9.9) the difference between the first-order rate constants was minimized, the enzymes proved to be less stable than at pH 8.6, and the choice of buffer and pH is thus a compromise. In the sodium pyrophosphate buffer at pH 8.6, the commercial preparation had a V_{max} for succinate semialdehyde dehydrogenase of 4.5 μmol/min/mg, and a K_m for succinate semialdehyde of 23 μ*M*; the GABA-transaminase had a V_{max} of 0.89 μmol/min/mg and a K_m for GABA of 475 μ*M*. Since the transaminase is obviously the limiting enzyme, the assay is designed to favor this activity. For a completion time of 3 min, approx 180 μg/mL are required. The activity of the SSADH present is in excess of that required to complete the reaction. The first-order rate constant for the amount of enzyme specified is 35 min^{-1}, fourfold greater than that of the transaminase.

Comment

Commercial Gabase adds an appreciable direct fluorescent blank, and the blank increases with incubation, presumably because of a contaminating enzyme or enzymes. The presence of reducing agent accelerates the blank formation, so when low levels of GABA are to be measured, the reducing agent can be omitted. When acid or alkali was heated in oil wells prior to addition of the GABA reagent, the blank value increased. This source of blank can be minimized by substituting mineral oil for the mineral-oil hexadecane mixture, reducing the vol of the first step, and heating 30 min at 60° rather than 20 min at 80°.

Glucose

$$\text{Glucose} + \text{ATP} \xrightarrow{\text{hexokinase}} \text{glucose-6-P} + \text{ADP}$$

[Re. 6-23]

$$\text{Glucose-6-P} + \text{NADP}^+ \xrightarrow[\text{dehydrogenase}]{\text{glucose-6-P}} \text{6-P-gluconolactone} + \text{NADPH} + \text{H}^+$$

Spectrophotometer, 15–200 nmol/mL

Reagent: Tris-HCl buffer, pH 8.1 (25 m*M* Tris base, 25 m*M* Tris-HCl); $MgCl_2$, 1 m*M*; ATP, 500 µ*M*; $NADP^+$, 500 µ*M*; hexokinase (yeast), 0.5 U/mL (2 µg/mL); and glucose-6-P dehydrogenase (yeast), 0.08 U/mL, equivalent to 0.2 µg/mL of crystalline enzyme.

Reaction time: Four to 6 min.

Conduct of the Assay

For standardization purposes, both enzymes can be incorporated in the reagent and the reaction started with glucose. Otherwise, hexokinase is omitted from the reagent and added last after an initial reading. If glucose-6-P is present the initial reading is not made until after time for it to react (2–4 min).

Fluorometer Direct Assay, 0.1–10 nmol/mL

Reagent: The buffer and $MgCl_2$ are the same as in Spectrophotometer. Dithiothreitol, 0.5 m*M*; ATP, 300 µ*M*; $NADP^+$, 30 µ*M*; hexokinase, 0.3 U/mL (1 µg/mL); and glucose-6-P dehydrogenase, 0.02 U/mL.

Reaction time: Five to 10 min.

Conduct of the Assay

Ordinarily, glucose-6-P dehydrogenase is incorporated into the reagent. The sample is added; after allowing a few minutes for any glucose-6-P to react, a reading is made and hexokinase is added. If glucose-6-P is to be measured in the same sample, both enzymes are omitted from the reagent. The sample is added first followed by the glucose-6-P dehydrogenase. Usually glucose-6-P levels are much lower than the glucose levels. Therefore, the readings before and after the dehydrogenase addition are made at a relatively sensitive setting. Before adding hexokinase, the sensitivity is reduced according to the glucose range expected and a new reading made, preferably with a stronger quinine standard. The glucose reaction is then initiated with hexokinase.

Cycling Assay, 1–10 pmol

Reagent for glucose reaction: Buffer, $MgCl_2$ and enzyme concentrations as in Fluorometric Direct Assay. ATP, 100 µ*M*; $NADP^+$, 5 µ*M*; and bovine plasma albumin, 0.02%.

Reaction vessels: Oil wells.

Procedure for Freeze-Dried Tissue Samples

Step 1: HCl, 1 µL of 0.02*M*, for samples and blanks, and 1 µL of 10 µ*M* glucose in 0.02*M* HCl for standards; 10 min at 60°.

Step 2: Glucose reagent; 5 µL; 20 min at room temperature.

Step 3: NaOH, 1 µL of 0.5*M*; 15 min at 60°.

Step 4: Transfer 5-µL aliquots to fluorometer tubes containing 100 µL of NADP cycling reagent with enzyme levels to give about 1500 cycles in 60 min at 38° (Chapter 5). Heat stop by 2 min in a 95–100° water bath.

Step 5: Add 1 mL of indicator reagent (Chapter 5), and read when reaction is complete.

Comment

The acid treatment is sufficient to destroy preformed NADPH and tissue enzymes, but too mild to hydrolyze an appreciable amount of glycogen. It may, however, partially break down UDP-glucose to glucose and UDP. For analysis of glucose samples in which enzyme destruction is unnecessary, the acid treatment step can of course be omitted.

Cycling Assay, 0.1–1 pmol

Reagent for glucose reaction: The same as in Cycling Assay, 1–10 pmol, except: ATP, 200 µ*M*; and $NADP^+$, 20 µ*M*.

Reaction vessels: Oil wells.

Procedure for Freeze-Dried Tissue Samples

Step 1: HCl, 0.05 µL of 0.02*M*, for samples and blanks, and 0.05 µL of 2–20 µ*M* glucose in 0.02*M* HCl for standards; 20 min in 80° oven.
Step 2: Glucose reagent; 0.5 µL; 20 min at room temperature.
Step 3: NaOH, 5 µL of 0.05*M*; 20 min in 80° oven.
Step 4: Transfer 4-µL aliquots to fluorometer tubes containing 50 µL of NADP cycling reagent with enzyme levels to give abut 15,000 cycles in 60 min at 38° (Chapter 5). Heat stop by 2 min in a 95–100° water bath.
Step 5: Add 1 mL of indicator reagent (Chapter 5), and read when reaction is complete.

Cycling Assay, 10–50 fmol

Reagent for glucose reaction: Buffer and $MgCl_2$ as in Fluorometric Direct Assay. ATP, 200 µ*M*; $NADP^+$, 20 µ*M*; hexokinase, 1.5 U/mL; glucose-6-P dehydrogenase, 0.14 U/mL; and bovine plasma albumin, 0.04%.
Reaction vessels: Oil wells.

Procedure

Step 1: As in Cycling Assay, 0.1–1 pmol, except the vol is 0.015 µL and standard concentrations are 1–4 µ*M*.
Steps 2 and 3: The same as in Cycling Assay except vol are reduced to 0.02 µL and 0.2 µL, respectively.
Step 4: Add to each oil well 5 µL of NADP cycling reagent with enzyme levels to give about 100,000 cycles overnight at 15° (Chapter 5). Stop with 2 µL of 0.5*M* NaOH, and heat 20 min in 80° oven.
Step 5: Transfer 5 µL to 1 mL of indicator reagent (Chapter 5), and read when reaction is complete.

Comment

Alkali is added in Step 5 to dissolve the cycling reagent enzymes. Otherwise, after heating, the coagulum would clog the transfer pipet.

Kinetics

The V_{max} for hexokinase is about 300 µmol/mg/min. The K_m for ATP with 1 m*M* glucose is 30 µ*M*; the K_m for glucose with 300 µ*M* ATP is 160 µ*M*. Thus, with low substrate levels, about five times more enzyme is required for measuring glucose than for ATP. With 1 µg/mL of hexokinase, $t_{1/2}$ (for glucose) = 0.7 × 160/300 = 0.37 min.

Glucose-1-Phosphate

$$\text{Glucose-1-P} \xrightarrow{\text{P-glucomutase}} \text{glucose-6-P}$$

[Re. 6-24]

$$\text{Glucose-6-P} + \text{NADP}^+ \xrightarrow[\text{dehydrogenase}]{\text{glucose-6-P}} \text{6-P-gluconolactone} + \text{NADPH} + \text{H}^+$$

Glucose-1-P levels in living material are almost invariably low compared to those of glucose-6-P because of the position of equilibrium between the two. Therefore, glucose-6-P must either be removed first or glucose-1-P measured as a small reading on top of a large reading from glucose-6-P *(see* Comment *below).*

Spectrophotometer, 15–200 nmol/mL

Reagent: Tris-HCl buffer, pH 8.1 (25 m*M* Tris base, 25 m*M* Tris-HCl); 1 m*M* $MgCl_2$; 0.1 m*M* EDTA; $NADP^+$, 500 μ*M*; glucose-1,6-P_2, 0.5 μ*M*; P-glucomutase (rabbit muscle), 0.05 U/mL (0.25 μg/mL); and glucose-6-P dehydrogenase (yeast), 0.04 U/mL, equivalent to 0.1 μg/mL of crystalline enzyme.
Reaction time: Four to 6 min.

Conduct of the Assay

If glucose-6-P is present, the dehydrogenase is added first and a reading is made after the reaction is complete (3–5 min). The mutase is then added to measure glucose-1-P. If substantially more glucose-6-P than glucose-1-P is present, it should be made certain that all the glucose-6-P has reacted. Because the G6PDH reaction is reversible, the reaction is not complete until the 6-P-gluconolactone has been hydrolyzed. This can result in a slow drift after most of the glucose-6-P has reacted. This makes it desirable to use a generous excess of $NADP^+$, and pH 8 rather than 7, to favorably affect the G6PDH equilibrium reaction and accelerate hydrolysis of the lactone.

Comment

P-Glucomutase is ordinarily partially phosphorylated; therefore, if more enzyme is used (1 U, 5 μg/mL) the coenzyme can be omitted. *(See*, however, comment on the fluorometric method *below*.) For biological samples, it may be desirable to add 0.5 μ*M* dithiothreitol as in the fluorometric method. If large amounts of glucose-6-P are present in the samples, they can be destroyed by heating in 0.1*M* NaOH for 15 min at 100° without any loss of glucose-1-P.

Fluorometer, Direct Assay, 0.1–10 nmol/mL

Reagent: The same reagent as in Spectrophotometer, except as follows: $NADP^+$, 100 μ*M* (30 μ*M* with lowest levels); dithiothreitol, 500 μ*M*; and P-glucomutase, 0.02 U/mL (0.1 μg/mL).

Reaction time: Four to 8 min.

Conduct of the Assay (See *Spectrophotometer*)

Comment

Sometimes glucose-1,6-P_2 preparations contain small amounts of glucose-6-P. If this is troublesome, the level of the coenzyme can be reduced to 0.1 μ*M* with only a modest decrease in rate. Alternatively, the coenzyme can be heated in alkali without loss, to destroy glucose-6-P (*see* comment under Spectrophotometer). Because glucose-1-P levels are usually very low compared to the other hexose phosphates, great care may have to be taken to obtain valid answers. A trace of P-glucoisomerase, for example, would lead to an erroneously high value. The best solution is to keep the amount of enzyme minimal and to control the reaction time.

If the sample contains a large amount of glucose-6-P, this can be removed with alkali, as described in Spectrophotometer. Alternatively, glucose-6-P can be allowed to react and the resultant NADPH destroyed by acidification. After neutralization, fresh glucose-6-P dehydrogenase is added, a reading taken at a more sensitive fluorometer setting, and P-glucomutase added to complete the reaction.

Kinetics

The kinetics are very favorable. The V_{max} of P-glucomutase is over 300 μmol/mg/min, but may be a third of this in commercial preparations. The K_m for glucose-1-P with saturating amounts of the coenzyme is 8 μ*M*. Assuming a V_{max} of 100 μmol/mg/min, the half-time for very low glucose-1-P levels would be only 0.06 min with 1 μg/mL of enzyme (0.7 × 8/100). Because of the low Michaelis constant, more enzyme is required for glucose-1-P levels encountered in the spectrophotometer. The apparent dissociation constant for the coenzyme, glucose-1,6-P_2, is 0.06 μ*M*. Its function is to phosphorylate the enzyme. When less than saturating levels of coenzyme are present, the Michaelis constant for glucose-1-P decreases. Consequently, the effect of using suboptimal concentrations of coenzyme is less noticeable in the measurement of low levels of glucose-1-P than of high levels.

Glucose-6-Phosphate

$$\text{Glucose-6-P} + \text{NADP}^+ \xrightarrow[\text{dehydrogenase}]{\text{glucose-6-P}} \text{6-P-gluconolactone} + \text{NADPH} \quad \text{[Re. 6-25]}$$

$$\text{6-P-gluconolactone} + H_2O \xrightarrow{\text{nonenzymatic}} \text{6-P-gluconate}$$

This is the indicator reaction for a family of analyses. The following protocols are designed to provide optimal conditions for measuring glucose-6-P itself. In other cases, the conditions are altered according to the demands of the other substances to be measured.

Spectrophotometer, 15–200 nmol/mL

Reagent: Tris buffer, pH 8.1 (50 m*M* Tris base, 50 m*M* Tris-HCl); NADP$^+$, 500 μ*M*; and glucose-6-P dehydrogenase (baker's yeast), 0.06 U/mL (0.15 μg of crystalline enzyme).
Reaction time: Three to 5 min.

Conduct of the Assay

Ordinarily, the enzyme is added separately to start the reaction after an initial reading has been made. For standardization purposes, the reaction could be started equally well with the glucose-6-P solution.

Fluorometer Direct Assay, 0.1–10 nmol/mL

Reagent: Buffer, the same as in Spectrophotometer, except only half as strong (50 m*M*). NADP$^+$, 100 μ*M* for 2–10 μ*M* glucose-6-P, 10 μ*M* for 0.1–2 μ*M* glucose-6-P; dithiothreitol, 0.1 m*M* (optional, *see* Comment *below)*; glucose-6-P dehydrogenase, 0.02 U/mL.
Reaction time: Two to 4 min.

Comment

The NADP$^+$ is kept as low as possible because of the likelihood of fluorescent trace impurities. However, an excess of at least threefold should be provided (see Kinetics *below*). Dithiothreitol is only required if the sample to be assayed contains oxidized glutathione, the enzyme is heavily contaminated with glutathione reductase, and the samples are not read promptly. In this case, some of the NADPH initially formed will be reoxidized. The danger is much greater when glucose-6-P dehydrogenase is used as the auxiliary enzyme in an assay that takes longer to complete and in which more glucose-6-P dehydrogenase is required.

In the assay of tissue or bacterial extracts, the blank fluorescence from flavins can be greatly reduced by changing the buffer to 100 or 200 m*M* imidazole, pH 7. The rate of reaction with low levels of glucose-6-P is practically unchanged; however, at least a tenfold excess of $NADP^+$ should be maintained to drive the reaction close to completion *(see* Kinetics *below).*

Cycling Assay, 1–10 pmol

Reagent for glucose-6-P reaction: The same as in Fluorometer Direct Assay (10 μ*M* $NADP^+$), but with 0.04% bovine plasma albumin.
Reaction vessels: Oil wells.

Procedure for Dry Tissue Samples

Step 1: HCl, 1 μL of 0.02*M*, for samples and blanks; and 1 μL of 2–10 μ*M* glucose-6-P in 0.02*M* HCl for standards; 10 min at 60°.
Step 2: Reagent, 5 μL; 20 min at room temperature.
Step 3: NaOH, 1 μL of 1*M*; 10 min at 60°.
Step 4: Transfer 5-μL aliquots to fluorometer tubes containing 100 μL of NADP cycling reagent with enzyme levels to give about 1500 cycles in 60 min at 38° (Chapter 5). Heat stop by 2 min in a 95–100° water bath.
Step 5: Add 1 mL of indicator reagent (Chapter 5), and read when reaction is complete.

Comment

This procedure and the two that follow are designed for measuring glucose-6-P in frozen-dried fragments from tissue sections. The first step is for the purpose of destroying enzymes and preformed NADPH. There may be other occasions to measure glucose-6-P with cycling amplifications when this is unnecessary. The samples may be acid extracts of tissue that either contain less glucose-6-P than can be measured directly in the fluorometer or are so fluorescent as to make direct measurement unsatisfactory. In either case, Step 1 would be unnecessary. If the amount of glucose-6-P is >10 pmol, Steps 2 and 3 can be carried out with larger volumes of reagent, in larger vessels, and 10-μL aliquots taken for Step 4.

Cycling Assay, 0.1–1 pmol

Glucose-6-P dehydroenase reagent: The same as in Cycling Assay, 1–10 pmol.
Reaction vessels: Oil wells.

Procedure

Step 1: HCl, 0.1 µL of 0.02*M*, for samples and blanks, and 0.1 µL of 1–10 µ*M* glucose-6-P in 0.02*M* HCl for standards; 20 min at 80°.
Step 2: Reagent, 0.5 µL; 20 min at room temperature.
Step 3: NaOH, 5 µL of 0.05*M*; 20 min at 80°.
Step 4: Transfer 4-µL aliquots into fluorometer tubes containing 100 µL of NADP cycling reagent with enzyme levels to give about 15,000 cycles in 60 min at 38° (Chapter 5). Heat stop by 2 min in a 95–100° water bath.
Step 5: Add 1 mL of indicator reagent (Chapter 5), and read when reaction is complete.

Cycling Assay, 10–50 fmol

Glucose-6-P dehydrogenase reagent: The same as in Cycling Assay, 1–10 pmol, except all components at double concentration.
Reagent vessels: Oil wells.

Procedure

Step 1: Volumes of 0.015 µL for samples and blanks, and 0.015 µL of 1–3 µ*M* glucose-6-P in 0.02*M* HCl for standards, 20 min at 80°.
Step 2: Reagent, 0.03 µL; 20 min at room temperature.
Step 3: NaOH, 0.02 µL of 0.1*M*; 20 min at 80°.
Step 4: Add to each oil well 5 µL of NADP cycling reagent with enzyme levels to give about 100,000 cycles overnight at 15° (Chapter 5). Stop with 2 µL of 0.5*M* NaOH, and heat 20 min in 80° oven.
Step 5: Transfer 5 µL to 1 mL of indicator reagent (Chapter 5), and read when the reaction is complete.

Kinetics

The kinetics of glucose-6-P dehydrogenase from baker's yeast are exceptionally favorable for measuring low levels of either glucose-6-P or $NADP^+$. At pH 8, the V_{max} is on the order of 400 µmol/mg/min (400 U/mg), the K_m for glucose-6-P is 10–20 µ*M*, and that for $NADP^+$ 2 µ*M*. This means that 0.1 µg/mL of crystalline enzyme will give a first-order rate constant as high as 4/min (0.1 × 400/10). Many assays using this enzyme as an auxiliary are conducted at pH 7 rather than 8. At pH 7, the lower V_{max} about 200 µmol/mg/min, is offset by even more favorable Michaelis constants, 2.5 µ*M* for glucose-6-P and 0.8 µ*M* for $NADP^+$.

Although the initial velocity at low glucose-6-P concentration is as rapid at pH 7 as 8, the equilibrium is less favorable and the lactone is more slowly hydrolyzed; therefore, a larger excess of $NADP^+$ is required to give stoichiometric reaction. A substantial $NADP^+$ excess is also desirable because of the strong inhibition by NADPH. This inhibition increases somewhat the amount of enzyme or time required for completion of the reaction. G6PDH from *Leuconostoc mesenteroides* has kinetic properties almost as favorable as those of baker's yeast and can be used interchangeably by doubling the units per milliliters.

Glucose-1,6-Bisphosphate

$$\text{P-Glucomutase + glucose-1,6-P}_2 \rightleftharpoons \text{phosphorylated P-glucomutase + glucose-1-P}$$

$$\text{Glucose-1-P} \xrightarrow[\text{P-glucomutase}]{\text{phosphorylated}} \text{glucose-6-P} \qquad \text{[Re. 6-26]}$$

$$\text{Glucose-6-P + NADP}^+ \xrightarrow[\text{dehydrogenase}]{\text{glucose-6-P}} \text{6-P-gluconolactone + NADPH + H}^+ \qquad \text{[Re. 6-27]}$$

P-Glucomutase must be phosphorylated by glucose-1,6-P_2 to be active. A small amount of glucose-1,6-diphosphate will induce the formation of a large amount of NADPH. Although the method was originally used in the spectrophotometer *(11),* only a fluorometric method is presented here. It is suitable for measuring 1–8 pmol of the coenzyme. Much greater sensitivity can be attained, as indicated at the end of this section.

Reagent: Imidazole buffer, pH 7 (30 m*M* imidazole base, 20 m*M* imidazole-HCl); $MgCl_2$, 2 m*M*; EDTA, 0.1 m*M*; bovine serum albumin, 0.01%; glucose-1-P, 15 µ*M*; $NADP^+$, 50 µ*M*; P-glucomutase (rabbit muscle), 0.002 U/mL (0.01 µg/mL); and glucose-6-P dehydrogenase (yeast), 0.125 U/mL (0.33 µg/mL of crystalline enzyme).

Conduct of the Assay

The reagent is prepared complete, except for the omission of glucose- 1-P, and is brought to room temperature. One milliliter of this reagent is pipeted into each of a series of fluorometer tubes at room temperature. The fluorometer is adjusted so that 5 μ*M* NADPH will read approx full scale. Standards are added to give 2, 5, and 8 n*M* concentrations (2, 5, and 8 μL of 1 μ*M* standard glucose-1,6-P_2). Samples are likewise prepared to make a set of 15 or 20 tubes. Several tubes are reserved as blanks. All tubes are read initially. The reaction is started by adding 5 μL of 3 m*M* glucose-1-P to each tube, including the blanks. The additions are made at intervals equal to the time it will take to read each tube. Readings are made at several exact intervals (e.g., 5, 10, 15, and 20 min) after glucose-1-P addition. Because the reaction has a high temperature coefficient (the Q_{10} is nearly 3), it is important that samples and standards be run at exactly the same temperature. The results are calculated from a standard curve.

Comment

The overall blank should not exceed 0.2–0.3 n*M* glucose-1,6-P_2. Part of the blank is contributed by the enzyme itself, which is present at about 0.1 n*M* concentration and is partly phosphorylated as purchased. It is for this reason that P-glucomutase concentration is kept low. The glucose-1-P level is also kept low, because preparations usually contain some of the coenzyme. Some preparations are so badly contaminated as to be unsuitable.

Increased Sensitivity

Much smaller quantities of glucose-1,6-P_2 could be easily measured by an indirect procedure. With the P-glucomutase level recommended above, the rate of NADPH formation is about 2000 mol/mol of coenzyme/h. This could be increased three or fourfold with four times the enzyme concentration without seriously increasing the blank. Glucose-1,6-P_2, 10 fmol in 10 μL of reagent (1 n*M* concentration) would yield at least 5×10^{-11} mol of NADPH, an amount easily measurable with accuracy by the strong alkali method.

Kinetics

The apparent dissociation constant for glucose-1,6-P_2 is very low, 60 n*M*. Therefore, in the analytical range recommended (1–10 n*M*) P-glucomutase is operating at 15% or less of its V_{max}. The K_m for glucose-1-P varies with coenzyme concentration *(28)*. At the low coenzyme levels encountered here, the K_m is <1 μ*M*. Thus, the level of glucose-1-P recommended (15 μ*M*) is nearly saturating. The V_{max} for commercial P-glucomutase is almost 200 µmol/mg/min under analytical conditions. The net result of these constants is that at the recommended enzyme level (0.01 mg/L) the velocity is 0.03–0.05 µ*M*/min with 1 n*M* glucose-1,6-P_2.

Glutamate

$$\text{Glutamate} + \text{NAD}^+ \xrightarrow[\text{dehydrogenase}]{\text{glutamate}} \alpha\text{-ketoglutarate} + \text{NH}_4^+ + \text{NADH} + \text{H}^+ \qquad \text{[Re. 6-28]}$$

The equilibrium of the analytical reaction is concentration-dependent (*see* Kinetics). At glutamate levels <10 µ*M* (fluorometric procedures), the reaction proceeds essentially to completion. At levels encountered in the spectrophotometer (100 µ*M*), the reaction would be incomplete with practicable levels of NAD^+. Therefore, H_2O_2 is added to oxidize the α-ketoglutarate and pull the reaction.

Spectrophotometer, 15–200 nmol/mL

Reagent: Tris-acetate buffer, pH 8.4 (33 m*M* Tris base, 17 m*M* Tris-acetate); NAD^+ 1 m*M*; ADP, 100 µ*M*; H_2O_2 10 m*M*; and glutamate dehydrogenase (beef liver), 4.5 U/mL (100 µg/mL) from a 50% glycerol solution.

Reaction time: Fifteen to 20 min.

Conduct of the Assay

The H_2O_2 is added to the reagent shortly before use. Since the 10 m*M* H_2O_2 slowly inactivates glutamate dehydrogenase, the enzyme is ordinarily added to the reagent in the spectrophotometer after the sample. Both NAD^+ and the enzyme contribute significant blank absorption at 340 nm, which may slowly change with time. Therefore, parallel blank samples are required. There is an initial rapid reaction as the equilibrium is attained. This is followed by a slower reaction brought about by α-ketoglutarate removal.

Comment

NH_4^+ ion should be low or absent, since it is one of the products and will therefore delay the reaction. The H_2O_2 will slowly destroy the NADH formed (2–3%/h). ADP is added to protect the enzyme from inactivation and to overcome NADH inhibition *(29)*.

Fluorometer Direct Assay, 0.1–10 nmol/mL

Reagent: The buffer and ADP concentrations are the same as in Spectrophotometer. NAD^+, 400 μ*M* (150 μ*M* with <2 μ*M* glutamate); and glutamate dehydrogenase, 2.25 U/mL (50 μg/mL).
Reaction time: Fifteen to 30 min.

Conduct of the Assay

The enzyme is added last, after the sample is in and an initial reading has been made.

Comment

No H_2O_2 is necessary if NH_4^+ concentration is kept low (<10 μ*M*). Otherwise, H_2O_2 is added as for the spectrophotometer. NAD^+ is reduced for lowest glutamate concentrations, because it contributes blank fluorescence. It can be reduced even further if longer reaction times are acceptable.

Cycling Assay, 2–10 pmol

Glutamate reagent: Tris-acetate buffer, pH 8.4 (66 m*M* Tris base, 34 m*M* Tris-acetate); bovine serum albumin, 0.02%; and NAD^+ 200 μ*M*; ADP, 100 μ*M*, liver glutamate dehydrogenase, 2.5 U/mL (70 μg/mL) from glycerol stock solution.
Reaction vessels: Oil wells.

Procedure

Step 1: Add standards (10–50 μ*M*) of glutamate in 0.2 μL of 0.02*M* HCl, and add tissue samples to 0.2 μL of 0.02*M* HCl; heat at 80° 20 min.
Step 2: Add 2 μL of reagent; 30–40 min at room temperature.
Step 3: Add 5 μL of 0.5*M* NaOH; 30 min in 80° oven.
Step 4: Transfer 5-μL aliquots to fluorometer tubes containing 100 μL of NAD cycling reagent with enzyme levels to give about 1500 cycles in 60 min at 25° (Chapter 5). Heat stop by 2 min in 95–100° water bath.
Step 5: Add 1 mL of indicator reagent (Chapter 5), and read when reaction is complete.

Cycling Assay, 0.5–2.5 pmol

Reagent: The same as for Cycling Assay, 2–10 pmol.
Reaction vessels: Oil wells.

Procedure

Step 1: Add standards of 10–50 μ*M* glutamate in 0.05 μL of 0.02*M* HCl, and add tissue samples to 0.05 μL of 0.02*M* HCl, heat at 80° 20 min.

Step 2: Add 0.5 µL of reagent; 30–40 min at room temperature.
Step 3: Add 0.2 µL of 0.2*M* NaOH; 30 min in 80° oven.
Step 4: To all oil wells, add 10 µL of NAD cycling reagent with enzyme levels to give about 5000 cycles in 60 min at 25° (Chapter 5). Stop with 2 µL of 1*M* NaOH plus 20 min in 80° oven.
Step 5: Transfer 10-µL aliquots to 1 mL of indicator reagent (Chapter 5). Read when reaction is complete.

Kinetics

The kinetic constants are not very favorable. Under analytical conditions, V_{max} for glutamate oxidation with 1 m*M* NAD^+ is about 12 µmol/mg/min. The Michaelis constant for each substrate is affected (favorably) by the other *(30)*. With 1 m*M* NAD^+, the K_m for glutamate is about 600 µ*M*; with 10 µ*M* glutamate, the K_m for NAD^+ is considerably >1 m*M*. Thus, if the back-reaction could be ignored (which it cannot) the half-time with 1 µg/mL of enzyme would be 0.7 × 600/12(NAD^+) = 35 min/(NAD^+), with (NAD^+) expressed as millimolar concentration. For example, with 50 µg/mL of enzyme and 0.4 m*M* NAD^+ the half-time would be 1.7 min ([3/50 × 0.4]). The back-reaction prolongs this in a complicated manner, so that roughly 10 half-times are required.

$$\frac{(\alpha\text{-KG})(\text{NADH})(\text{NH}_4^+)}{(\text{GLUT})(\text{NAD}^+)} = 12.5\ \mu\text{M} \qquad \text{[Eq. 6-2]}$$

This expresses the fact that, because two reactants are forming three products, the lower the initial glutamate concentration, the more nearly complete will be the reaction, *even if* NAD^+ *is reduced in proportion.* The dilution effect is even more marked if NAD^+ is held constant. For example, according to this equation, with 1 m*M* NAD^+ and 100 µ*M* glutamate (Spectrophotometer), the reaction will be only 70% complete at equilibrium, whereas with 20 µ*M* glutamate, it will be 97% complete. With 10 µ*M* glutamate and 400 µ*M* NAD^+ (Fluorometer Direct Assay) in spite of the lower NAD^+, the reaction will be 98% complete. This shows why H_2O_2 is required only for glutamate concentrations >10 or 20 µ*M*.

For the reaction between α-ketoglutarate and H_2O_2, the first-order rate constant is $k = 0.06\ \text{min}^{-1}\ [H_2O_2]$, with H_2O_2 expressed as millimolar concentration; thus $k = 0.6\ \text{min}^{-1}$ with 10 m*M* H_2O_2 as recommended.

This means a rate of 60%/min of the α-ketoglutarate present at any instant, not of the glutamate present. Consequently, no matter how much enzyme is added, the rate of reaction with H_2O_2 will depend on how far the reaction is pushed by NAD^+ in the first place. Thus, when H_2O_2 has to be used, a high NAD^+ level is required not only to satisfy kinetic requirements, but also to shift the equilibrium toward α-ketoglutarate.

Glutamine and Glutamate

$$\text{Glutamine} \xrightarrow{\text{glutaminase}} \text{glutamate} + NH_4^+$$

[Re. 6-29]

$$\text{Glutamate} + NAD^+ \xrightarrow[\text{dehydrogenase}]{\text{glutamate}} \alpha\text{-ketoglutarate} + NH_4^+ + NADH + H^+$$

The method is based on the conversion of glutamine to glutamate by glutaminase, and the subsequent oxidative deamination of glutamate with glutamate dehydrogenase to form NADH (*see* Glutamate Assay). The system measures both glutamate and glutamine in the same sample. The glutamine is either determined by difference or is determined separately after destroying the NADH from the glutamate assay.

Fluorometer Direct Assay, 1–10 nmol of Glutamine plus Glutamate

Glutaminase reagent: Acetate buffer, pH 4.9 (16 m*M* acetic acid, 34 m*M* sodium acetate; EDTA, 0.2 m*M*; and *E. coli* glutaminase, 10 μg/mL [1 U/mL])

Glutamate reagent: Tris-acetate buffer (33 m*M* Tris-base, 17 m*M* Tris-acetate; NAD^+, 400 μ*M*; ADP, 100 μ*M*; and glutamate dehydrogenase, 2 U/mL [50 μg/mL]).

Conduct of the Assay

Sample extracts in volumes of <10 μL are added to 50 μL of glutaminase reagent in 10 × 75 mm tubes; standards contain 1–5 nmol of glutamate and 1–3 nmol of glutamine. A duplicate set of tubes without the glutaminase measures only the glutamate, and the glutamine is determined by difference. Incubate 30–60 min at room temperature. Add 1 mL of glutamate reagent and read when reaction is complete, about 30 min at room temperature. Greater sensitivity may be achieved by reading the tubes first at the appropriate setting, adding the glutamate dehydrogenase, and reading when the reaction is complete.

Cycling Assay, 0.2–1.5 pmol of Glutamate and/or Glutamine

Glutaminase reagent: The same as for Fluorometer Direct Assay, but all components double-strength.

Glutamate reagent: Tris-HCl buffer, pH 8.4 (70 m*M* Tris base, 30 m*M* Tris-HCl); NAD^+, 1 m*M*; ADP, 0.6 m*M*; bovine serum albumin; and 0.1%, glutamate dehydrogenase, 4.5 U/mL (100 µg/mL).

Reaction vessels: Oil wells.

Procedure for Glutamate Assay

Step 1: HCl, 0.25 µL of 5 m*M* for samples and blanks, and 0.25 µL of 1–5 µ*M* glutamate in 5 m*M* HCl for standards. Heat 20 min at 75°.

Step 2: To the cooled tubes, add 0.25 µL of glutamate reagent; 30 min at room temperature.

Step 3: Transfer 0.15 µL of the sample to 0.5 µL of 0.05*M* NaOH under oil in another rack. Heat 20 min at 75°.

Step 4: Transfer 0.5 µL of each sample to 100 µL of NAD cycling reagent in 10 × 75 mm tubes with enzyme levels to provide about 2500 cycles in 60 min at 25° (Chapter 5). Heat stop by 2 min in 95–100° in water bath.

Step 5: Add 1 mL of indicator reagent (Chapter 5). Read when reaction is complete.

Procedure for Glutamine Assay

Step 1: To the sample remaining at Step 3 above (0.35 µL) add 0.05 µL of 0.5*M* HCl (to destroy the NADH from glutamate).

Step 2: Add 0.4 µL of glutaminase reagent; 30 min at room temperature.

Step 3: Add 0.4 µL of glutamate reagent at pH 9.4; 30 min at room temperature.

Step 4: Add 4 µL of 0.05*M* NaOH; 30 min at 75°.

Step 5: Transfer 4 µL to 100 µL of NAD cycling reagent in 10 × 75 mm tubes with enzyme levels to provide about 25,000 cycles in 60 min at 25° (Chapter 5). Heat stop by 2 min in 95–100° water bath.

Step 6: Add 1 mL of indicator reagent (Chapter 5). Read when reaction is complete.

Glycerol

$$\text{Glycerol} + \text{ATP} \xrightarrow{\text{glycerokinase}} \text{glycero-P} + \text{ADP}$$

[Re. 6-30]

$$\text{Glycero-P} + \text{NAD}^+ \xrightarrow[\text{dehydrogenase}]{\text{glycero-P}} \text{dihydroxyacetone-P} + \text{NADH} + \text{H}^+$$

$$\text{Dihydroxyacetone-P} + \text{hydrazine} \rightarrow \text{dihydroxyacetone-P hydrazone}$$

Spectrophotometer, 15–200 nmol/mL

Reagent: Hydrazine buffer, pH 9.2 (350 m*M* hydrazine base, 50 m*M* hydrazine HCl); NAD^+, 2 m*M*; glycero-P-dehydrogenase from rabbit muscle, 3.4 U/mL (20 μg/mL); and glycerokinase from *Candida mycoderma,* 0.085 U/mL (1 μg/mL).
Reaction times: Glycero-P, 15 min; glycerol, 20 min.

Conduct of the Assay

If glycero-P is to be measured, the sample is added, a reading is made, and the glycero-P-dehydrogenase added. When the reaction is over, glycerokinase is added to measure the glycerol.

Comment

Because glassware and glycero-P-dehydrogenase were frequently found to be contaminated with glycerol, and glycerokinase as purchased may be in a glycerol solution to stabilize the enzymatic activity, care is taken to avoid touching glassware, and the enzymes are sedimented by centrifugation and washed with 5 vol of 2.5*M* $(NH_4)_2SO_4$. The enzymes are then dissolved in 50 m*M* Tris-HCl buffer, pH 8.0, before use.

Fluorometer Direct Assay, 0.1–10 nmol/mL

Reagent: The reagent is the same as Spectrophotometer, except as follows: NAD^+, 0.2 m*M*; glycero-P-dehydrogenase. 0.34 U/mL (2 μg/mL); and glycerokinase 0.085 U/ml (1 μg/mL).

Conduct of the Assay

See Spectrophotometer *above.*

Comment

The method for glycerophosphate in this chapter outlines the precautions concerning the fluorescent enhancement of NADH by glycero-P-dehydrogenase, as well as the kinetics of the reaction.

α-Glycerophosphate: Method I

$$\text{Glycero-P+NAD}^+ \xrightarrow[\text{dehydrogenase}]{\text{glycero-P-}} \text{dihydroxyacetone-P+NADH+H+}$$

[Re. 6-31]

$$\text{Dihydroxyacetone-P+hydrazine} \rightarrow \text{DHAP-hydrazone}$$

Because of its greater simplicity, this will usually be the preferred method for direct assays, both in the spectrophotometer and in the fluorometer. Only one enzyme is required. For indirect assays, the second method, which avoids the use of hydrazine, has distinct advantages.

Spectrophotometer, 15–200 nmol/mL

Reagent: Hydrazine buffer, pH 9.2 (350 m*M* hydrazine; 50 m*M* hydrazine-HCl); NAD^+, 2 m*M*; and rabbit muscle glycero-P-dehydrogenase, 3.4 U/mL (20 μg/mL).

Reaction time: Ten to 20 min.

Conduct of the Assay

The reaction is ordinarily started with the enzyme after a preliminary reading.

Comment

A blank sample should be run concurrently because of a small increase in absorption with time.

Fluorometer Direct Assay, 0.2–10 nmol/mL

Reagent: Hydrazine buffer as for Spectrophotometer. NAD, at least 50 times the concentration of glycero-P expected; and glycero-P-dehydrogenase; 0.85 U/mL (5 μg/mL).

Reaction time: Ten to 20 min.

Conduct of the Assay

The enzyme is added last after an initial reading.

Comment

There is significant fluorescence enhancement with this amount of glycero-P-dehydrogenase. Lower concentrations of NADH are enhanced disproportionately; therefore, standards should be included at several levels in the range of samples.

With 5 μg/mL of glycero-P-dehydrogenase, the fluorescence enhancement at pH 9 is 18, 14, and 8%, with 2, 4, and 10 μ*M* NADH, respectively.

Kinetics

The time required depends largely on the ratio of NAD^+ to glycero-P. The rate of reaction of dihydroxyacetone-P with hydrazine is rather slow and depends on the concentration of both reactants. With high enzyme levels, the dihydroxyacetone-P concentration depends in turn on the equilibrium constant for the ratio of NAD^+ to glycero-P. The reaction proceeds in the kinetically unfavorable direction. At pH 9.4:

$$\text{Keq} = \frac{[\text{glycero–P}]\,[\text{NAD}^+]}{[\text{dihydroxyacetone-P}][\text{NADH}]} = 50 \qquad \text{[Eq. 6-3]}$$

It is therefore desirable to provide a high ratio of NAD^+ to glycero-P. The NAD^+ levels chosen are a compromise between this consideration and the wish to avoid excessive initial blank values.

Approximate kinetic constants at pH 9 are V_{max} = 80 µmol/mg/min, $K_{glycero\text{-}P}$ = 250 µ*M* (with 850 µ*M* NAD^+), and K_{NAD+} = 350 µ*M* (with 40 µ*M* glycero-P). With these constants, in the presence of 500 µ*M* NAD^+, the *initial* rate would be about 20%/min (and essentially independent of dihydroxyacetone-P concentration) with only 1 µg/mL of enzyme [i.e., k = (80/250) (500/850) min^{-1}]. However, because of the back-reaction and the low K_m for NADH, much more enzyme is needed.

α-Glycerophosphate: Method II

$$\text{Glycero-P} + \text{NAD}^+ \xrightarrow[\text{dehydrogenase}]{\text{glycero-P-}} \text{dihydroxyacetone-P} + \text{NADH} + \text{H}^+$$

$$\text{Dihydroxyacetone-P} \xrightarrow[\text{isomerase}]{\text{triose}} \text{glyceraldehyde-P} \qquad \text{[Re. 6-32]}$$

$$\text{Glyceraldehyde-P} + \text{NAD}^+ \xrightarrow[\text{arsenate}]{\text{GAPDH}} \text{3-P-glycerate} + \text{NADH} + \text{H}^+$$

This analytical sequence proposed by Matschinsky *(31)* avoids the use of hydrazine. It is therefore of advantage for indirect procedures. Also in its favor is the fact that 2 mol of NADH are produced/mol of glycero-P. The method has the disadvantage that two additional enzymes are required, and that triose phosphates must be removed, measured first, or ignored. Fortunately, in most cases, glycero-P levels are much greater than those of the triose phosphates.

Spectrophotometer, 8–100 nmol/mL

Reagent: 2-Amino-2-methylpropanediol, pH 9.0 (30 m*M* as the free base, 20 m*M* as the hydrochloride); NAD$^+$ 2 m*M*; mercaptoethanol, 2 m*M*; Na_2HAsO_4, 1 m*M*; glycero-P-dehydrogenase, 3.4 U/mL (20 µg/mL); triose-P isomerase, 24 U/mL (10 µg/mL); and glyceraldehyde-P dehydrogenase, 3.6 U/mL (100 µg/mL) (all rabbit muscle enzymes). Stock enzyme suspensions in $(NH_4)_2SO_4$ are centrifuged and dissolved in Tris buffer, pH 8, to remove most of the sulfate.

Reaction time: Fifteen to 20 min.

Conduct of the Assay

For standardization purposes, the glycero-P can be added last. In this case, the three enzymes can be incorporated in the reagent where they are sufficiently stable for at least an hour. For biological samples, the reaction is started with glycero-P-dehydrogenase after first allowing 5 or 10 min for triose phosphates, if present, to react.

Comment

Sulfate is removed to a large extent because of its inhibitory effect on all three enzymes. The level should not exceed 2 or 3 m*M*.

Fluorometer Direct Assay, 0.1–5 nmol/mL

Reagent: Buffer, mercaptoethanol, and arsenate as for the spectrophotometer. NAD$^+$, 1 m*M* (100 µ*M* with 1 µ*M* glycero-P or less); glycero-P-dehydrogenase, 1.7 U/mL (10 µg/mL); triose-P isomerase 12 U/mL (5 µg/mL); and glyceraldehyde-P dehydrogenase, 1.44 U/mL (40 µg/mL).

Reaction time: Fifteen to 20 min.

Conduct of the Assay

The sample is added. If triose phosphates are present, glyceraldehyde-P dehydrogenase and triose-P isomerase are added, and a reading taken 5 or 10 min later, after which the glycero-P-dehydrogenase is added.

Comment

In principle, glyceraldehyde-P, dihydroxyacetone-P, and glycero-P can all be measured on the same sample by stepwise addition of the three enzymes. In practice, this will usually not be satisfactory because of the relatively low levels of the triose phosphates, particularly glyceraldehyde-P. It is ordinarily much better to measure the triose phosphates on a separate sample with a somewhat different reagent *(see* Method II for dihydroxyacetone-P, and so forth).

Because of fluorescence enhancement by glycero-P-dehydrogenase, standards at several levels may be required, particularly at the lowest concentrations. With 10 µg/mL (2 U/mL) of glycero-P-dehydrogenase, enhancement at this pH is 35, 22, and 12%, for 2, 4, and 10 µ*M* NADH, respectively.

Kinetics

The reaction sequence involves two reversible steps followed by a nonreversible one. The equilibria of the reversible steps are unfavorable. The triose isomerase equilibrium constant favors dihydroxyacetone-P by a factor of 25:1. For glycero-P-dehydrogenase at pH 9:

$$K_{eq} = \frac{[\text{glycero-P}][\text{NAD}^+]}{[\text{dihydroxyacetone-P}][\text{NADH}]} = 125 \qquad \text{[Eq 6-4]}$$

As a result of these two equilibria, with an excess of glycero-P-dehydrogenase, triose isomerase, and a tenfold excess of NAD^+, only 1% of the substrate would be present as glyceraldehyde-P (2% with a tenfold excess). It is this situation that demands the relatively high concentration of glyceraldehyde-P dehydrogenase. The kinetics of this enzyme were not thoroughly studied under analytical conditions. However, the essential parameter is the first-order rate constant for glyceraldehyde-P, which is about 2/min with 1 µg/mL of enzyme. (The rate is essentially the same for 100–2000 µ*M* NAD^+.) This means that, if only 1% of substrate is present as glyceraldehyde-P, the apparent rate constant for glycero-P could not exceed 0.02/min with 1 µg/mL of glyceraldehyde-P dehydrogenase. For the reaction to be complete in 20 min, the minimal rate constant would need to be at least 10 times this (i.e., at least 10 µg/mL of this enzyme is needed, no matter how much of other enzymes is used). In practice, even more is needed because reasonable levels of the other enzymes cannot keep the two unfavorable steps at equilibrium.

Glycogen

$$(\text{Glc})_n \xrightarrow{\text{amylo-}\alpha\text{-1,4-}\alpha\text{-1,6-glucosidase}} \text{n glucose}$$

$$\text{Glucose + ATP} \xrightarrow{\text{hexokinase}} \text{glucose-6-P + ADP} \qquad \text{[Re. 6-33]}$$

$$\text{Glucose-6-P + NADP}^+ \xrightarrow[\text{dehydrogenase}]{\text{glucose-6-P}} \text{6-P-gluconolactone + NADPH + H}^+$$

This assay must be conducted in two steps, since the amyloglucosidase requires an acid pH, and the glucose measurement requires a neutral or slightly alkaline pH. Furthermore, the final product to be measured, NADPH, is not stable at the optimum pH for the glucosidase. The procedure is based on that of Passonneau and Lauderdale *(32)* and Lust et al. *(33)*. Extracts containing both glucose and glycogen can be analyzed in the absence and presence of glucosidase, and thus, glucose and glucose plus glycogen are measured. If the presence of glucose is troublesome, it can be destroyed by heating the extract or sample in 0.1*M* NaOH for 20 min at 80° or 5 min at 100°. If the sample volume is large, the proper pH for enzyme Step 1 can be attained by adding an appropriate amount of acetic acid. If, for example, the sample is 50 µL of 0.1*M* NaOH, an equal vol of 0.2*M* acetic acid could be added to produce the desired pH. The enzyme would then be added separately. In order to measure glucose and glycogen separately, duplicate samples should be included, and the amyloglucosidase should be omitted from one.

Spectrophotometer, 15–200 nmol of Glucosyl U/mL

Reagent for glycogen hydrolysis: Sodium acetate buffer, pH 4.7 (50 m*M* acetic acid, 50 m*M* Na acetate), and glucosidase, 0.007 U/mL (0.5 µg/mL).

Reagent for glucose measurement: Tris-HCl buffer, pH 8.8 (85 m*M* Tris base, 15 m*M* Tris-HCl); $MgCl_2$, 5 m*M*; dithiothreitol, 0.1 m*M*; ATP, 500 μ*M*; $NADP^+$, 400 μ*M*; hexokinase, 0.7 U/mL (5 μg/mL); and glucose-6-P-dehydrogenase, 0.2 U/mL (0.5 μg/mL).

Reaction time: Thirty minutes for glycogen hydrolysis and 10 min for glucose measurement.

Conduct of the Assay

The initial incubation of glycogen with glucosidase can be carried out in 50–100 μL vol. Glycogen standards are added in volumes of 5–10 μL (15–200 nmol). After the initial 30-min incubation, 1 mL of the glucose reagent is added. The hexokinase is omitted from the reagent and added after an initial reading. Samples without amyloglucosidase will measure glucose alone; the second set of samples with the glucosidase will measure both glucose and glycogen.

Fluorometer Direct Assay, 0.1–10 nmol of Glucosyl U/mL

Reagent for glycogen hydrolysis: The same as in Spectrophotometer.

Reagent for glucose measurement: The same as in Spectrophotometer, except ATP, 150 μ*M*; and $NADP^+$, 40 μ*M*.

Cycling Assay, 1–10 pmol of Glucosyl Units

Reagent for glycogen hydrolysis: Tris-acetate buffer, pH 4.5 (50 m*M* Tris base, 100 m*M* acetic acid); amyloglucosidase, 0.04 U/mL (3 μg/mL); and bovine plasma albumin, 0.04%.

Reagent for glucose measurement: Tris-HCl buffer (85 m*M* Tris base, 15 m*M* Tris-HCl); $MgCl_2$, 5 m*M*; dithiothreitol, 0.1 m*M*; ATP, 500 μ*M*; $NADP^+$, 100 μ*M*; and hexokinase, 0.7 U/mL (5 μg/mL), glucose-6-P dehydrogenase, 0.2 U/mL (0.5 μg/mL).

Reaction vessels: Oil wells.

Procedure

Step 1: Samples, blanks, and glycogen standards are all introduced into the oil wells in 1 μL of 0.02*M* NaOH, and heated 30 min in a 90° oven to destroy glucose.

Step 2: 2 μL of glycogen hydrolysis reagent; 30 min at room temperature.

Step 3: 2 μL of glucose reagent; 20 min at room temperature.

Step 4: 2 μL of 0.25*M* NaOH; 20 min at 80°.

Step 5: Transfer 5-μL aliquots to fluorometer tubes containing 100 μL of NADP cycling reagent with enzyme levels to give about 1500 cycles in 60 min at 38° (Chapter 5). Heat stop by 2 min in a 95–100° water bath.

Step 6: Add 1 mL of indicator reagent (Chapter 5), and read when the reaction is complete.

Guanine Nucleotides

$$\text{GTP + succinate + coenzyme A} \xrightarrow{\text{succinyl Co A synthetase}} \text{GDP + succinyl CoA + P}_i$$

$$\text{GDP + P-pyruvate} \xrightarrow{\text{pyruvate kinase}} \text{GTP + pyruvate}$$

$$\text{Pyruvate + NADH + H}^+ \xrightarrow{\text{lactate dehydrogenase}} \text{NAD}^+ \text{ + lactate} \qquad \text{[Re. 6-34]}$$

$$\text{GMP + ATP} \xrightarrow{\text{guanylate kinase}} \text{GDP + ADP}$$

$$\text{GDP + ADP + 2 P-pyruvate} \xrightarrow{\text{pyruvate kinase}} \text{ATP + GTP + 2 pyruvate}$$

$$\text{2 Pyruvate + 2 NADH + 2 H}^+ \xrightarrow{\text{lactate dehydrogenase}} \text{2 NAD}^+ \text{ + 2 lactate} \qquad \text{[Re. 6-35]}$$

The enzymes used in the analytic sequence are not specific for guanine nucleotides. Succinyl CoA synthetase from pig heart and guanylate kinase from pig brain react with inosine nucleotides as well. However, since the inosine nucleotide levels in tissue are relatively low (no more than 25% of the guanine nucleotides in brain, for example), the error is rather small. A more serious problem is that all of the guanine nucleotides are ultimately measured with pyruvate kinase, and this enzyme reacts with all of the purine and pyrimidine dinucleotides *(18,34,35)*. Since ADP is relatively more abundant than GDP and GMP, the error is large if ADP is included.

Therefore, when GDP is to be measured, the ADP in the sample is converted to ATP by the addition of P-creatine and creatine kinase. When GMP is to be measured, greater accuracy is achieved by converting both ADP and GDP to the respective trinucleotides with a larger amount of creatine kinase *(see Kinetics)*.

Fluorometer Direct Analysis, 0.1–10 nmol

Reagent for conversion of GTP to GDP: Tris-succinate buffer, pH 7.7 (200 m*M* Tris base, 100 m*M* Tris-succinate); $MgCl_2$, 5 m*M*; dithiothreitol, 0.5 m*M*; CoA, 1 m*M*; P-creatine, 10 m*M*; hydrazine, 10 m*M*; EDTA, 0.1 m*M*; and AP_5A, 10 µ*M*. For conversion of ADP to ATP, add creatine kinase, 0.35 U/mL (1 µg/mL). For conversion of GDP to GTP, add creatine kinase, 35 U/mL (100 µg/mL). For conversion of GTP to GDP, add succinyl CoA synthetase (pig heart), 0.05 U/mL (5 µg/mL).

Conduct of the Assay

When all three nucleotides are to be measured, three sets of samples are prepared as follows:

1. Reagent containing creatine kinase, 1 U/mL (3 µg/mL);
2. Reagent containing creatine kinase, 1 U/mL (3 µg/mL) and succinyl CoA synthetase, 0.05 U/mL (5 µg/mL); and
3. Reagent containing creatine kinase, 35 U/mL (100 µg/mL).

The first set of tubes is used for the measurement of GDP, the second set for GTP and GDP, and the third for GMP.

Step 1: Samples and standards are added in vol of 50 µL to an equal vol of reagent in 5 × 50 mm test tubes. The reaction is stopped by the addition of 10 µL of 3*M* perchloric acid, and the tubes centrifuged. Alternatively, this step can be carried out in 10 × 75 mm fluorometer tubes and the reaction stopped by heating at 100° for 3 min. This procedure eliminates centrifugation, sample transfer, and pH adjustment, but the turbidity of the solutions may interfere with subsequent fluorescence measurements at high sensitivity.

Step 2: Measurement of GDP: The GDP in the samples, either as a tissue constituent or as the product of the succinyl CoA synthetase reaction, is measured with pyruvate kinase and lactate dehydrogenase. If the first step was conducted in 10 × 75 mm tubes, 1 mL of the GDP reagent is added. If the samples were treated with perchloric acid and centrifuged, 100-µL aliquots are removed to fluorometer tubes and neutralized with 10 µL of 2.5*M* NaOH, and 1 mL of the GDP reagent added.

Reagent: Tris-HCl buffer, pH 7.5 (12 m*M* Tris base, 38 m*M* Tris-HCl); $MgCl_2$, 2 m*M*; KCl, 75 m*M*; hydrazine, 10 m*M*; P-pyruvate, 40 µ*M*; EDTA, 100 µ*M*; NADH, 2 µ*M*; ATP, 20 µ*M* (for GMP assay); beef heart lactate dehydrogenase, 0.5 U/mL (2 µg/mL); pyruvate kinase, 4 U/mL (20 µg/mL); and guanylate kinase, 0.02 U/mL (2 µg/mL) (for GMP assay). The tubes are read in the fluorometer, and pyruvate kinase added to all tubes. After the reaction is complete, a second reading is made.

Step 3: Measurement of GMP: To the reagent for GDP is added 20 μ*M* ATP, and 1 mL is added to the third set of tubes. An initial reading is taken, and the guanylate kinase is added. A second reading is made when the reaction is complete.

Reaction times: Step 1: Thirty minutes for conversion of GTP to GDP and removal of ADP. Step 2: Five minutes for measurement of GDP. Step 3: Twenty minutes for measurement of GMP.

Kinetics

The reaction must be conducted in two steps, since the GTP formed in the reaction with pyruvate kinase reacts with the succinyl CoA and a cycling reaction ensues. The conversion of ADP to ATP without conversion of GDP to GTP by treatment with creatine kinase and P-creatine is kinetically favorable, under the conditions described. The K_m for ADP in the reagent is 0.15 m*M* and the V_{max} 40 μmol/min/mg, whereas the corresponding numbers for GDP are 4 m*M* and 10 μmol/min/mg. The apparent first-order rate constants (V_{max}/K_m)are the 0.27 min^{-1} and 0.0025 min^{-1}. The half-time for conversion of ADP with 1 μg/mL creatine kinase is 3 min for ADP and 300 min for GDP. After a 30-min incubation, the destruction of ADP is virtually complete (<0.1% left), whereas only about 6% of the GDP would react. In practice, standard ADP added to the incubation mixture was undetectable, and there was no measureable loss of GDP.

The treatment of tissue to remove both ADP and GDP is not so critical, since the aim here is to remove all the dinucleotides and only GMP is to be measured. The calculated half-time with 35 U/mL (100 μg/mL) of creatine kinase is 3 min.

Succinyl CoA Synthetase Reaction

The succinyl CoA reaction is reversible, and conditions were chosen to ensure the conversion of GTP to GDP was essentially complete. The equilibrium for the reaction has been determined *(36)*.

$$K = \frac{(\text{succinate})(\text{CoA})(\text{GTP})}{(\text{succinyl CoA})(\text{GDP})(P_i)} = 3.7 \qquad \text{[Eq. 6-5]}$$

When the reaction is complete, the final concentrations of succinyl CoA, GDP, and P_i will approximate those of the initial concentration of (GTP) = $(GTP)_i$. Making these substitutions:

$$(GTP)_f = \frac{3.7\ (GTP)_i^3}{(succinate)(CoA)} \qquad [Eq. 6-6]$$

where $(GTP)_f$ is the final concentration of GTP. For the reaction to be 99% complete $(GTP)_f = 0.01\ (GTP)_i$. The equation then becomes $(GTP)_i^2\ 370$ = (succinate)(CoA). With GTP concentrations of 10, 25, or 100 μ*M* in the reaction, and for the reaction to proceed to 99%, the product of (succinate)(CoA) must be ≥0.04, 0.23, and 3.7 mM^2, respectively. High succinate concentrations presented no disadvantage, whereas high concentrations of CoA may contribute to the fluorescent blank. Succinate buffer at 100 m*M* concentration was used with CoA at a relatively low concentration.

The amount of succinyl CoA synthetase needed for a half-time of 3 min for 50 μ*M* GTP based on a K_m of 10 μ*M* (*37*) and a V_{max} of 10 μmol/min/mg protein and in the presence of saturating succinate would be:

$$V_{max} = \frac{4 \times K_m + 50}{t_{0.98}} = \frac{4 \times 10 + 50}{20\ min} = 4.5\ \mu M/min \qquad [Eq. 6-7]$$

In practice, because of the complex back-reaction, it was necessary to use a concentration of 5 μg/mL of the enzyme, equivalent to 50 μ*M*/min.

The K_m of pyruvate kinase for GDP was determined to be 280 μ*M* in the reagent used, with a V_{max} of 28 μmol/min/mg protein. The amount needed to complete the conversion of GDP in 5 min was calculated to be 10 μg/mL; as a margin of safety, 20 μg/mL were used routinely.

Cycling Assay, 1–3 pmol

The measurement of GMP at this sensitivity has not been attempted.

Reagent for conversion of GTP to GDP: Imidazole-HCl buffer, pH 7.0 (60 m*M* imidazole base, 40 m*M* imidazole-HCl); $MgCl_2$, 2 m*M*; succinate, 10 m*M*; dithiothreitol, 0.5 m*M*; coenzyme A, 0.04 m*M*, hydrazine, 10 m*M*; bovine serum albumin, 0.02%; P-creatine, 1.5 m*M*; creatine kinase, 0.35 U/mL (1 μg/mL); and succinyl CoA synthetase, 0.05 U/mL (5 μg/mL) when present.

Reagent for measurement of GDP: Imidazole-HCl buffer, pH 7.0 (60 m*M* imidazole base, 40 m*M* imidazole HCl); $MgCl_2$, 2 m*M*; KCl, 75 m*M*; P-pyruvate, 10 μ*M*; hydrazine, 100 μ*M*; ascorbate, 2 m*M*; NADH, 2–5 μ*M*; beef heart lactate dehydrogenase, 0.5 U/mL (2 μg/mL); and pyruvate kinase, 4 U/mL (20 μg/mL).

Reaction vessels: Oil wells.

Procedure

Step 1: NaOH, 0.05 µL of 0.1*M* for samples and blanks, or 0.1*M* NaOH containing 20–50 µ*M* GDP and/or GTP.

Step 2: To all samples, add 2 µL of the reagent without succinyl CoA synthetase (Rack 1). One-microliter portions removed and added to another oil well to 1-µL droplets containing the reagent with succinyl CoA synthetase; 30 min at room temperature, and 20 min at 80° to destroy the enzymes.

Step 3: To all samples, add 4 µL of the reagent for the measurement of GDP; 10 min at room temperature.

Step 4: To all samples, add 5 µL of 0.2*M* HCl.

Step 5: Transfer a 5-µL aliquot of the samples to fluorometer tubes containing 100 µL of NAD cycling reagent with enzyme levels to give 3000 cycles in 60 min at 25° (Chapter 5). Heat stop by 2 min in a 95–100° water bath.

Step 6: Add 1 mL of the indicator reagent (Chapter 5), and read when reaction has stopped.

Isocitrate

$$\text{Isocitrate} + \text{NADP}^+ \xrightarrow[\text{dehydrogenase}]{\text{isocitrate}} \alpha\text{-ketoglutarate} + \text{CO}_2 + \text{NADPH} + \text{H}^+$$

[Re. 6-36]

Spectrophotometer, 15–200 nmol/mL

Reagent: Tris-HCl, pH 8.1 (25 m*M* Tris base, 25 m*M* Tris-HCl); $MnCl_2$, 100 µ*M*; $NADP^+$, 300 µ*M*; and isocitrate dehydrogenase ($NADP^+$-specific from pig heart) 0.4 U/mL (10 µg/mL). Stock enzyme dilutions are made in 50% glycerol in Tris buffer (*see* Comment).

Reaction time: Ten minutes.

Conduct of the Assay

The enzyme is added last after the sample is in and the initial reading made.

Comment

Isocitrate dehydrogenase is unstable when highly diluted in buffer, especially at dilutions >5 µg/mL. It is stabilized by isocitrate, which protects it during the assay. It is also stabilized by bovine plasma albumin, although this does not appear to be owing to the albumin itself. Therefore, if the enzyme must be added to the reagent before the sample, at least 0.01% bovine plasma albumin should also be present. The enzyme is also stabilized by 50% glycerol. Therefore, this is used for preparing dilute stock solutions.

Fluorometer Direct Assay, 0.1–10 nmol/mL

Reagent: The same buffer and $MnCl_2$ concentrations are used as in Spectrophotometer. $NADP^+$, 35 μ*M*; bovine plasma albumin, 0.01%; and isocitrate dehydrogenase, 0.08 U/mL (2 μg/mL).

Reaction time: Five to 10 min.

Conduct of the Assay

The enzyme is added last.

Comment

For isocitrate concentrations of 1 μ*M* or less, $NADP^+$ can be reduced to 5 μ*M* and the enzyme reduced to 1 μg/mL. Greater dilution of the enzyme is not recommended because of increased instability. For very low levels of isocitrate, the albumin can be omitted, but it may be necessary to increase the amount of enzyme or prolong the assay time.

Isocitrate levels are usually very low in living cells, because the equilibrium ratio between citrate and isocitrate favors citrate by a factor of 15. Indirect cycling procedures may be required to obtain accurate estimations. We have had no experience with such assays, but no difficulty is anticipated if the specific reagent is prepared with at least 0.02% bovine plasma albumin, the isocitrate dehydrogenase concentration is not <1 or 2 μg/mL, and the reagent is used promptly after enzyme addition. The possibility of trace contamination with aconitase should be carefully checked.

Kinetics

The V_{max} is low, about 40 μmol/mg/min. This is compensated by very low Michaelis constants for the substrates, 0.3 μ*M* for $NADP^+$, 0.4 μ*M* for isocitrate. From this, the calculated amount of enzyme needed for an assay time of 10 min with 1, 10, and 100 μ*M* isocitrate would be 0.005, 0.025, and 0.25 μg/mL (Chapter 2). Because of enzyme instability at high dilution, somewhat larger amounts of enzymes are required in practice.

The K_m for Mn^{2+} is 7 μ*M*. Levels higher than 100 or 200 μ*M* should be avoided because of possible development of significant absorption in the region of NADPH fluorescence. Mg^{2+} can be substituted for Mn^{2+}. Its K_m is 48 μ*M*. The V_{max} is about the same, but the K_m for isocitrate is four times larger. Therefore, Mn^{2+} is distinctly preferable with low substrate levels.

α-Ketoglutarate

$$\alpha\text{-ketoglutarate} + NH_4^+ + NADH + H^+ \xrightarrow[\text{dehydrogenase}]{\text{glutamate}} \text{glutamate} + NAD^+ \quad \text{[Re. 6-37]}$$

Spectrophotometer, 15–200 nmol/mL

Reagent: Imidazole acetate buffer (20 m*M* imidazole base: 20 m*M* imidazole acetate); ammonium acetate, 25 m*M*; ADP, 100 μM; NADH, 100 μM; and beef liver glutamate dehydrogenase, 0.15 U/mL (4 μg/mL).
Reaction time: Ten minutes.

Conduct of the Assay

For samples that may contribute significant absorption, the enzyme is omitted until after the sample is added and an initial reading made. The enzyme is then added in a 5- or 10-μL vol.

Fluorometer Direct Assay, 2–8 nmol/mL

Reagent: The same as for Spectrophotometer, except NADH is reduced to 10 μ*M* and the enzyme is omitted.

Conduct of the Assay

Ordinarily, the sample is added first. After an initial reading, the enzyme is added.

Reaction time: Ten to 15 min.

Fluorometer Indirect Assay, 0.2–1 nmol

Reagent: The same as for Spectrophotometer, except NADH is reduced to 10 μ*M*.

Procedure

Step 1: Samples (*see* Comment *below*) and standards in 10-μL vol or less are added to 100 μL of reagent in fluorometer tubes.
Step 2: After 15 min, 10 μL of 0.5*M* HCl is added (to remove excess NADH).
Step 3: After at least 10 min, 1 mL of 6*M* NaOH containing 10 m*M* imidazole base is added rapidly with immediate mixing.
Step 4: The tubes are heated 10 min in a 60° water bath, and read exactly at room temperature.

Cycling Assay, 1–10 pmol

Reagent: Imidazole-acetate buffer (60 m*M* imidazole base, 20 m*M* imidazole acetate); ammonium acetate, 50 m*M*, ADP, 200 μ*M*; NADH, 20 μ*M*; and glutamate dehydrogenase, 0.3 U/mL (8 μg/mL).

Reaction vessels: Tubes, 6 × 50 mm.

Procedure

Step 1: Samples, standards, and blanks in 10-μL vol are added to 10 μL of reagent.

Step 2: After 15 min at room temperature, add 2 μL 1*M* HCl.

Step 3: After at least 10 min, transfer 10-μL aliquots to fluorometer tubes containing 100 μL of NAD cycling reagent with enzyme levels to give about 2000 cycles/h at 38° (Chapter 5). Heat stop by 2 min in a 95–100° water bath.

Step 4: Add 1 mL of indicator reagent (Chapter 5), and read when reaction is complete.

Comment

All tissue samples for Fluorometer, Indirect Assay, and Cycling Assay require pretreatment in 0.02*M*, or stronger, NaOH for 20 min at 60° to remove preformed NAD^+ and inactivate disturbing tissue enzymes.

Equilibrium Constant

According to Olson and Anfinsen *(38)* at pH 7.

$$K_{eq} = \frac{[\alpha KG][NH_4^+][NADH]}{[Glut][NAD^+]} = 0.45 \times 10^{-6}\ M = 0.45\ \mu M \qquad [Eq.\ 6\text{-}8]$$

(where αKG is α-ketoglutarate, and Glut is glutamate). At pH 8, K_{eq} = 4.5 μ*M* (i.e., tenfold less favorable).

Since (NH_4^+) is ordinarily used at millimolar concentrations, the equilibrium position is very favorable at pH 7 for reduction of α-ketoglutarate. Therefore, the fact that this is a reversible reaction would not be of analytical concern if it were not that glutamate levels in tissues are very high, and NAD^+ levels may be at least as high as those of α-ketoglutarate.

A useful way to arrange the equilibrium equation is:

$$\frac{[\alpha KG]}{[Glut]} = \frac{K_{eq}\ [NAD^+]}{[NH_4^+]\ [NADH]} \qquad [Eq.\ 6\text{-}9]$$

Table 1
Extent of Glutamate Dehydrogenase Reaction
Under Specified Conditions

	Initial μM	Final, pH 7 μM
a-Ketoglutarate	0.5	0.013
Glutamate	50	50.5
NAD^+	1.0	1.49
NADH	0.75	0.263
ΔNADH	—	0.487 (97%)

At pH 7, if all concentrations are micromolar and $[NH_4^+]$ is set at 10,000 μM:

$$\frac{[\alpha KG]}{[Glut]} = \frac{0.45[NAD^+]}{10{,}000[NADH]} = \frac{[NAD^+]}{22{,}000[NADH]} \quad \text{[Eq. 6-10]}$$

From this equation, it is simple to calculate how nearly complete the analytical reaction will be. Take as an example a tissue that contains per kilogram 0.1 mmol of α-ketoglutarate, 10 mmol of glutamate, and 0.2 mmol of NAD^+. (Endogenous NADH is destroyed in making an acid tissue extract.) Let us suppose that during the enzymatic assay the overall tissue dilution is 200-fold, and a 50% excess of NADH is provided. Table 1 shows the initial and the equilibrium concentrations that would result during the assay (i.e., the reaction would be 97% complete).

The equilibrium is even more favorable below pH 7, but NADH instability is a deterrent. Another way to favor the reaction would be to increase NH_4^+ concentration to 25 m*M* or even 50 m*M*. Levels above 50 m*M*, although favorable to the equilibrium, would slow the rate.

Reaction Kinetics

Another, and usually more important, reason for keeping the pH low is that it has a very favorable effect on the reaction rate at low substrate levels. This is not owing to any effect on the V_{max}, which is no greater at pH 7 than at pH 8. The favorable effect of low pH is on the Michaelis constants for both α-ketoglutarate and NADH. The K_m for α-ketoglutarate is nearly 10 times smaller at pH 7 than at pH 8. Although the reaction is even faster below pH 7, the advantage is not great enough to compensate for the beginning instability of NADH.

In addition to pH, the type of buffer and its concentration have profound influence on the Michaelis constants for both substrates. With NADH and α-ketoglutarate at low levels (5 and 3 μ*M*, respectively), comparisons were made at pH 7 between a number of buffers and buffer strengths. With 20 m*M* imidazole-acetate buffer taken as 100%, rates were 50% with 100 m*M* imidazole-acetate, 20% with 100 m*M* imidazole-HCl, and 10% with 100 m*M* phosphate buffers. As in the case of pH, the V_{max} values were only moderately affected by the changes in buffer and buffer concentration. As a result of these tests imidazole-acetate is recommended at pH 6.9 and a concentration of 40 m*M*. A lower concentration would give a somewhat faster rate but the higher buffering capacity is ordinarily desirable.

When measuring low levels of α-ketoglutarate directly in the fluorometer, the NADH levels, as has been said, are of necessity also low. This means that both substrates may be well below their respective Michaelis constants which are approx 100 μ*M* for α-ketoglutarate and 4 μ*M* for NADH under recommended conditions. In this event, the reaction becomes truly second order. This has two consequences. For a given percentage excess of NADH more enzyme will be needed as the concentration of NADH is decreased. In addition, with a given initial NADH concentration, the time for completion will vary with the amount of α-ketoglutarate present. This general situation has been described fully in Chapter 2, but a specific illustration is included here. For example, if α-ketoglutarate is either 80 or 40% of the initial NADH concentration, at the higher level, the half-time is increased by 16%, but the time for 98% completion is more than double. Whereas a first-order reaction is 98% complete in 5.6 half-times, the second-order reactions, to be 98% complete, require 7.3 and 13 half-times at the 40 and 80% levels, respectively. There is obviously a need for caution when the analytical reaction is second order and most of the pyridine nucleotide is used up.

Lactate: Method I

$$\text{Lactate} + \text{NAD}^+ \xrightarrow[\text{dehydrogenase}]{\text{lactate}} \text{pyruvate} + \text{NADH} + \text{H}^+ \quad \text{[Re. 6-38a]}$$

$$\text{Pyruvate + glutamate} \xrightarrow[\text{transaminase}]{\text{glut-pyr}} \text{alanine} + \alpha\text{-ketoglutarate} \qquad \text{[Re. 6-38b]}$$

Although tissue lactate levels are relatively high, lactate measurement may cause difficulty because of the ease of contamination *(see* Final Comments) and because of the unfavorable equilibrium *(see* Kinetics *below).* This first method, initially described by Noll *(15),* has several advantages over the alternative method with hydrazine that compensate for the fact that an additional enzyme is required. Particularly in the spectrophotometer, blanks are easier to control, and there are fewer complications in adapting to high sensitivity.

Spectrophotometer, 15–200 nmol/mL

Reagent: 2-Amino-2-methylpropanol buffer, pH 9.9 (50 m*M* in the free base, 50 m*M* in the hydrochloride); NAD^+, 1.5 m*M*; glutamate, 50 m*M* (pH 9.8–10, *see* Comment); lactate dehydrogenase (beef heart), 20 U/mL (100 μg/mL); and glutamate-pyruvate transaminase (pig heart), 8 U/mL (100 μg/mL).

Reaction time: Eight to 12 min.

Conduct of the Assay

For standardization purposes, the lactate can be added last after making a preliminary reading. In this case, both enzymes should be in the reagent for 10 or 15 min to permit any contaminating lactate to react. For analyzing biological samples, the lactate dehydrogenase is ordinarily added last to permit an initial reading of the reagent plus sample. In this case, it is necessary to conduct a separate blank. Because the enzymes are somewhat unstable at this alkaline pH, neither enzyme should be added to the reagent more than 30 min before starting the reaction.

Comment

The glutamate required is conveniently prepared from a stock 1*M* solution of monosodium glutamate, which is brought to a pH of about 10 by incorporating 0.9*M* NaOH. (The third pK_a of glutamic acid is about 9.) The reaction is carried out at pH 9.8–10. The addition of $(NH_4)_2SO_4$ with the enzymes can lower the pH; otherwise a 10-m*M* concentration of this salt does not materially affect the rate or end point.

Fluorometer Direct Assay, 0.2–10 nmol/mL

Reagent: The buffer concentration is reduced to 50 m*M*; NAD^+, 500 μ*M* (200 μ*M* with 2 μ*M* lactate or less); glutamate, 2 m*M* (0.5 m*M* with 2 μ*M* lactate or less); lactate dehydrogenase, 10 U/mL (50 μg/mL); and glutamate-pyruvate transaminase, 4 U/mL (50 μg/mL).
Reaction time: Fifteen to 20 min.

Conduct of the Assay

The transaminase can be incorporated in the reagent, but not more than 30 min ahead of time. The reaction is ordinarily started with lactate dehydrogenase.

Comment

Because of the lower lactate levels in the fluorometer, the glutamate concentration is reduced; less is needed to drive the reaction to completion. The level used in the spectophotometer is actually inhibitory. As pyruvate levels are decreased, the glutamate required for maximal activity is also decreased *(see* Kinetics *below).* Because the lactate concentrations are lower than in the spectrophotometer procedure, contamination with lactate is more of a problem. The tubes must be handled carefully to avoid contamination by lactate from the fingers. It is best not to touch them near the top. It may be necessary to wear rubber gloves or to pick up the tubes with forceps.

Fluorometer Indirect Assay, 0.1–1 nmol

Because of the danger of lactate contamination, there is advantage in carrying out the analysis in two steps with the first step at a reduced volume.

Reagent: Buffer and transaminase as in B; NAD^+, 100 μ*M*; glutamate, 5 m*M*; and lactate dehydrogenase, 20 U/mL (100 μg/mL).

Procedure

Step 1: Fifty microliters of complete reagent plus 1–10 μL of sample incubated 30 min in a tube 7 × 70 mm.
Step 2: Add 50 μL of pH 12 phosphate buffer (0.25*M* Na_3PO_4, 0.25*M* K_2HPO_4); heat 15 min at 60°.
Step 3: Add a 50-μL aliquot to 1 mL of 6*M* NaOH containing 10 m*M* imidazole and 0.03% H_2O_2; heat 15 min at 60°; read fluorescence.

Cycling Assay, 1–10 pmol

Reagent for lactate reaction: 2-Amino-2-methyl-1-propanol buffer, pH 9.9 (50 m*M* free base, 50 m*M* hydrochloride); NAD^+, 50 μ*M*; glutamate, 0.5 m*M*; bovine plasma albumin, 0.02%; lactate dehydrogenase, 20 U/mL (100 μg/mL); and glutamate-pyruvate transaminase, 4 U/mL (50 μg/mL).
Reaction vessels: Oil wells.

Procedure for Freeze-Dried Samples

Step 1: NaOH, 1 μL of 0.02*M* for samples and blanks, and 1 μL of 1–10 μ*M* lactate in 0.02*M* NaOH for standards; heated 10 min at 60°.
Step 2: Reagent, 5 μL; 30 min at room temperature.
Step 3: NaOH, 5 μL of 0.2*M*; 10 min at 60°.
Step 4: Transfer 5-μL aliquots to fluorometer tubes containing 100 μL of NAD cycling reagent with enzyme levels to give about 2,000 cycles in 60 min at 25° (Chapter 5). Heat stop by 2 min in a 95–100° water bath.
Step 5: Add 1 mL of indicator reagent (Chapter 5), and read when reaction is complete.

Cycling Assay, 0.1–1 pmol

Reagent for lactate reaction: The same as in Fluorometer Indirect Assay.
Reagent vessels: Oil wells.

Procedure

Step 1: NaOH, 0.1 μL of 0.02*M*, for samples and blanks; and 0.1 μL of 2–10 μ*M* lactate in 0.02*M* NaOH for standards; 20 min at 80°.
Step 2: Reagent, 0.5 μL; room temperature 30 min.
Step 3: NaOH, 5 μL of 0.05*M*; 20 min at 80°.
Step 4: Transfer 4-μL aliquots to fluorometer tubes containing 100 μL of NAD cycling reagent with enzyme levels to give about 15,000 cycles in 60 min at 25° (Chapter 5). Heat stop by 2 min in a 95–100° water bath.
Step 5: Add 1 mL of indicator reagent and read when reaction is complete.

Cycling Assay, 20–100 fmol

Reagent for lactate reaction: The same as in Fluorometer Indirect Assay, but all components at double concentration.
Reaction vessels: Oil wells.

Procedure

Step 1: NaOH, 0.015 μL of 0.02*M*, for samples and blanks; and 0.015 μL of 2–5 μ*M* lactate in 0.02*M* NaOH for standards; 20 min at 80°.
Step 2: Reagent, 0.015 μL; 30 min at room temperature.

Step 3: NaOH, 0.2 µL of 0.1*M*; 20 min at 80°.

Step 4: Add to each oil well 5 µL of NAD cycling reagent with enzyme levels to give about 50,000 cycles overnight at 15° (Chapter 5). Stop with 2 µL of 0.5*M* NaOH, and heat 20 min in an 80° oven.

Step 5: Transfer 5 µL to 1 mL of indicator reagent (Chapter 5), and read when reaction is complete.

Final Comments

As indicated, lactate is an exceedingiy troublesome contaminant in the laboratory. All glassware used for reagents and analysis should be repeatedly rinsed, and it is recommended that gloves be worn during the rinsing process. The lactate contamination of the reagent can be reduced to 0.5 µ*M* or less if proper precautions are observed. Skeletal muscle lactate dehydrogenase is less satisfactory than the heart enzyme for the first enzyme step, because it is much less stable at alkaline pH.

Kinetics

The kinetics of lactate dehydrogenase are complicated and rather unfavorable. The Michaelis constants are large, and the equilibrium lies on the side of pyruvate reduction. There is major kinetic interaction between the two substrates. Under analytical conditions, the apparent K_m for NAD^+ is about 200 µ*M*, whereas that for lacate is nearly 2 m*M*. The V_{max} at pH 10 is about 100 µmol/mg/min or about half of that in the opposite direction at pH 7. The analytical reaction is carried out at as high a pH as possible without enzyme inactivation, because this shifts the equilibrium in the desired direction. It is also important to keep the NAD^+ concentration high, not only because of the large Michaelis constant but, because this also shifts the equilibrium toward pyruvate formation.

At pH 9.9, the lactic dehydrogenase equilibrium is:

$$\frac{[\text{lac}][\text{NAD}^+]}{[\text{pyr}][\text{NADH}]} = 60 \qquad \text{[Eq. 6-11]}$$

The transaminase equilibrium at pH 9.9 is:

$$\frac{[\text{glut}][\text{pyr}]}{[\alpha\text{-KG}][\text{ala}]} = 1.6 \qquad \text{[Eq. 6-12]}$$

For the overall reaction (Reactions [6-38a and 6-38b]) to proceed to virtual completion, the final concentrations of α-ketoglutarate, alanine, and NADH will be essentially equal to the initial level of lactate, $[lac]_o$. Making these substitutions and combining (Equations [6-11 and 6-12]):

$$[lac]_f = \frac{96[lac]_o^3}{[glut][NAD^+]} \qquad \text{[Eq. 6-13]}$$

where $[lac]_f$ is the final lactate concentration.

For the reaction to be 99% complete, $[lac]_f = 0.01[lac]_o$. Making this substitution in Eq. [Eq. 6-13] and rearranging:

$$[glut][NAD^+] = 9600[lac]_o^2 \qquad \text{[Eq. 6-14]}$$

This equation defines the concentrations of glutamate and NAD^+ that will drive the reaction 99% to completion with different levels of lactate. For example, with 0.1, 0.01, and 0.001 m*M* lactate, the product of glutamate and NAD^+ concentrations (also millimolar) would need to be at least 100, 1, and 0.01 mM^2, respectively. This has governed in part the choice of levels for the reagents above.

The kinetics of glutamate-pyruvate transaminase have not been well studied at pH 10, but the following suffices for the present purpose. The apparent K_m for glutamate is about 1.5 m*M* with 40 µ*M* pyruvate (i.e., at initial spectrophotometer levels), but is much lower (0.15 m*M*) with 5 µ*M* pyruvate. The apparent Michaelis constant for pyruvate is similarly affected by the glutamate and <5 µ*M* with 0.1 m*M* glutamate. These kinetic properties make it unnecessary to use high glutamate concentrations except for high lactate levels in the spectrophotometer. High levels of glutamate are actually inhibitory. The V_{max} at pH 10 was not measured, but with 1–5 m*M* glutamate, the apparent first-order rate constant with low pyruvate levels is about 0.04/min with an enzyme concentration of 1 µg/mL.

Lactate: Method II

$$\text{Lactate} + NAD^+ \xrightarrow[\text{dehydrogenase}]{\text{lactate}} \text{pyruvate} + NADH + H^+$$

[Re. 6-39]

$$\text{Pyruvate} + \text{hydrazine} \rightarrow \text{pyruvate hydrazone}$$

Spectrophotometer, 15–200 nmol/mL

Reagent: Hydrazine buffer, pH 9.6 (190 m*M* hydrazine, 10 m*M* HCl); NAD^+, 2 m*M*; and lactate dehydrogenase (heart), 20 U/mL (100 µg/mL). The lactate dehydrogenase stock suspension is centrifuged and dissolved in 0.02*M* Tris buffer, after removing as much of the original supernatant fluid as possible (to eliminate most of the NH_4^+).

Reaction time: Ten to 20 min (depending on lactate level).

Conduct of the Assay

The enzyme is added separately to start the reaction after the sample has been added and an initial reading has been made. Concomitant blank samples are essential.

Comment

Hydrazine is the only buffer used, because the combination of hydrazine, NAD^+, and buffers tried (glycine, aminomethylpropanol) increases the blank absorbance. Hydrazine is a poor buffer at the best pH reaction, 9.7–10.

Care must be taken that the pH is not shifted by the sample or by other additions. This is the reason for removal of most of the NH_4^+ from the enzyme preparation.

Fluorometer, Direct Assay, 0.5–10 nmol/mL

Reagent: 2-Amino-2-methyl-1-propanol buffer, 100 m*M*, pH 10 (50 m*M* free base, 50 m*M* hydrochloride); hydrazine, 100 m*M*; NAD, 200 µ*M* (for 2–10 µ*M* lactate), 50 µ*M* (for 0.5–2 µ*M* lactate); and lactate dehydrogenase, 8 U/mL (40 µg/mL).

Reaction time: Ten to 20 min.

Conduct of the Assay

As in Spectrophotometer.

Comment

The reagent is better buffered than for the spectrophotometric assay. It is therefore, unnecessary to remove NH_4^+ from the enzyme preparation if the final NH_4^+ concentration does not exceed 10 m*M*. If the buffer adds an appreciable blank, as it may, glycine buffer, pH 10, can be used instead, or the hydrazine buffer used in the spectrophotometer can be substituted. It is important that the ratio of NAD^+ to lactate does not fall below 20:1. (For more sensitive procedures for lactate, *see* Method I).

Kinetics

For the kinetics of lactate dehydrogenase at pH 10, *see* the first lactate method. The hydrazine reaction is relatively slow, and is accelerated in proportion to the concentration of free pyruvate and the hydrazine concentration. At pH 10, the equilibrium constant is about 50 for:

$$\frac{[\text{lactate}][\text{NAD}^+]}{[\text{pyruvate}][\text{NADH}]} \qquad \text{[Eq. 6-15]}$$

This means that with an initial lactate:NAD^+ ratio of 20:1, the reaction will proceed only about 45% to equilibrium without the benefit of hydrazine.

Malate: Method I

$$\text{Malate} + \text{NAD}^+ \xrightarrow[\text{dehydrogenase}]{\text{malate}} \text{oxaloacetate} + \text{NADH} + \text{H}^+$$

[Re. 6-40]

$$\text{Oxaloacetate} + \text{glutamate} \xrightarrow[\text{transaminase}]{\text{aspartate}} \text{aspartate} + \alpha\text{-ketoglutarate}$$

This method is analogous to the first lactate method. It is superior in several respects to the alternative procedure with hydrazine, even though it requires an additional enzyme. It can be made faster, there is less problem with blanks, particularly in the spectrophotometer, and it is much easier to adapt to high-sensitivity procedures.

Spectrophotometer, 15–200 nmol/mL

Reagent: 2-Amino-2-methylpropanol buffer, pH 9.9 (50 m*M* in free base, 50 m*M* in the hydrochloride); NAD^+, 2 m*M*; glutamate, 40 m*M*, pH 9.9 (10% monosodium salt, 90% disodium salt); malate dehydrogenase (pig heart), 3.5 U/mL (5 µg/mL); and aspartate transaminase (pig heart), 0.9 U/mL (5 µg/mL).

Reaction time: Four to 8 min.

Conduct of the Assay

For standardization purposes, the malate can be added last after making a preliminary reading. For other purposes, if the sample may add an appreciable absorption blank, it is more satisfactory to add malate dehydrogenase last. In neither case should the enzyme remain in the reagent for more than 10 or 15 min because of slight instability at this alkaline pH.

Comment

The NH_4^+ in the enzyme preparations will lower the pH slightly. If the amounts to be added would exceed 10 m*M*, most of the $(NH_4)_2SO_4$ should be removed by centrifugation before use. Otherwise, the assay will be prolonged and the reaction may be incomplete.

Fluorometer Direct Assay, 0.1–10 nmol/mL

Reagent: The buffer concentration is reduced to 50 m*M*. NAD^+, 200 µ*M* (50 µ*M* with 1 µ*M* malate or less); glutamate, 10 m*M* (1 m*M* with 1 µ*M* malate or less); malate dehydrogenase, 3.5 U/mL (7 U/mL with 50 µ*M* NAD^+); and aspartate transaminase, 0.4 U/mL.
Reaction time: Three to 6 min.

Conduct of the Assay

Ordinarily, the sample is added first. After an initial, reading the enzymes are added together as a 2.5: 1 mixture.

Comment

Because of the weaker buffer used to reduce the fluorescence blank, the enzymes will ordinarily need to be centrifuged to remove most of the NH_4^+ *(see Spectrophotometer, Comment).* With malate levels of 1 µ*M* or less, glutamate and NAD^+ are decreased to reduce further the fluorescence blank. At lowest levels, it may be desirable to reduce the buffer concentration to 20 m*M*.

Analysis with <0.1 pmol

We have not included specific protocols for measuring malate at very low levels, but have not encountered difficulties with levels down to the fmol range. *See*, for example, Hintz et al. *(39).*

Kinetics

The analytical reaction depends on two equilibria, one unfavorable, and one slightly favorable:

$$\frac{[\text{oxaloacetate}][\text{NADH}][\text{H}^+]}{[\text{malate}][\text{NAD}^+]} = 8 \times 10^{-13} \qquad [\text{Eq. 6-16}]$$

At pH 9.9, this becomes $6.4 \times 10^{-3} = 1/156$. Under conditions of the assay, it was observed that:

$$\frac{[\alpha\text{-ketoglutarate}]\ [\text{aspartate}]}{[\text{glutamate}]\ [\text{oxaloacetate}]} = 2.7 \qquad \text{[Eq. 6-17]}$$

Others have found at neutral pH a value of about 7 *(40)*. Combining the two equilibrium constants:

$$\frac{[\text{malate}]\ [\text{glutamate}]\ [\text{NAD}^+]}{[\alpha\text{-ketoglutarate}]\ [\text{aspartate}]\ [\text{NADH}]} = 58 \qquad \text{[Eq. 6-18]}$$

If the overall reaction is to approach completeness, the final concentrations of NADH, α-ketoglutarate, and aspartate will have to be nearly equal to the initial malate concentration $[\text{malate}]_o$. Substituting and rearranging Eq. [6 - 18] (and ignoring the decrease in glutamate and NAD^+ concentrations):

$$\frac{[\text{malate}]}{[\text{malate}]_o} = \frac{58[\text{malate}]_o^2}{[\text{glutamate}]\ [\text{NAD}^+]} \qquad \text{[Eq. 6-19]}$$

From this, the minimal concentrations of glutamate and NAD^+ can be calculated. For the reaction to be 99% complete, i.e., $[\text{malate}]/[\text{malate}]_o = 0.01$, with 0.001, 0.01, and 0.1 m*M* malate, requires the product of the millimolar concentrations of glutamate and NAD^+ to be 0.006, 0.6, and 58, respectively. Because NAD^+ has more absorption at 340 nm and more fluorescence than glutamate, it is an advantage to keep glutamate high and NAD^+ low.

The kinetic constants for both enzymes are quite favorable. Malate dehydrogenase has a V_{max} of about 100 µmol/mg/min at pH 10, with Michaelis constants of about 200 µ*M* for both malate and NAD^+. Consequently, when NAD^+ is reduced to the extent it is in the fluorometric procedures, there is a need to increase malate dehydrogenase. Transaminase kinetics have not been well studied under assay conditions, but the V_{max} is at least 50 µmol/mg/min. The K_m for oxaloacetate is probably about 15 µ*M* (much greater than at pH 7), and the K_m for glutamate is <100 µ*M* (much smaller than at pH 7). Consequently, the required concentration of transaminase is decreased at low malate levels.

Malate: Method II

$$\text{Malate} + \text{NAD}^+ \xrightarrow[\text{dehydrogenase}]{\text{malate}} \text{oxaloacetate} + \text{NADH} + \text{H}^+ \quad \text{[Re. 6-41]}$$

$$\text{Oxaloacetate} + \text{hydrazine} \rightarrow \text{oxaloacetate-hydrazone}$$

Spectrophotometer, 15–200 nmol/mL

Reagent: Hydrazine pH 9.3 (180 m*M* hydrazine base, 20 m*M* hydrazine-HCl); NAD^+, 2 m*M*; and malate dehydrogenase (pig heart), 2.1 U/mL (3 µg/mL).

Reaction time: Thirty to 50 min.

Procedure

The reaction can be started with either the enzyme or the sample after an initial reading. It is important to have concomitant blanks.

Comment

The hydrazine serves as the only buffer. Inclusion of other buffers has given higher blank values, which tend to increase with time. It is important that neither the sample nor the enzyme lower the pH significantly. For this reason, NH_4^+ from the enzyme preparation should not be >5 m*M*.

Fluorometer Direct Assay, 0.1–10 nmol/mL

Reagent: 2-Amino-2-methylpropanol, pH 9.9 (50 m*M* free base, 50 m*M* hydrochloride); hydrazine, 200 m*M*; NAD^+, 100 times the highest expected malate concentration. Malate dehydrogenase, 0.7 U/mL (1.4 U/mL and 2.1 U/mL with 200 and 50 µ*M* NAD^+).

Reaction time: Fifteen to 30 min.

Procedure

The reaction is started with the enzyme after a preliminary reading.

Comment

The rate is very dependent on the ratio of NAD^+ to malate. This governs the steady-state proportion of oxaloacetate, which in turn governs the rate or reaction with hydrazine, which is rate-limiting. At low absolute NAD^+ levels, the malate dehydrogenase is increased to compensate for the fact that the K_m is about 200 µ*M*.

Indirect Assays

For malate amounts of $<0.2 \times 10^{-9}$ mol, the alternative method with transaminase as auxiliary enzyme is strongly preferred.

Kinetics

See Method I.

Oxaloacetate

$$\text{Oxaloacetate} + \text{NADH} + \text{H}^+ \xrightarrow[\text{dehydrogenase}]{\text{malate}} \text{malate} + \text{NAD}^+ \quad \text{[Re. 6-42]}$$

Spectrophotometer, 15–120 nmol/mL

Reagent: Tris-HCl buffer, pH 7.5 (10 m*M* Tris base, 40 m*M* Tris-HCl); NADH, 50–150 μ*M* (at least a 20% excess); and malate dehydrogenase (pig heart), 0.07 U/mL (0.1 μg/mL).
Reaction time: Two to 4 min.

Conduct of the Assay

Either the enzyme or sample can be added last.

Comment

Oxaloacetate is unstable except in strong acid (0.5*M* HCl or stronger), in which it can be stored frozen for long periods without loss. It is actually less stable in slightly acid solutions (pH 3–5) than it is in neutral or slightly alkaline solution. The loss at 25° and pH 7 is 5%/h.

Fluorometer Direct Assay, 0.1–8 nmol/mL

Reagent: The same buffer and enzyme level as in Spectrophotometer. NADH, 0.2–10 μ*M* (20–75% excess).
Reaction time: Less than 3 min.

Conduct of the Assay

The enzyme is added last after a preliminary reading. The samples should be kept frozen until shortly before analysis and should not be allowed to remain more than 30 min in the fluorometer tube before enzyme addition.

Comment

Although only 0.1 μg/mL of enzyme is recommended because of the extremely favorable kinetics, it is a generous amount and could be reduced if necessary, except with the very lowest levels. Oxaloacetate has been found to be very low in tissues. It may be necessary to use as much as 500 μL of tissue extract to obtain reliable results. In this case, the reagent can be made at twice the stated concentration and used in a volume equal to the extract.

Kinetics

The V_{max} is over 700 μmol/mg/min. Under assay conditions the K_m for NADH is about 8 μ*M*, and that for oxaloacetate about 2 μ*M*. This means that with 0.1 μg/mL of enzyme (V_{max} = 70 pmol/min), 10 μ*M* NADH, and 5 μ*M* oxaloacetate, the reaction should be 98% complete in 1 min (Chapter 2).

Phosphate (Inorganic)

$$\text{Glycogen} + P_i \xrightarrow{\text{phosphorylase}} \text{glucose-1-P}$$

$$\text{Glucose-1-P} \xrightarrow{\text{P-glucomutase}} \text{glucose-6-P}$$

[Re. 6-43]

$$\text{Glucose-6-P} + NADP^+ \xrightarrow[\text{dehydrogenase}]{\text{glucose-6-P}} \text{6-P-gluconolactone} + NADPH + H^+$$

The assay for P_i by this three-enzyme sequence is attributed to Fawaz et al. *(41)*. P_i is a ubiquitous substance that is often very difficult to avoid. It is present in ordinary dust, and a component of many soaps and cleaning powders. Some, but not all $NADP^+$ preparations have contained disturbing amounts of P_i. The same is true for some of the analytical enzyme preparations. In this case, removal is possible by dialysis or "washing" with $(NH_4)_2SO_4$ solutions strong enough not to dissolve the enzymes.

Spectrophotometer, 15–200 nmol/mL

Reagent: Imidazole buffer, pH 7.0 (25 m*M* imidazole base, 25 m*M* imidazole-HCl); glycogen, 0.8 mg/mL; 5'AMP, 10 μ*M*; glucose-1,6-P_2, 0.5 μ*M*; magnesium acetate, 0.5 m*M*; EDTA, 1 m*M*; dithiothreitol, 0.5 m*M*; $NADP^+$, 1 m*M*; glucose-6-P dehydrogenase, 0.3 U/mL; rabbit muscle phosphoglucomutase, 2 U/mL (10 μg/mL); and rabbit muscle phosphorylase *a*, 0.8 U/mL (50 μg/mL).

Reaction time: Ten to 20 min.

Conduct of the Assay

If the samples contribute significant glucose-6-P or absorption at 340 μ*M*, the phosphorylase can be omitted from the reagent and a preliminary reading made before adding the phosphorylase in a relatively small volume.

Fluorometer Direct Assay, 1–10 nmol/mL

Reagent: The same as for Spectrophotometer, except $NADP^+$ is reduced to 50 μ*M*, P-glucomutase to 0.6 U/mL, and glucose-6-P dehydrogenase to 0.1 U/mL.

Reaction time: Fifteen to 25 min.

Conduct of the Assay

Unless the samples contribute significant fluorescence or glucose-6-P, the phosphorylase is incorporated in the reagent. This permits any P_i in the reagent or contaminating the fluorometer tube to react ahead of time. Each tube can then be read before adding the sample to give a complete reagent blank.

Fluorometer Indirect Assay, 1–10 nmol

Reagent: The same as for Fluorometer Direct Assay with all three enzymes present.

Procedure

Step 1: Samples in 5 μL or less are added to 100 μL of complete reagent in a fluorometer tube.

Step 2: After 30 min at room temperature, add 10 μL of 0.5*M* NaOH. Heat 20 min in 60° water bath.

Step 3. Add 1 mL of 6*M* NaOH containing 10 m*M* imidazole. Heat again 20 min at 60°, cool exactly to room temperature, and read.

Enzymatic Cycling, 1–10 pmol

Reagent: Imidazole buffer, pH about 6.5 (30 m*M* imidazole base, 70 m*M* imidazole-HCl); glycogen, 1.5 mg/mL; 5'-AMP, 20 μ*M*; glucose-1,6-P_2, 1 μ*M*; magnesium acetate, 1 m*M*; EDTA, 2 m*M*; dithiothreitol, 1 m*M*; $NADP^+$, 100 μ*M*; glucose-6-P dehydrogenase, 0.6 U/mL; P-phosphoglucomutase, 4 U/mL; and phosphorylase, 1.5 U/mL.

Reaction vessels: Oil wells.

Procedure

Step 1: Add freeze-dried tissue samples to 1 µL of 0.02*M* NaOH. Heat 20 min in 80° oven.

Step 2: Add 1 µL of reagent. Incubate 30 min.

Step 3: Add 5 µL of 0.5*M* NaOH. Heat 20 min in 80° oven.

Step 4: Transfer 5 µL to fluorometer tube containing 100 µL of NADP cycling reagent with enzyme levels to give about 1500 cycles in 60 min at 38° (Chapter 5). Heat stop by 3 min in 95–100° water bath.

Step 5: Add 1 mL of indicator reagent (Chapter 5), and read when reaction is complete.

Comment

5'AMP is included in the reagent for its favorable effects on the kinetics of phosphorylase *(see below)*. The 5'-AMP is kept low, however, because high levels may cause a progressively increasing blank. This is the result of release of P_i by certain phosphorylase preparations and is presumably owing to a contaminating phosphatase. The glycogen level recommended is 20 times the Michaelis constant, which is about 0.25 m*M* (calculated as glucosyl units) at the low levels of P_i measured, and is therefore, not critical.

Glutathione reductase is a common contaminant of glucose-6-P dehydrogenase, which in combination with oxidized glutathione from tissues can reoxidize NADPH. Dithiothreitol is added to prevent this by keeping glutathione in the reduced state. (It also ensures the maintenance of phosphorylase in the reduced state.)

The assay is designed for measurement of P_i in the presence of organic phosphate metabolites, including ATP. EDTA is added in excess over Mg^{2+} concentration in order to prevent possible ATPase action. Commercial preparations of phosphorylase have been found to be seriously contaminated with ATPase. In addition, ATPase is difficult to destroy completely in tissue samples. The excess of EDTA over Mg^{2+} lowers the free Mg^{2+} below that required for ATPase action and yet leaves a sufficient level for P-glucomutase in the reagent.

Kinetic Considerations

The most critical enzyme in the assay is glycogen phosphorylase *a*. Its kinetics are somewhat complicated in that the two substates, glycogen and P_i, show strong cooperativity and the Michaelis constants are affected (favorably) by the presence of 5'-AMP. Glycogen can be used at a high level, and P_i will always be at a relatively low level. Consequently, the significant Michaelis constants are that for P_i with glycogen saturating, and that for glycogen in the absence of P_i. In the presence of 5'-AMP these constants for muscle phosphorylase *a* are 1 and 0.25 m*M* (in glucosyl units), respectively. (In the absence of 5'-AMP, they are both much higher, 7 and 1 m*M*, respectively.) The high K_m for P_i (even with AMP present) sets a high requirement for phosphorylase, but means that the same enzyme concentration will suffice for 1 μ*M* or 200 μ*M* P_i concentrations.

Much lower concentrations of the other two enzymes are sufficient. P-Glucomutase has a high specific activity and low Michaelis constant (5 μ*M*), and would be used in very small amounts if it were not for the handicap imposed by the low level of free Mg^{2+}. The low Michaelis constant means that substantially more enzyme is required at the spectrophotometric level of assay than at fluorometer levels.

Glucose-6-P dehydrogenase is used in higher concentrations than in the method for glucose-6-P, and at least a four- or fivefold excess of $NADP^+$ is provided. At pH 7, the dehydrogenase activity is less than at pH 8, and there is greater slowing as the reaction proceeds owing to the less favorable equilibrium position for the reaction *(see Glucose-6-P Methods)* and the fact that NADPH inhibits more strongly at this pH. Keeping the dehydrogenase reaction relatively fast also compensates a little for the delay occasioned by the other two enzymes.

Both auxiliary enzymes are inhibited by sulfate; this is particularly true of the dehydrogenase, which is 50 and 80% inhibited by 5 and 16 m*M* sulfate, respectively. Therefore, an increase in the amounts of the auxiliary enzymes beyond those recommended might actually decrease the rate of reaction, unless most of the $(NH_4)_2SO_4$ used in the suspending medium is removed by centrifugation and resuspension of the enzyme in sulfate-free buffer.

6-P-Gluconate

$$\text{6-P-Gluconate} + NADP^+ \xrightarrow[\text{dehydrogenase}]{\text{6-P-gluconate}} \text{ribulose-5-P} + CO_2 + NADPH + H^+ \qquad \text{[Re. 6-44]}$$

Spectrophotometer, 15–200 nmol/mL

Reagent: Imidazole acetate, pH 7.0 (30 m*M* imidazole base, 20 m*M* imidazole acetate); $MgCl_2$, 5 m*M*; dithiothreitol, 0.5 m*M*; ammonium acetate,

30 m*M*; EDTA, 1 m*M*; NADP$^+$, 500 μ*M*; and 6-P-gluconate dehydrogenase (yeast), 0.06 U/mL (5 μg/mL).

Reaction time: Eight to 10 min.

Conduct of the Assay

It may be preferable to add the enzyme last after an initial reading has been made. The enzyme tends to lose part of its activity at high dilution.

Comment

The reaction is faster with higher NADP$^+$ levels, but, owing to impurities usually present, extra NADP$^+$ will increase the initial blank reading more than is desirable.

Fluorometric Direct Assay, 0.1–10 nmol/mL

Reagent: The ingredients are the same as in Spectrophotometer, except as follows: NADP$^+$, 100 μ*M* (may be reduced to 20 μ*M* for 1 μ*M* 6-P-gluconate or less); and 6-P-gluconate dehydrogenase, 0.024 U/mL (2 μg/mL).

Reaction time: Five to 10 min.

Conduct of the Assay

The enzyme is added last.

Comment

Tissue levels are apt to be very low. It may be desirable to increase the imidazole buffer strength to 0.1*M* to quench flavin fluorescence. We have not had experience with the application of cycling amplification to 6-P-gluconate measurement, but there would be a distinct advantage even when the amount of tissue is not limiting. No difficulty is foreseen in such adaptation. *See*, for example, the glucose-6-P methods, this chapter.

Kinetics

The kinetics are favorable, in spite of the low V_{max} (12 μmol/mg/min), because of the low Michaelis constants for both substrates. The K_m for 6-P-gluconate is affected by reagent composition. As shown by Pontremoli et al. *(42)*, NH_4^+ and Mg^{2+} each have a favorable effect by lowering this K_m, which under assay conditions is about 20 μ*M*. The additives do not have much effect on the V_{max}. The K_m for NADP$^+$ is 2–4 μ*M*. In spite of this low value, when measuring higher levels of 6-

P-gluconate, it is necessary to add high levels of $NADP^+$, because NADPH is an exceedingly potent inhibitor, competitive with NADP. In the spectrophotometer under assay conditions, because of this inhibition, the reaction follows almost a first-order rate curve, whereas one would expect an almost linear course with both substrates so much higher than their Michaelis constants.

3-P-Glycerate: Method I

$$\text{3-P-Glycerate} + \text{ATP} \xrightarrow[\text{kinase}]{\text{P-glycerate}} \text{1,3-diphosphoglycerate} + \text{ADP}$$

[Re. 6-45]

$$\text{1,3-Diphosphoglycerate} + \text{NADH} + H^+ \xrightarrow[\text{dehydrogenase}]{\text{glyceraldehyde-P}} \text{glyceraldehyde-P} + NAD^+ + P_i$$

This is the simpler of the two methods offered, but has the disadvantage for direct analyses that it is based on NADH disappearance. Therefore, the NADH concentration must be chosen with care. For indirect analyses, NAD^+ appearance can be measured; this method is therefore preferable in this case, except possibly with extremely small volumes *(see Comment)*.

Spectrophotometer, 15–200 nmol/mL

Reagent: Imidazole-HCl buffer, pH 7.1 (30 m*M* imidazole base, 20 m*M* imidazole-HCl); mercaptoethanol, 2 m*M*; $MgCl_2$, 1 m*M*; NaCl, 20 m*M*; ATP, 1 m*M*; NADH, 100–150 µ*M* (at least a 20% excess); P-glycerate kinase (rabbit muscle), 0.36 U/mL (2 µg/mL); and glyceraldehyde-P dehydrogenase (rabbit muscle), 0.18 U/mL (5 µg/mL). Most of the $(NH_4)_2SO_4$ should be removed from the enzymes by centrifugation.

Reaction time: Three to 5 min.

Conduct of the Assay

For standardization purposes, the 3-P-glycerate can be added last, eliminating the need for a separate blank. Otherwise, the sample is added first. The enzymes can be added in either order or together as a mixture. In this case, 1,3-diphosphoglycerate would be included in the assay; however, the levels in biological samples are exceedingly low and, moreover, 1,3-diphosphoglycerate is extremely unstable and would disappear from extracts without special precautions.

Fluorometer Direct Assay, 0.2-8 nmol/mL

Reagent: This is the same as for the spectrophotometer, except for the following changes: NADH, 0.5–10 µ*M* (a 20–80% excess); and ATP, 300 µ*M*.
Reaction time: Five to 10 min.

Conduct of the Assay

The enzymes are ordinarily added last after the sample. Otherwise, if glyceraldehyde-P dehydrogenase is added to the reagent before the sample, it may be desirable to add 0.01% bovine serum albumin to protect it from partial inactivation.

Comment

Sulfate increases the K_m for glyceraldehyde-P and should be kept below 1 m*M*. Indirect procedures have not been presented, but would be of advantage at low levels of substrate. After allowing the reaction to proceed in a vol of 200 µL or less, excess NADH would be destroyed, and the NAD^+ measured by the strong alkali procedure. This would permit greater leeway in regard to excess NADH and would diminish the significance of the blank contribution from the biological material. For very small volumes, where cycling is required, NADH oxidation in the blanks becomes a problem. If ascorbic acid does not provide a solution (Chapter 1), it may be preferable to use Method II, in which NADH rather than NAD^+ is the product to be measured.

Kinetics

The kinetics of P-glycerate kinase are quite favorable. Under analytical conditions, the V_{max} is about 400 µmol/mg/min. The kinetic relationships are somewhat complicated. Half-maximal velocity is obtained with about 500 µ*M* ATP and 500 µ*M* 3-P-glycerate; however, the rate with 20 µ*M* 3-P-glycerate is 20% of the maximum, rather than 4% as expected from "normal" kinetics. NaCl, as well as other salts, has a favorable effect by lowering the K_m for 3-P-glycerate. With 300 µ*M* ATP, as recommended for the fluorometric method, the half-time with low levels of 3-P-glycerate would be 15 s for 1 µg/mL of enzyme.

The kinetics of glyceraldehyde-P dehydrogenase are not quite so favorable. The V_{max} is moderately high (probably about 150 µmol/mg/min), the K_m for NADH is about 2 µ*M*, and that for 1,3-diphosphoglycerate is also low (15 or 20 µ*M*). The problem is that the equilibrium constant for [ATP][3-PGA]/[ADP][DPGA] is about 2000 at pH 7 (i.e., very unfavorable). With the 1 m*M* ATP, the 3-P-glycerate level required to give half-maximal activity (glyceraldehyde-P dehydrogenase rate-limiting) is about 200 μ*M*. With 300 μ*M* ATP, it is nearer 1 m*M*. Offsetting this is the fact that, with ATP held constant, the diphosphoglycerate equilibrium level is roughly the square root of the 3-P-glycerate level. Consequently, with 300 μ*M* ATP, reduction of 3-P-glycerate from 800 to 3 μ*M* (266-fold) only reduces the rate about 30-fold.

3-P-Glycerate: Method II

$$\text{3-P-Glycerate} + \text{ATP} \xrightarrow[\text{kinase}]{\text{P-glycerate}} \text{1,3-diphosphoglycerate} + \text{ADP}$$

$$\text{1,3-Diphosphoglycerate} + \text{NADH} + \text{H}^+ \xrightarrow[\text{dehydrogenase}]{\text{glyceraldehyde-P}} \text{glyceraldehyde-P} + \text{NAD}^+ + \text{P}_i$$

Destruction of NADH with acid [Re. 6-46]

$$\text{Dihydroxyacetone-P} \xrightarrow[\text{isomerase}]{\text{triose-P}} \text{glyceraldehyde-P}$$

$$\text{Glyceraldehyde-P} + \text{NAD}^+ + \text{arsenate} \xrightarrow[\text{GAP dehydrogenase}]{} \text{3-P-glycerate} + \text{NADH} + \text{H}^+$$

This is more complicated than the alternative method. It has the advantage that readings increase with increased substrate, and it is consequently more flexible and more accurate with lower substrate levels.

Current preparations of P-glycerate kinase are badly contaminated with triose-P isomerase. This means that at the time of acidification, most of the 3-P-glycerate has been converted to dihydroxyacetone-P. Consequently, it is necessary to include triose-P isomerase in the reagent after acidification. In spite of the complexity of this method, it has given highly reproducible results without difficulty. No spectrophotometric assay is presented, because it is simpler and completely satisfactory to standardize 3-P-glycerate solutions with the first method.

Fluorometer, Direct Measurement of Final Steps, 0.1–10 nmol/mL

Reagent before acidification: Imidazole-HCl buffer, pH 7 (18 m*M* imidazole base, 12 m*M* imidazole-HCl); $MgCl_2$, 1 m*M*; mercaptoethanol, 5 m*M*; ATP, 100 μ*M*; NaCl, 20 m*M*; NADH, 20 μ*M* (5 μ*M* with 1 μ*M* 3-P-glycerate or less); P-glycerate kinase (rabbit muscle), 0.18 U/mL (1 μg/mL); and glyceraldehyde-P-dehydrogenase (rabbit muscle), 0.36 U/mL (10 μg/mL). Centrifuge enzyme preparations to remove most of the $(NH_4)_2SO_4$ if present.

Reagent composition during final step: Imidazole-HCl buffer, pH 7.0 (78 m*M* imidazole base, 52 m*M* imidazole-HCl); $MgCl_2$, mercaptoethanol, ATP, and NaCl from the first step essentially unchanged; NAD^+, 100 μ*M*; Na_2HAsO_4, 1 m*M*; triose-P isomerase (rabbit muscle), 2.4 U/mL (1 μg/mL); and glyceraldehyde-P dehydrogenase 0.72 U/mL (20 μg/mL).

Procedure

Step 1: The sample is added to 1 mL of reagent in a fluorometer tube. The P-glycerate kinase can be incorporated in the reagent; however, it is preferable to add the dehydrogenase last after the samples, because otherwise it loses part of its activity. After 5 min at room temperature, 10 μL of 4*M* HCl are added, and the tube is allowed to stand for at least 10 min at room temperature. Twenty microliters of 5 m*M* imidazole base are then added, followed by 10 μL of a mixture of 10 m*M* NAD^+ in 100 m*M* Na_2HAsO_4.

Step 2: After making a reading, 4 μL of a mixture containing 0.5% glyceraldehyde-P dehydrogenase and 0.02% triose-P isomerase are added. A final reading is made when the reaction is over (5–10 min).

Comment

Most of the $(NH_4)_2SO_4$ is removed from the glyceraldehyde-P dehydrogenase, because it is inhibitory to triose-P isomerase and the glyceraldehyde-3-P dehydrogenase. Imidazole-HCl rather than the acetate is used as the buffer to decrease the amount of HCl required for acidification. NaCl is added because of its favorable effect on P-glycerate kinase kinetics (*see* Kinetics in Method I). The first-step reaction is expected to be finished in 1 or 2 min. It is desirable to check this directly in the fluorometer each day with an extra tube to which 3-P-glycerate has been added equal to 50–70% of the NADH level. Otherwise, no readings are necessary in Step 1; however, readings on one tube in the set can be used to check on the time required for complete NADH destruction (expected to be <5 min).

The assay as described measures the sum of 3-P-glycerate and the triose phosphates. As a rule, 3-P-glycerate is likely to be present in greater amounts than the triose phosphates. When necessary, correction can be made by a separate measurement of the triose phosphates. Alternatively, they can be destroyed beforehand by heating 10 min at 60° in 0.02*M* NaOH; 3-glycerate is extremely stable in alkali.

Kinetics

The kinetics of the reactions in Step 1 are given in Method I. The kinetics of the reactions in Step 2 are described in Method II for dihydroxyacetone-P.

P-Pyruvate

$$\text{P-Pyruvate} + \text{ADP} \xrightarrow[\text{kinase}]{\text{pyruvate}} \text{pyruvate} + \text{ATP}$$

[Re. 6-47]

$$\text{Pyruvate} + \text{NADH} + \text{H}^+ \xrightarrow[\text{dehydrogenase}]{\text{lactate}} \text{lactate} + \text{NAD}^+$$

Levels of P-pyruvate are low in most tissues. Because lactate dehydrogenase reacts with pyruvate and other keto acids present in tissue extracts, precision is greatly improved if these keto acids are removed with hydrazine prior to the assay. During the assay, lactate dehydrogenase is used in sufficient amount to reduce pyruvate before it has a chance to react with the hydrazine. Because the spectrophotometric assay will ordinarily be used only for standardization, the hydrazine step is omitted.

Spectrophotometer, 15–120 nmol/mL

Reagent: Phosphate buffer, pH 7 (30 m*M* K_2HPO_4:20 m*M* NaH_2PO_4); $MgCl_2$, 2 m*M*; ADP, 200 µ*M*; NADH 50–150 µ*M* (at least a 20% excess); pyruvate kinase, 0.15 U/mL (1 µg/mL); and lactate dehydrogenase (beef heart), 0.2 U/mL (1 µg/mL).

Reaction time: Two to 4 min.

Conduct of the Assay

Unless pyruvate is also to be measured, the reagent may be prepared with lactate dehydrogenase added, but without the pyruvate kinase. The sample is added. A reading is made after 1 or 2 min to allow time for possible presence of pyruvate. Pyruvate kinase is then added.

Comment

For standardization of P-pyruvate solutions, there is no reason to pretreat the sample with hydrazine. If, however, P-pyruvate is to be measured spectophotometrically in the presence of a larger amount of pyruvate, the pretreatment may be carried out exactly as described in the next protocol and with the same reagent (enzymes omitted). Notice the increase in lactate dehydrogenase. Because the hydrazine pretreatment takes 20 or 30 min, this would be conveniently carried out in test tubes ahead of time. The sample would then be transferred to the spectrophotometer cell, lactate dehydrogenase added, a reading made, and the reaction started with pyruvate kinase.

Fluorometer Direct Assay, 0.1–8 nmol/mL

Reagent: The same as in Spectrophotometer, except: NADH 0.2–10 μ*M* (to provide a 20–75% excess); and hydrazine-HCl, 10 m*M*; lactate dehydrogenase, 0.5 U/mL (increased to 1 U/mL with NADH <1 μ*M*).
Reaction time: Two to 5 min.

Conduct of the Assay

The reagent is prepared without either enzyme. The sample is added to reagent in the fluorometer tube. After 20–30 min at room temperature, lactate dehydrogenase is added and a reading made (1- or 2-min delay is sufficient to make sure that any trace of pyruvate in sample or reagent has reacted). The analytical reaction is then started with pyruvate kinase. The final reading should be made within 30 min (5 min should be sufficient), because the pyruvate reaction with hydrazine is slowly reversible.

Imidazole can be substituted for phosphate to diminish fluorescence of tissue flavins. With 200 m*M* imidazole, the pyruvate kinase and lactate dehydrogenase should both be doubled. If imidazole is used, KCl is added at a concentration of 75 m*M* to take care of the K^+ requirement of pyruvate kinase.

Kinetics

The approximate kinetic constants for pyruvate kinase at pH 7 in 50 m*M* phosphate buffer are V_{max}, 100 μmol/mg/min (100 U/mg); K_m for P-pyruvate, 30 μ*M*; and K_m for ADP, 200 μ*M*. From these data, the calculated halftime for 10 μ*M* substrate or less with 200 μ*M* ADP and 1 μg/mL of enzyme would be 0.4 min [0.7(30/100)(400/200)]. In 200 m*M* imidazole at the same pH, the velocity with low P-pyruvate is a third to half as fast, probably because of an increase in the Michaelis constant.

Pyruvate kinase requires both Mg^{2+} and K^+. The reaction of pyruvate with the large excess of hydrazine proceeds at a rate that can be considered as a first-order reaction for pyruvate. The rate constant for this reaction at 25° with 10 m*M* hydrazine is about 0.2/min (0.02/min multiplied by the millimolar concentration of hydrazine), giving a half-time of 3 or 4 min. For 99% of liberated pyruvate to react with NADH, the lactate dehydrogenase apparent first-order rate constant for pyruvate must be 100 times faster than that with hydrazine, or 20/min. Even with 100 μ*M* P-pyruvate, the free pyruvate concentration will never exceed first-order concentrations; therefore, the same lactate dehydrogenase concentration is appropriate for high and low P-pyruvate concentrations. However, at very low levels, if the NADH concentration is reduced below 1 μ*M*, an increase in lactate dehydrogenase is indicated.

Potassium

The method for potassium, adapted from Outlaw and Lowry (*43*), is based on the fact that pyruvate kinase requires both a divalent and a monovalent cation for activity, and that potassium is 40-fold more effective than sodium. The rate of the reaction is measured by a coupled enzymatic step. *See* final Blank Values in regard to the great danger of contamination of reagents with monovalent cations and their avoidance.

$$\text{ADP} + \text{P-pyruvate} \xrightarrow[Mg^{2+},\ K^+]{\text{pyruvate kinase}} \text{pyruvate} + \text{ATP}$$

[Re. 6-48]

$$\text{NADH} + \text{pyruvate} + H^+ \xrightarrow{\text{lactate dehydrogenase}} \text{NAD}^+ + \text{lactate}$$

Fluorometric Direct Assay, 10–100 nmol

Reagent: Imidazole-HCl buffer pH 7.2 (15 m*M* imidazole base, 10 m*M* imidazole HCl); $MgCl_2$, 1 m*M*; ADP, 0.2 m*M*; P-pyruvate, 0.1 m*M*; bovine plasma albumin, 0.02%; beef heart lactate dehydrogenase, 1.5 U/mL (3 µg/mL); and rabbit muscle pyruvate kinase, 2.5 U/mL (5 µg/mL).

Conduct of the Assay

Aliquots of standards or extracts are added to 1 mL of basic reagent containing 10 µ*M* NADH. An initial fluorescence reading is made after 5 min (to allow any pyruvate to react) and a final reading is made after 30 min. Total volume should not vary more than 10 µL.

Indirect Assay for Tissue Extracts, 1–10 nmol

Reagent: The reagent is the same as in Fluorometric Direct Assay, but with 12 U/mL (25 µg/mL) pyruvate kinase.

Procedure

Step 1: Aliquots of standards or extracts are added to 100 µL of basic reagent in 6 × 50 mm serological tubes at measured time intervals. Total volume should not vary more than 2 µL.

Step 2: After 60 min at room temperature, add 1 µL of 10 m*M* NADH to each tube in the same time sequence as Step 1.

Step 3: After 2–5 min add 1 µL of 5*M* HCl.

Step 4: After at least 5 min, transfer 50-µL aliquots to fluorometer tubes containing 1 mL of 6*M* NaOH with 10 m*M* imidazole, 10 min at 60°.

Step 5: After cooling in a water bath to room temperature, the tubes are wiped and the fluorescence read. To increase precision, the fluorescence of each tube may be read before and after the addition of the sample.

Indirect Assay for Tissue Sections, 0.2–2 nmol

Reagent: Imidazole-HCl buffer, pH 7.6 (40 m*M* imidazole base, 10 m*M* imidazole HCl); $MgCl_2$, 2 m*M*; ADP, 0.4 m*M*; P-pyruvate, 0.2 m*M*; bovine serum albumin, 0.04%; beef heart lactate dehydrogenase, 3 U/mL (6 µg/mL); and pyruvate kinase, 25 U/mL (50 µg/mL).

Procedure

Step 1: Add 50 µL of 0.01*M* HCl to 6 × 50 mm serological tubes. Samples are added to the acid. Standards are 2–20 µ*M* K^+ in 0.01*M* HCl. Heat 20 min in 60° water bath.

Step 2: Add 50 µL of reagent and mix. Total vol should not vary more than 2 µL.

Step 3: After 60 min at room temperature, add 5 µL of 0.3 m*M* NADH in 25 m*M* imidazole buffer, pH 7.2.

Step 4: After 5 min, add 50 µL 0.2*M* HCl to each tube.

Step 5: Transfer 100 µL to fluorometer tubes containing 1 mL 6*M* NaOH with 10 m*M* imidazole; 10 min at 60°.

Step 6: The tubes are cooled and read.

Cycling Assay, 10–100 pmol

Reagent: The same as for Indirect Assay for Tissue Sections.
Reaction vessels: Oil wells.

Procedure

Step 1: Add 1 µL of 0.01*M* HCl to each well. Samples are added to the acid. Standards are 10–100 µ*M* K^+ in 0.01*M* HCl. Heat 20 min at 80°.

Step 2: Add 1 µL of reagent: Incubate 60 min. Total volume should not vary more than 0.1 µL.

Step 3: After 60 min, add 1 µL of 100 µ*M* NADH in 25 m*M* imidazole buffer, pH 7.2.

Step 4: After 5 min, add 1 µL of 0.5*M* HCl.

Step 5: Transfer 2 µL to fluorometer tubes containing 100 µL of NAD cycling reagent with enzyme levels to give about 500 cycles in 60 min at 25° (Chapter 5). Heat 3 min in water bath at 95–100°.

Step 6: Add 1 mL of indicator reagent (Chapter 5), and read when reaction is complete.

Cycling Assay, 1–10 pmol

Reagent: The same as for Indirect Assay for Tissue Sections.
Reactions vessels: Oil wells.

Procedure

Steps 1 and 2: These steps are identical to Cycling Assay, except the volumes are reduced to 0.1 µL.

Step 3: After 60 min, add 0.1 µL of 100 µ*M* NADH.

Step 4: After 5 min, add 0.1 µL of 0.2*M* HCl to each droplet.

Step 5: After at least 2 min, add 10 µL of NAD cycling reagent with enzyme levels to give about 2500 cycles in 60 min at 25° (Chapter 5).

Step 6: Add 1 µL of 1*M* NaOH at the same intervals as the cycling reagent. Heat 20 min at 80°.

Step 7: Transfer 10 µL from the oil well to 1 mL of indicator reagent (Chapter 5), and read when reaction is complete.

Comment

With 12 U (25 μg) of pyruvate kinase, the expected yield is about 0.5 mol of NAD^+/mol of K^+/h. It may be necessary to check each new preparation of pyruvate kinase. Higher yields may be obtained by increasing the temperature. There is a 65% higher yield at 35° than 25°. This is an advantage, because it decreases the significance of tissue blanks, such as the endogenous NAD^+. The yield of NAD^+ was virtually unchanged when Mg^{2+} was increased from 1–2 m*M*; thus, Mg^{2+} from the tissues should not distort the results.

Blank Values

There are two components of the blank. There are those that add a constant amount to the signal, such as pyruvate in the P-pyruvate, nonenzymatic oxidation of NADH, and fluorescence of the reagents. Neither pyruvate kinase nor lactate dehydrogenase preparations have contained enough NAD^+ to be troublesome in the assay.

The other component is owing to cation contamination, which causes stimulation of pyruvate kinase and increases the blanks in parallel with the yield from K^+. This component is the most troublesome. The worst offender is NH_4^+, became it is equivalent to K^+ as a cofactor for pyruvate kinase, is a common laboratory component, and NH_3 is volatile and absorbed to glass. To minimize this problem, the following precautions are recommended. Pyruvate kinase and lactate dehydrogenase, which are supplied in $(NH_4)_2SO_4$ suspensions, are centrifuged, the supernatant fluid discarded, and the enzymes dissolved in the same vol of 25 m*M* Tris-HCl, pH 8.1, containing 0.02% bovine serum albumin. The enzymes are then dialyzed against 500 vol of the same buffer for 4 h, the buffer replaced, and dialysis continued for 4 h.

The glassware was rinsed immediately before use with 5 m*M* NaOH, followed by water and handled with latex surgical gloves. The blank pyruvate kinase activity can be reduced to the equivalent of 2 μ*M* K^+ in Fluorometric Direct Assay, and 6–10 μ*M* K^+ in the indirect methods.

The overall constant component of the blanks can be kept equivalent to 3 μ*M* NAD^+. This would correspond to 6 μ*M* K^+ for a 0.5:1 NAD^+ to K^+ yield. About 30% of this can be attributed to the pyruvate kinase step reagents. The remainder of the constant blank is primarily the result of the fluorescence of strong NaOH (Indirect Assay for Tissue Extracts and Indirect Assay for Tissue Section) or the indicator reagent (Cycling Assays). The contribution from the cycling reagent should be negligible.

The NADH is added for a minimal period of time to decrease the contribution of nonenzymatic oxidation. In 0.1 µL of reagent, NADH was oxidized 5% in 1 h.

The overall blanks should never exceed the equivalent of 15 µ*M* K^+. The analyst is referred to the original paper *(43)* for further suggestions about removing disturbing monovalent cations from reagents.

Pyrophosphate, Method I

$$PP_i + \text{UDPglucose} \xrightarrow[\text{pyrophosphorylase}]{\text{UDPglucose}} \text{UTP} + \text{glucose-1-P}$$

$$\text{Glucose-1-P} \xrightarrow{\text{P-glucomutase}} \text{glucose-6-P} \qquad \text{[Re. 6-49]}$$

$$\text{Glucose-6-P} + NADP^+ \xrightarrow[\text{dehydrogenase}]{\text{glucose-6-P}} \text{6-P-gluconate} + NADPH + H^+$$

Pyrophosphate is present in low concentrations in most tissues, approximately that of glucose-1-P. Since glucose-6-P and glucose-1-P are at near equilibrium in most tissues, the glucose-6-P is present at 10–20-fold the concentration of PP_i, making the accurate measurement of glucose-1-P or PP_i very difficult. To obviate the problem of measuring a small increment of NADPH absorbance or fluorescence, the NADPH formed by the oxidation of glucose-6-P can be destroyed with acid. The pyrophosphate can then be measured by the addition of UDP-glucose (UDPG) pyrophosphorylase in the presence of the auxiliary enzymes P-glucomutase and glucose-6-P dehydrogenase.

A second option described below is to destroy the glucose-1-P and glucose-6-P with acid and alkali, respectively. Glucose-1-P can be hydrolyzed to glucose and inorganic phosphate by heating in 0.1*M* HCl for 10 min at 80°; the recovery of PP_i under these conditions is over 90%. Suitable PP_i standards should be incorporated in each assay to determine the recovery. Glucose-6-P can be destroyed by heating in 0.1*M* NaOH for 20 min at 80°. No loss of PP_i occurs under these alkaline conditions, but it is recommended that standards treated with alkali be included when tissue levels are determined. An alternative is to treat tissues only with alkali to destroy the glucose-6-P; the samples can then be analyzed with and without the addition of UDPG pyrophosphorylase. Such analyses are described in the cycling section and have proven to be satisfactory. (The spectrophotometer method is not recommended for unfractionated tissue, because the required amount of tissue extract would cause problems owing to turbidity and UV absorption.)

Spectrophotometer, 15–120 nmol/mL

Reagent: Tris-HCl buffer, pH 8.0 (25 m*M* Tris base, 25 m*M* Tris-HCl); $NADP^+$, 500 μ*M*; UDPG, 1 m*M*; glucose-1,6-P_2, 0.5 μ*M*; EDTA, 0.1 m*M*; $MgCl_2$, 1 m*M*; glucose-6-P dehydrogenase, 0.25 U/mL (0.5 μg/mL); P-glucomutase, 0.4 U/mL (2 μg/mL); and UDPG pyrophosphorylase (beef liver), 0.2 U/mL (2 μg/mL).

Reaction time: Two to 3 min for glucose-6-P, 4–6 min for glucose-1-P, and 30 min for PP_i.

Conduct of the Assay

For standardizing PP_i solutions, all three enzymes can be included in the reagent and the sample added after an initial reading. For samples that may contribute absorption at 340 nm or will necessitate considerable dilution of the sample, it is better to add the pyrophosphorylase last.

Fluorometer Direct Analysis, 0.1–9 nmol/mL

Reagent: The reagent is the same as in Spectrophotometer, except: $NADP^+$, 50 μM, and omission of pyrophosphorylase.

Procedure

Step 1: Add samples to 100 μL of 0.1*M* NaOH; heat 20 min at 80°.

Step 2: Add 10 μL of 1*M* HCl.

Step 3: Add 1 mL of reagent. After 5 min for glucose-1-P, if present to react, read and add 0.2 U of UDPG pyrophosphorylase.

Step 4: Read after reaction is complete, about 30 min.

Conduct of the Assay

Tissue samples can be treated before analysis to destroy the glucose-6-P, which would be the most disturbing reaction. A small vol of concentrated NaOH can be added to bring the final concentration to

0.1*M* NaOH, and the sample heated at 80° for 20 min. Standards of glucose-6-P and PP_i are treated in the same fashion. The sample is then neutralized with an equivalent amount of HCl, and an aliquot taken for analysis. The samples in this case should be analyzed in the absence and presence of UDPG pyrophosphorylase. The PP_i concentration can be determined by the difference.

Cycling, 0.05–0.25 pmol

Reagent: The same as for Spectrophotometer, except: $NADP^+$, 10 µ*M*, and bovine serum albumin, 0.02%.
Reaction vessels: Oil wells.

Procedure

Step 1: NaOH, 0.05 µL of 0.1*M* for samples, blanks, and standards (0.05 µL of 1–5 µ*M* PP_i, glucose-1-P, and glucose-6-P).
Step 2: Add 1 µL of pyrophosphate reagent; 40 min at room temperature.
Step 3: Add 2 µL 0.2*M* NaOH; 30 min at 80°.
Step 4: Transfer 2 µL into fluorometer tubes containing 100 µL of NADP cycling reagent with enzyme levels to give about 25,000 cycles in 2 h at 38° (Chapter 5). Incubate 1 h at 38°, then 2 min at 100°.
Step 5: Add 1 mL of indicator reagent (Chapter 5), and read when reaction is complete.

Kinetics

The UDPG pyphosphorylase of liver has a K_m for UDPG of about 50 µ*M* under the conditions described, and that for PP_i is near 150 µ*M*. The Michaelis constants of the yeast enzyme are somewhat more favorable for the measurement of PP_i (the K_ms for UDPG and PP_i are 80 and 30 µ*M*, respectively, in the reagent given). However, the yeast enzyme that was tested had disturbing side reactions that led to the formation of NADPH in excess of the the PP_i present. Although the problems were not investigated exhaustively, it appeared that the yeast enzyme was contaminated with 6-P-gluconate dehydrogenase. The liver enzyme contributes a direct fluorescent blank, equivalent to 0.1 µ*M* NADPH when used at thc concentration recommended. The amount of enzyme was kept low in order to minimize the blank, resulting in a reaction time of 30 min. When the measurement of PP_i was made by enzymatic cycling, the enzyme at the level recommended contributed a blank equivalent to 0.04 µ*M* at the specific enzyme step.

Pyrophosphate, Method II

An alternative method for measuring pyrophosphate utilizes the pyrophosphate-dependent fructose-6-P kinase from a bacterial source.

$$PP_i + \text{fructose-6-P} \xrightarrow[\text{kinase}]{PP_i\text{-dependent fructose-6-P}} \text{fructose-1,6-P}_2 + P_i$$

$$\text{Fructose-1,6-P}_2 \xrightarrow{\text{aldolase}} \text{dihydroxyacetone-P} + \text{glyceraldehyde-P}$$

$$\text{Glyceraldehyde-3-P} \xrightarrow{\text{triose-P isomerase}} \text{dihydroxyacetone-P} \qquad \text{[Re. 6-50]}$$

$$\text{2 Dihydroxyacetone-P} + 2\,\text{NADH} + 2\text{H}^+ \xrightarrow[\text{dehydrogenase}]{\alpha\text{-glycero-P}} 2\ \alpha\text{-glycero-P} + 2\ \text{NAD}^+$$

Spectrophotometer, 15–120 nmol/mL

Reagent: Imidazole buffer pH 7.4 (60 m*M* imidazole base, 40 m*M* imidazole HCl); $MgCl_2$, 2 m*M*; fructose-6-P, 1 m*M*; NADH, 100 µ*M*; aldolase 0.02 U/mL (2 µg/mL); triose-P-isomerase 0.2 U/mL (0.4 µg/mL); α-glycero-P-dehydrogenase, 0.7 U/mL (4 µg/mL); and pyrophosphate-dependent fructose-6-P kinase from *Proprionibacterium freudenreichii,* 0.1 U/mL.
Reaction time: Three to 5 min.

Conduct of the Assay

The enzymes can be added stepwise, if it is desired to measure dihydroxyacetone-P, glyceraldehyde-P, fructose 1,6-P_2, and pyrophosphate. Otherwise, all the enzymes, except pyrophosphate-dependent fructose-P kinase, can be incorporated into the reagent, and an initial reading taken after the three reactions have occurred. The fructose-6-P kinase is then added, and a second reading taken when the reaction is complete.

Fluorometer, Direct Assay 0.1–4 nmol/mL

Reagent: The reagent is the same as Spectrophotometer, except: NADH, 0.2–10 µ*M* (20–70% excess); aldolase, 0.01 U/mL; and glycero-P-dehydrogenase, 0.2 U/mL.
Reaction time: Three to 5 min.

Conduct of the Assay

As in the spectrophotometric assay, the metabolites may be measured individually by the sequential addition of the enzymes. However, the level of fructose-1,6-P_2 in some tissues (brain, for example) is an order of magnitude higher in concentration than the pyrophosphate. In this case, all of the enzymes may be incorporated into the reagent, along with sufficient NADH to permit the measurement of pyrophosphate after the first three reactions are complete. The remaining NADH is set to read full-scale, and a reading is made. The fructose-6-P kinase is then added to determine the pyrophosphate concentration. The yield of NADH oxidation is twice the pyrophosphate concentration, which is an advantage over the preceding measurement, since pyrophosphate concentrations in tissues are on the order of 10–20 µmol/kg wet wt *(44)*.

Comment

The fructose-6-P kinase is supplied as a lyophilized powder. The powder containing about 5% protein (50 µg/10 U of enzyme activity) is dissolved in 1 mL of 50% glycerol. The preparation is stable for at least a week at 4°. Extensive tests of stability were not made.

Pyruvate

$$\text{Pyruvate} + \text{NADH} + \text{H}^+ \xrightarrow[\text{dehydrogenase}]{\text{lactate}} \text{lactate} + \text{NAD}^+ \qquad \text{[Re. 6-51]}$$

Although pyruvate is not as abundant a contaminant as lactate, it is nevertheless widespread, and because tissue levels are much lower than in the case of lactate, the danger from contamination is almost as great.

Spectrophotometer, 15–120 nmol/mL

Reagent: Phosphate buffer, pH 7 (30 m*M* K_2HPO_4, 20 m*M* NaH_2PO_4); NADH, 50–150 µ*M* (at least a 20% excess); and lactate dehydrogenase (beef heart), 0.1 U/mL (0.5 µg/mL).

Reaction time: Two to 4 min.

Conduct of Assay

Ordinarily, the sample is added first. After an initial reading, the reaction is started with enzyme added in small volume. For standardizing pure solutions, the enzyme can be incorporated in the reagent and the reaction started with the sample, eliminating the need for a blank.

Fluorometer Direct Assay 0.1–8 nmol/mL

Reagent: Phosphate buffer as in spectrophotometer; NADH, 0.2–10 μ*M* (20–75% excess over expected pyruvate); and lactate dehydrogenase, 0.04 U/mL (0.2 U/mL with <1 μ*M* NADH).
Reaction time: Three to 10 min.

Conduct of the Assay

The reaction is ordinarily started by addition of the enzyme after the sample is in and an initial reading made.

Comment

Pyruvate stock solutions are not indefinitely stable even frozen, because of a tendency to polymerize. Therefore, when used as fluorometer standards, they should be occasionally restandardized in the spectrophotometer. Heart lactate dehydrogenase can strongly enhance the fluorescence of NADH. At the enzyme level recommended, this is scarcely noticeable. This is handled by running suitable standards and blanks. This is one reason for not using more enzyme than necessary. A more important reason is that lactate dehydrogenase reacts with many keto acids besides pyruvate, although at a much slower rate. If a large excess of enzyme is used, certain tissues (brain, for example) will give an erroneously high result. If tissue blank fluorescence from flavins is a problem, this can be greatly reduced by substituting 100 or 200 m*M* imidazole buffer, pH 7, for the phosphate. With the higher imidazole concentration, lactate dehydrogenase should be increased threefold or the reaction time prolonged *(see* Kinetics *below).*

Fluorometer Indirect Assay 10–100 pmol

Reagent: Imidazole buffer, pH 7 (30 m*M* imidazole, 20 m*M* imidazole-HCl); ascorbic acid, 2m*M*; NADH 2 μ*M*; bovine plasma albumin, 0.01%; and lactate dehydrogenase, 0.04 U/mL (0.2 μg/mL). The reagent should be prepared within an hour of use and kept on ice. The NADH is added after the ascorbic acid and should be as free as possible of NAD^+ *(see* Chapter 1). The ascorbic acid is added from a 100-m*M* stock solution stored at –20° or below.
Reaction vessels: Fluorometer tubes.

Procedure

Step 1: One hundred microliters of reagent at 0°. Standards, 10 μL of 1–10 μ*M* pyruvate. Samples in volumes of 20 μL or less.

Step 2: Twenty minutes at room temperature (water bath), and return to ice bath.
Step 3: HCl, 20 µL of 2*M*; room temperature at least 5 min.
Step 4: NaOH, 1 mL of 6*M* containing 10 m*M* imidazole; 10 min 60°.
Step 5: Fluorescence reading exactly at room temperature.

Comment

This indirect assay is based on measurement of NAD^+. In general, living cells contain more NAD^+ than pyruvate and often much more. Therefore, in analyzing tissues or microorganisms with this indirect analysis, it would usually be necessary to destroy most of the NAD^+. This can be accomplished by heating in alkali as in Cycling Assay below. Pyruvate is not indefinitely stable in alkali; therefore, the alkaline treatment should not be excessive *(see* Chapter 1 for minimal conditions for NAD^+ destruction). The buffer is changed from phosphate to imidazole, because phosphate buffers cause appreciable NADH oxidation.

Cycling Assay, 1–5 pmol

Pyruvate reagent: The same buffer and NADH concentratian as in Fluorometer Indirect Assay. Ascorbic acid, 2 m*M*; bovine plasma albumin, 0.02%; and lactate dehydrogenase 0.02 U/mL. The ascorbic acid is added before the NADH from a stock 9% (50 m*M*) solution that is stored at –20°.
Reaction vessels: Oil wells.

Procedure for Frozen-Dried Tissue Samples

Step 1: Use 1 µL of 0.02*M* NaOH for samples and blanks, and 1 µL of 2 to 5 µ*M* pyruvate in 0.02*M* NaOH for standards; heat 20 min at 80°.
Step 2: Add 5 µL of pyruvate reagent; incubate 20 min at room temperature.
Step 3: Add 1 µL of 1*M* HCl. Allow to stand at least 3 min.
Step 4: Transfer 5-µL aliquots to fluorometer tubes containing 100 µL of NAD cycling reagent with enzyme levels to give about 3000 cycles in 60 min at 25° (Chapter 5). Heat stop by 2 min in a 95–100° water bath.
Step 5: Add 1 mL of indicator reagent (Chapter 5), and read when reaction is complete.

Cycling Assay, 0.1–0.5 pmol

Reagents: The same as in Cycling Assay, 1–5 pmol.
Reaction vessels: Oil well.

Procedure

Steps 1–3: The same as in Cycling Assay, 1–5 pmol, but all volumes reduced tenfold.

Step 4: Add to each oil well 10 µL of NAD cycling reagent with enzyme levels to give about 25,000 cycles in 2 h at 25° (Chapter 5). Stop with 2 µL of 0.5*M* NaOH and heat 20 min in 80° oven.

Step 5: Transfer 10 µL to 1 mL of indicator reagent (Chapter 5), and read when reaction is complete.

Kinetics

The kinetic situation is favorable. The V_{max} is large, 200–250 µmol/mg/min, and the Michaelis constants are small. Increase in one substrate increases the apparent Michaelis constant for the other. At pH 7 in 50 m*M* phosphate buffer, the apparent K_m for pyruvate is 30 µ*M* with 100 µ*M* NADH, but only 6 µ*M* with 0.25 µ*M* NADH. Conversely, the K_m for NADH is about 2.5 µ*M* with 400 µ*M* pyruvate and only 1 µ*M* with 3 µ*M* pyruvate. The practical result is that the amount of lactate dehydrogenase does not need to be increased as much as otherwise expected to measure small concentrations of pyruvate with low levels of NADH.

The equilibrium is very favorable at pH 7. Starting with equal concentrations of pyruvate and NADH, the reaction would proceed until only 0.5% of the initial pyruvate remained.

Because of the use of lactate dehydrogenase as an auxiliary enzyme in a number of analyses, its kinetics are of interest under several conditions. At pH 8 in 50 m*M* Tris or glycylglycine buffers, the V_{max} is at least as great as at pH 7, and the K_m for NADH is if anything more favorable, but the K_m for pyruvate is increased eightfold or tenfold. This means that more enzyme must be used. Dilute imidazole (50 m*M*) at pH 7 is about as favorable as phosphate, but if the concentration is increased to 200 m*M*, three times as much enzyme is required because of an increase in the K_m for pyruvate.

Skeletal muscle lactate dehydrogenase has less favorable kinetics than the heart enzyme for measuring pyruvate. Although the V_{max} is 50% greater, the K_m for pyruvate is five or six times larger. The K_m for NADH is about the same for both enzymes.

Sucrose

$$\text{Sucrose} \xrightarrow{\text{invertase}} \text{glucose + fructose}$$

$$\text{Glucose + fructose + 2 ATP} \xrightarrow{\text{hexokinase}} \text{glucose-6-P + fructose-6-P + 2 ADP}$$

$$\text{Fructose-6-P} \xrightarrow{\text{P-glucoisomerase}} \text{glucose-6-P} \qquad \text{[Re. 6-52]}$$

$$\text{2 Glucose-6-P + 2 NADP}^+ \xrightarrow[\text{dehydrogenase}]{\text{glucose-6-P}} \text{2 6-P-gluconolactone + 2 NADPH + 2 H}^+$$

Sucrose can be measured separately from existing glucose, fructose, and their phosphorylated forms by first destroying these with alkali. Alternatively, invertase can be added after the free glucose and fructose in the sample have been measured. The method has been described by Jones, et al. *(45)*.

Spectrophotometer, 8–100 nmol/mL

Reagent: Imidazole-HCl buffer, pH 6.9 (40 m*M* imidazole base, 60 m*M* imidazole HCl); $NADP^+$, 0.4 m*M*; ATP, 1 m*M*; $MgCl_2$, 5 m*M*; bovine serum albumin, 0.02%; yeast invertase, 6 U/mL (20 µg/mL); yeast P-glucoisomerase, 0.35 U/mL (1 µg/mL); yeast glucose-6-P dehydrogenase, 0.35 U/mL (1 µg/mL); and yeast hexokinase 0.15 U/mL (2 µg/mL).

Procedure

Step 1: The tissue extract (at least 50 µ*M* in sucrose) is made 0.02*M* in NaOH, and heated at 90° for 3 min along with appropriate sucrose standards.

Step 2: An aliquot of the extract, 200 µL or less, is added to 1 mL of the reagent.

Step 3: Read after 30 min at 20–25°.

Fluorometer Direct Assay, 0.1–5 nmol/mL

Reagent: The same as for Spectrophotometer, but the buffer is 50 m*M*; ATP, 0.2 m*M*; NADP, 0.1 m*M* (0.03 m*M* in the range of 0.1–0.5 nmol/mL).

Fluorometer Indirect Assay, 0.01–1 nmol/mL

Procedure

Reagent: The same as for Fluorometer Direct Assay.

Step 1: If the volume is small, heating in 0.02*M* NaOH should be done in sealed tubes or under oil.

Step 2: The sample in a vol of 10 µL or less is added to 50 µL of the reagent in 10 × 75 mm tubes.

Step 3: After 30 min at 20–25°, 50 µL of phosphate buffer (0.25*M* Na_3PO_4, 0.25*M* K_2HPO_4) are added, mixed and the tubes heated 15 min at 60°.

Step 4: To each tube is added 1 mL of 6*M* NaOH containing 10 m*M* imidazole and 10 m*M* H_2O_2. Heat 60 min for 15 min.

Step 5: Cool to room temperature, wipe dry, and read.

Cycling Assay, 1–10 pmol

Reagent: The same as for Fluorometer Direct Assay, except the buffer is 80 m*M* imidazole base, 20 m*M* imidazole-HCl, pH 7.4, and $NADP^+$ is 50 µ*M*. All other components are double-strength.

Reaction vessels: Oil wells.

Procedure

Step 1: Blanks of 0.02*M* NaOH, and standards of 1–10 µ*M* sucrose in the same medium are introduced into oil wells in 1 µL vol. Freeze-dried samples of tissues are added to 1 µL of 0.02*M* NaOH. The rack is heated at 90° for 30 min.

Step 2: One µL of the specific reagent is added to each droplet, 30 min at 20–25°.

Step 3: Ten µL of 0.06*M* NaOH are added to each droplet, heat at 80° for 20 min.

Step 4: Transfer 5-µL aliquots to fluorometer tubes containing 100 µL of NADP cycling reagent with enzyme levels to give about 1200 cycles in 60 min at 38° (Chapter 5). Heat stop by 2 min in 95–100° water bath.

Step 5: Add 1 mL of indicator reagent (Chapter 5), and read when reaction is complete.

Cycling Assay, 0.1–1 pmol

Reagent: The reagent is the same as for Cycling Assay.

Reaction vessels: Oil wells.

Procedure

Steps 1 and 2: The same as for Cycling Assay, 1–10 pmol, but the vol are each reduced to 0.1 µL.

Step 3: Add 0.5 µL of 0.06 *M* NaOH.

Step 4: To all oil wells, add 10 µL of NADP cycling reagent with enzyme levels to give about 6000 cycles in 60 min at 38° (Chapter 5). Stop reaction with 2 µL of 1*M* NaOH plus 20 min in 80° oven.
Step 5: Transfer 10-µL aliquots into 1 mL of indicator reagent (Chapter 5), and read when reaction is complete.

Cycling Assay, 10–100 fmol

Reagent: The same as for Cycling Assay, 10–100 pmol.
Reaction vessels: Oil wells.

Procedure

Step 1: Add samples and standards in a vol of 0.02 µL, as in protocol Cycling Assay, 0.1–1 pmol.
Step 2: Add 0.02 µL of reagent, 20–25 min at 30°.
Step 3: Add 0.1 µL of 0.06*M* NaOH, heat at 80°, 20 min.
Step 4: To oil wells, add 10 µL of NADP cycling reagent with enzyme levels to give 50,000 cycles in 2 h at 38° C (Chapter 5). Stop reaction with 2 µL of 1*M* NaOH plus 20 min in 80° oven.
Step 5: Transfer 10-µL aliquots into 1 mL of indicator reagent (Chapter 5), read when reaction is complete.

Total Nucleotide Triphosphates

$$\text{XTP} + \text{Fructose-6-P} \xrightarrow{\text{P-fructokinase}} \text{XDP} + \text{fructose-1,6-P}_2$$

$$\text{Fructose-1,6-P}_2 \xrightarrow{\text{aldolase}} \text{dihydroxyacetone-P} + \text{glyceraldehyde-P}$$

$$\text{Glyceraldehyde-P} \xrightarrow[\text{isomerase}]{\text{triose-P}} \text{dihydroxyacetone-P}$$

[Re. 6-53]

$$2\ \text{Dihydroxyacetone-P} + 2\ \text{NADH} \xrightarrow[\text{dehydrogenase}]{\text{glycero-P-}} 2\ \text{glycero-P} + \text{NAD}^+ + 2\text{H}^+$$

The procedure is adapted from an unpublished method of Stanley R. Nelson; we wish to thank him for allowing us to include it. P-fructokinase reacts with all purine and pyrimidine nucleotide triphosphates tested (ATP, ITP, GTP, CTP, UTP, TTP).

Spectrophotometer, 8–60 nmol/mL

Reagent: Imidazole-HCl buffer, pH 7.0 (25 m*M* imidazole base, 25 m*M* imidazole-HCl), fructose-6-P, 300 μ*M*; $MgCl_2$, 2 m*M*; K_2HPO_4, 5 m*M*; NADH, 50–150 μ*M* (20–50% excess); P-fructokinase (rabbit muscle), 0.2 U/mL; aldolase (rabbit muscle), 0.15 U/mL (1 μg/mL); triose-P-isomerase (rabbit muscle), 1.2 U/mL (0.5 μg/mL); and glycero-P-dehydrogenase (rabbit muscle), 0.16 U/mL (4 μg/mL).

Reaction time: Four to 10 min, depending on the nucleotide.

Conduct of the Assay

Unless it is necessary to measure fructose-1,6-P_2 in the same sample, all enzymes except P-fructokinase can be incorporated in the reagent. The sample is added, and after taking an initial reading, P-fructokinase is added.

Comment

It is desirable to test the system with UTP, which is one of the slower reacting ribonucleotides. In the case of deoxyribonucleotides, it may be necessary to increase the amount of P-fructokinase. TTP, at least, reacts rather slowly. P-fructokinase is very unstable in simple buffers, particularly at pH 7 or below. Dilutions of stock enzyme preparations should be made in pH 8 Tris buffer containing 10 m*M* K_2HPO_4. Phosphate is a potent stabilizer and is included in the reagent itself for this reason.

Fluorometer Direct Assay, 0.05–4 nmol/mL

Reagent: The reagent is the same as in Spectrophotometer, except as follows: NADH, 0.2–10 μ*M* (20–75% excess); P-fructokinase, 0.06 U/mL; and glycero-P-dehydrogenase, 0.04 U/mL.

Reaction time: Four to 8 min.

Conduct of the Assay

See Spectrophotometer *above.*

Comment

In the case of animal tissues, it will seldom be desirable to measure fructose-1,6-P_2 in the same sample as the total triphosphates because of the large discrepancy in levels. In bacterial extracts, this may not always be true, in which case a stepwise addition of the enzymes can be made. (*See also* Comments under Spectrophotometer.)

Kinetics

For P-fructokinase with specific activity of 200 μmol/mg/min (200 U/mg), the half-times for substrate concentrations of 10 μ*M* or less and 1 μg/mL of P-fructokinase, would be 0.2–0.4 min for ATP and GTP and 0.4–1 min for ITP, CTP, and UTP. This refers to the P-fructokinase step only. The K_m for fructose-6-P is <50 μ*M*.

Uridine-5'-Diphosphoglucose

$$\text{UDP-glucose} + 2\ \text{NAD}^+ \xrightarrow[\text{dehydrogenase}]{\text{UDPG}} \text{UDP-glucuronate} + 2\ \text{NADH} + 2\ \text{H}^+$$

[Re. 6-54]

Spectrophotometer, 8–100 nmol/mL

Reagent: Tris-HCl buffer, pH 8.1 (25 m*M* Tris base, 25 m*M* Tris-HCl); $MgCl_2$, 2 m*M*; NAD, 1 m*M*; and UDP-glucose dehydrogenase 0.008 U/mL (15 μg/mL).
Reaction time: Ten to 20 min.

Conduct of the Assay

For standardization purposes, it is preferable to add the UDP-glucose last, because the enzyme may add appreciably to the absorption at 340 nm.

Comment

The enzyme solution loses activity at 4°, but keeps well frozen at –40°.

Fluorometer Direct Assay, 0.05–5 nmol/mL

Reagent: The buffer and $MgCl_2$ are the same as in Spectrophotometer. NAD^+, 100 μ*M* (20 μ*M* with 1 μ*M* UDPG or less), and UDP-glucose dehydrogenase, 0.004 U/mL (0.002 U/mL with 1 μ*M* UDPG or less).
Reaction time: Ten to 20 min.

Conduct of the Assay

Ordinarily, the enzyme is added last. Because of the amount of enzyme needed with current preparations, there is likely to be an appeciable contribution of fluorescence. In some instances, a small drift in the blank reading has also been observed.

Kinetics

The K_m for UDP-glucose is of the order of 5 μM, and that of NAD^+ about 10 μM under analytical conditions. NADH is a very strong competitive inhibitor of NAD^+. Consequently the reaction rate falls off even when both substrates are far above their K_ms. When UDP-glucose is in excess and NAD^+ is in the 50 μM range, it is virtually impossible to drive the reaction to completion with NAD^+ because of this inhibition. In the measurement of UDP-glucose, it is advisable to provide a tenfold excess of NAD^+.

Chapter 7

A Collection of Enzyme Assays

To measure an enzyme is in some ways easier, but in many ways harder than to measure a metabolite. It is easier because an enzyme is a catalyst and can usually be made to generate a great deal more of its product than is present in the tissue of origin. For example, muscle lactate dehydrogenase can generate in an hour 100,000 times more lactate than is ever present in resting muscle. It is harder to measure an enzyme because, unlike metabolites, which always have identical properties wherever they are found in the biological world, enzymes having the same function can differ greatly. Each enzyme that catalyzes a particular reaction will more often than not differ from species to species, from cell type to cell type, and even from one kind of organelle to another. The differences may concern kinetic properties, optimum pH, sensitivity to activators and inhibitors, and so forth. In addition, the activity of every enzyme is affected by temperature, pH, and usually by ionic strength, type of buffer, and so forth. Finally, an enzyme is in a real sense alive, and most enzymes can easily die or be killed. All of this presents a challenge to the analyst, especially when the protocol has not been designed for the particular cell type under examination.

Expression of Enzyme Activities

The International Unit (IU or simply U) is defined as the activity that forms 1 μmol of product/min under specified conditions. Standard practice is to express activity of isolated enzymes as U/mg of enzyme protein. Tissue activities are more often expressed as mol, mmol, and so on/min or h/unit of tissue wet wt, dry wt, or protein, at a given temperature. Here we have usually expressed examples of tissue activities as mol (or mmol)/kg wet or dry wt at 20°C. Another standard unit, rarely used in biological studies, is the "katal" or "kat," which is defined as the activity that will increase the rate of reaction by 1 mol/s.

The 44 methods presented are based on the decrease in NADH or the formation of NAD^+ in 20 cases, the formation of NADPH in seventeen cases, the formation of NADH in seven cases, and the formation of $NADP^+$ in one case. The options presented usually call for changes in NADH or NADPH to be measured by changes in fluorescence or UV absorption. Changes in NAD^+ are measured by the fluorescence produced with strong NaOH or in a few cases, by converting the NAD^+ back to NADH.

Variations in Assays for Any Given Enzyme

1. *Direct assay in the spectrophotometer,* i.e., there is an increase or decrease in NADH or NADPH that can be followed directly in the instrument. This has an advantage in simplicity, but a disadvantage in that, when measuring enzymes of low activity, the tissue homogenates cannot be diluted enough to avoid problems of turbidity and of disturbing side reactions caused by other enzymes present. The low dilution may make control of effects of other tissue enzymes difficult.
2. *Direct assay in the fluorometer.* This eliminates some of the disadvantages with direct assays in the spectrophotometer. Because of the much greater sensitivity, tissue homogenates can usually be diluted to a point where turbidity is not a problem; and because the reaction is carried out in inexpensive tubes, large numbers of samples can be analyzed simultaneously with repeated readings made over periods as long as an hour or more. There is, however, one limitation to direct fluorometric assays that is illustrated by an example. The K_m for NADH with glutamate dehydrogenase is 4 μM. In a spectrophotometer, the NADH concentration can be kept far enough above this level throughout the assay to keep the velocity at 95% or more of the V_{max}. In the fluorometer, if the initial NADH concentration is 10 μM (i.e., about the upper limit for fluorometer linearity), the initial rate will be only 70% of V_{max} and would fall to 55% when the NADH falls to 5 μM. However, this limitation for direct assays does not apply if the NADH oxidation is made by an auxiliary enzyme catalyzing a sequential reaction. For example, in the direct fluorometric assay for aspartate aminotransferase, the oxaloacetate generated by the transaminase is rapidly converted to malate with NADH at an initial concentration of only 10 μM plus a relatively high level of malate dehydrogenase. Note that with both spectrophotometric and fluorometric direct assays, if the NADH or NADPH is generated or removed by an auxiliary enzyme or enzymes, there may be a significant initial lag until the auxiliary enzyme(s) catch up. Therefore, it is important to take a series of readings to obtain linear (steady-state) reaction data *(see* Chapter 2).

3. Indirect assay in the fluorometer. The specific action is allowed to proceed for a fixed time (usually an hour), then stopped and the product measured in a variety of ways depending on whether the first step involved a pyridine nucleotide or whether the pyridine nucleotide reaction took place in the auxiliary second step. Ordinarily, the entire procedure is carried out in the final fluorometer tube.
4. Indirect assay with enzymatic cycling. This differs from the above only in that the pyridine product is finally amplified to whatever degree is necessary. Usually, the first step is carried out in a smaller vessel than a fluorometer tube: serological tube or oil well.

Because it was out of the question to provide protocols for all circumstances, it is hoped that the analyst, with the aid of the many examples given, will easily adapt the procedures to his or her particular needs.

Adenylate Deaminase (EC 3.5.4.6) *(1)*

There are two separate enzymatic steps:

1) **AMP → IMP + NH_4^+**

[Re. 7-1]

2) 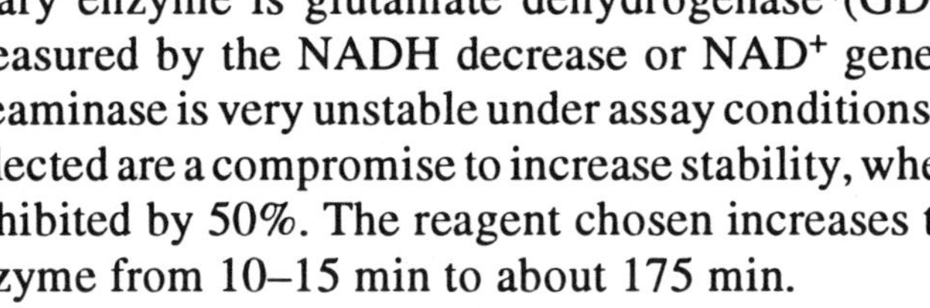

The auxiliary enzyme is glutamate dehydrogenase (GDH). The activity is measured by the NADH decrease or NAD^+ generated in Step 2. The deaminase is very unstable under assay conditions, and the conditions selected are a compromise to increase stability, whereas the reaction is inhibited by 50%. The reagent chosen increases the half-life of the enzyme from 10–15 min to about 175 min.

Sample tissue activities (mol/kg dry wt/h, 20°): kidney, 0.6; rabbit soleus muscle, 3; and rabbit tibialis anterior, 20.

Spectrophotometer (0.3–1.5 nmol/min, 20–90 nmol of Product)

Reagent 1: Imidazole buffer, pH 6.5 (15 m*M* imidazole base, 35 m*M* imidazole acetate); KCl, 0.1*M*; glycerol, 30%; ATP, 1 m*M*; 5'AMP, 5 m*M*; dithiothreitol, 1 m*M*; and bovine serum albumin, 0.05%.

Reagent 2: Tris-acetate buffer, pH 8.8 (85 m*M* Tris base, 15 m*M* Tris acetate); α-ketoglutarate, 2 m*M*; NADH, 0.1 m*M*; ADP, 100 μ*M*; bovine serum albumin, 0.04%; and beef liver glutamate dehydrogenase, 36 U/mL (300 μg/mL) from lyophilized powder or in glycerol to avoid NH_4^+.

To further minimize ammonia contamination, the buffers are prepared just before use from the buffer bases, plus sufficient acetic anhydride to provide the acid equivalents. This procedure (devised by Stephen H. Herrman, personal communication) converts any ammonia present to acetamide, and the acetic anhydride is converted within 10 min to acetic acid. The water, buffer base, glycerol, and KCl are combined, the acetic anyhydride added, and after 10 min, the other components are added. Pure acetic anhydride is 9.5*M*, equivalent to 19 mol of acetic acid/L.

Procedure

Step 1: To 0.1 mL of reagent in a 10 × 75 mm tube is added an appropriate amount of tissue to be analyzed (e.g., 10 μg of fast-twitch muscle, wet wt) or 10 μL of 3–5 m*M* $(NH_4)_2SO_4$ standard.
Step 2: Incubate at 25°C for 1 h.
Step 3: Add 5 μL of 0.7*M* HCl; heat at 60° 20 min to stop the enzyme activity.
Step 4: Add 1 mL of reagent 2, and read in spectrophotometer when the reaction is complete, approx 20 min.

Fluorometer Direct Assay (0.03–0.12 nmol/min, 2–8 nmol of Product)

Reagent 1: The same as for Spectrophotometer.
Reagent 2: The same as for Spectrophotometer, except NADH, 10 μ*M*.

Procedure

Step 1: To 0.1 mL of reagent 1 is added the appropriate amount of tissue (equivalent to 1–2 μg of fast-twitch muscle, wet wt). Standards are 5 μL of 1.5 m*M* $(NH_4)_2SO_4$. Incubate at 25° for 60 min.
Step 2: Add 5 μL of 0.7*M* HCl; heat at 60° for 10 min.
Step 3: Add 1 mL of reagent 2. Read when reaction is complete, about 20 min.

Fluorometer Indirect Assay (3–15 pmol/min, 0.2–1 nmol of Product)

Reagent 1: The same as for Spectrophotometer.
Reagent 2: The same as for Spectrophotometer.

Procedure

Step 1: One to 3 µL of tissue homogenate (e.g., 40–200 ng of muscle wet wt) are added to 50 µL of reagent 1. Standards are 60–120 µ*M* $(NH_4)_2SO_4$ added in the same volume. Incubate 60 min at room temperature.
Step 2: Add 5 µL of 0.7*M* HCl; 60° for 10 min.
Step 3: Add 0.1 mL reagent 2; incubate 20 min at room temperature.
Step 4: Add 20 µL 1*M* HCl; room temperature for 10 min.
Step 5: Transfer 50 µL to 1 mL of 6*M* NaOH containing 10 m*M* imidazole; heat 10 min at 60°. Cool and read on the fluorometer.

Adenylate Kinase (EC 2.7.4.3) *(2,3)*

The four enzymatic reactions are combined. The AMP is removed to prevent product inhibition.

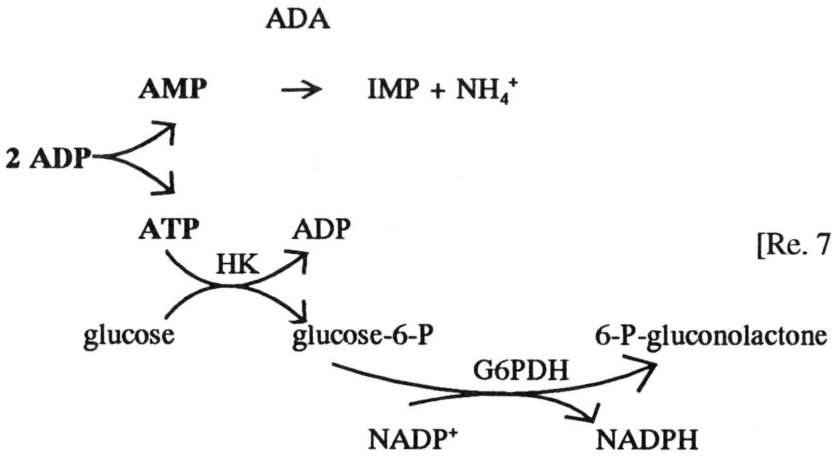

[Re. 7-2]

The auxiliary enzymes are AMP deaminase, (ADA), hexokinase (HK), and glucose-6-P dehydrogenase (G6PDH).

Sample tissue activities (mol/kg dry wt/h, 20°): rat kidney, 0.4–0.8; rat soleus muscle, 50; rat tibialis anterior muscle, 250; and mouse brain, 18.

Spectrophotometer (5–10 nmol/min, 50–100 nmol Product)

Reagent: Imidazole-HCl, pH 7.0 (60 m*M* imidazole base, 40 m*M* imidazole HCl); ADP, 3 m*M*; $MgCl_2$, 2 m*M*; glucose, 2 m*M*; bovine serum albumin, 0.025%; yeast hexokinase, 1.4 U/mL (10 µg/mL); rabbit muscle AMP deaminase, 1 U/mL (20 µg/mL); $NADP^+$, 1.5 m*M*; and glucose-6-P dehydrogenase from *Leuconostoc mesenteroides*, 0.28 U/mL (0.5 µg/mL).

Procedure

One milliliter of reagent is added to cuvets, and a reading is taken. Tissue homogenate is added to give an OD change of 0.030–0.060/min (5–10 μg of muscle, 80–150 μg of brain, wet wt). Readings are taken every 1 or 2 min.

Fluorometer (0.02–0.2 nmol/min, 1–10 nmol Product)

Reagent 1: The reagent is the same as for Spectrophotometer, except the glucose-6-P dehydrogenase and $NADP^+$ are omitted.

Reagent 2: Tris-HCl buffer, pH 8.0 (25 m*M* Tris base, 25 m*M* Tris HCl), $NADP^+$, 50 μ*M*; EDTA 1 m*M*; glucose-6-P dehydrogenase, 0.28 U/mL (0.5 μg/mL); and rabbit muscle P-glucoisomerase, 0.5 U/mL (1 μg/mL).

Procedure

Step 1: Samples (e.g., 50 μg kidney, 0.1 μg muscle, wet wt) are added in a vol of 1 μL into 50 μL of reagent 1 in 10 × 75 mm tubes. Standards are 50 μL of 0.1 and 0.2 m*M* ATP in reagent 1, 60 min at 25°.

Step 2: A vol of 5 μL of 1*M* HCl is added to stop the reaction; 10 min at room temperature.

Step 3: Add 1 mL of reagent 2; 10–15 min at room temperature. Read on fluorometer.

Alanine Transaminase (EC 2.6.1.2) *(1,4)*

1) Alanine + α-ketoglutarate → pyruvate + glutamate

$$\text{2) Pyruvate + NADH} \xrightarrow{\text{LDH}} \text{lactate + NAD}^+ \qquad \text{[Re. 7-3]}$$

The auxiliary enzyme is lactate dehydrogenase (LDH). The two enzyme steps are carried out separately. This permits destruction after enzyme Step 1 of potentially disturbing side reactions caused by NADH oxidase and glutamate dehydrogenase in tissue samples. For the same reason, a direct assay combining the two reactions above is impractical.

Rather than measure the decrease in NADH in Step 2 above, the NADH is destroyed with HCl, and the NAD^+ is converted to its highly fluorescent strong alkali product. This extends the analytical range almost tenfold and makes it unnecessary to predict carefully the amount of NADH to use in Step 2. When still more sensitivity is required, the NAD^+ is measured by enzymatic cycling.

Sample tissue activities (mol/kg dry wt/h, 20°): rat kidney, 0.03–0.16; rabbit tibialis anterior muscle, 0.4; and rabbit soleus muscle, 1.2. Activities are about 40% higher at 25°.

Fluorometer Indirect Assay (0.2–10 nmol of Product)

Reagent 1: Tris-HCl buffer, pH 8.1 (25 m*M* Tris base, 25 m*M* Tris-HCl); L-alanine, 50 m*M*; monosodium α-ketoglutarate, 0.5 m*M*; pyridoxal phosphate, 5 μ*M*; and bovine serum albumin, 0.02%.

Reagent 2: Imidazole buffer, pH 7 (50 m*M* imidazole-HCl, 50 m*M* imidazole base), NADH, 200 μ*M*, and beef heart lactate dehydrogenase, 0.5 U/mL (2 μg/mL). Standards are 5 μL of 0.2–2 m*M* pyruvate carried throughout the procedure.

Procedure

Step 1: To 100 μL of reagent 1 in 10 × 75 mm fluorometer tubes are added 5 μL of an appropriate tissue dilution (10–50 μg of kidney, 2–25 μg of soleus muscle). Incubate 1 h at 25°, followed by 2 min at 100°.

Step 2: Add 100 μL of reagent 2; room temperature 15 min.

Step 3: Add 20 μL 1*M* HCl; room temperature at least 15 min.

Step 4: Add 1 mL 6*M* NaOH containing 10 m*M* imidazole; 60° for 20 min. Cool and read on the fluorometer.

Comment

The K_m for alanine is very high, about 50 m*M*. This means the observed activity with the recommended reagent is only about half the V_{max}. It also means that it is essential to test the purity of the L-alanine, since some batches have been contaminated with L-aspartate or pyruvate. Either contaminant would give erroneously high values.

Aldolase *(5,6)* (Fructose-1, 6-bisphosphate, EC 4.1.2.13)

The three enzymatic steps are combined (note that 2 mol of NAD^+ are generated/mol of fructose bisphosphate consumed):

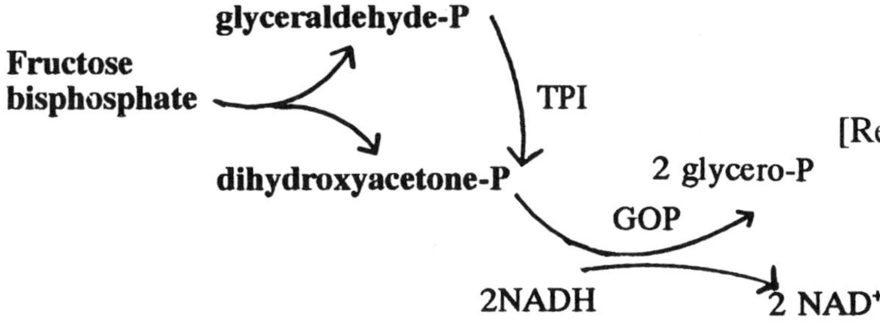

[Re. 7-4]

The auxiliary enzymes are triose-P isomerase (TPI) and glycero-P dehydrogenase (GOP). (The spectrophotometric method is not recommended for unfractionated tissue in general, because the amount of tissue required in most cases would cause problems owing to turbidity and competing tissue enzymes.)

Sample tissue activities (mol/kg dry wt/h, 20°): rat brain, 1.2; mouse brain, 0.6; rat kidney cortex, 0.4; and liver, 0.2.

Spectrophotometer (5–10 nmol/min, 50–100 nmol of Product)

Reagent: Tris buffer, pH 7.7 (17 m*M* Tris base, 33 m*M* Tris-HCl); fructose-1,6-bisphosphate, 0.3 m*M*; NADH, 150 μ*M*; bovine serum albumin, 0.02%; triose-P isomerase 10 U/mL; glycero-P dehydrogenase, 0.7 U/mL; and sodium amytal, 2 m*M*.

Conduct of the Assay

The reaction is carried out in 1 mL of reagent in a cuvet. The sample is added, and the rate of oxidation of NADH monitored for 10 or 20 min.

Fluorometer Direct Assay (0.1–1 nmol/min, 1–8 nmol of Product)

Reagent: The reagent is the same as for Spectrophotometer, except NADH, 10 μ*M*.

Conduct of the Assay

The reaction is carried out in 1 mL of reagent, following the disappearance of NADH. Appropriate tissue amounts are 50–200 μg of kidney, and 15–50 μg of brain, wet wt.

Fluorometer Indirect Assay (2–100 pmol/min, 0.1–5 nmol of Product)

Reagent: The reagent is the same as for Spectrophotometer, except NADH, 1 m*M*; and nicotinamide, 20 m*M*.

Procedure

Step 1: To 50 μL of reagent in 10 × 75 mm tubes are added 5 μL of sample containing, for example, 10 μg of liver or brain. Standards are 5 μL of 2 m*M* NAD^+ or dihydroxyacetone-P in the complete reagent. Incubate 60 min at 25°.

Step 2: Add 0.5 μL of 5*M* HCl at the same rate as the sample addition. Wait at least 10 min.

Step 3: Add 1 mL of 6*M* NaOH containing 10 m*M* imidazole; 30 min at 38° or 10 min at 60°. Cool and read.

Comment

The nicotinamide prevents the enzymatic destruction of the product NAD^+. Tissue blanks with the substrate omitted are needed because of the likelihood of NADH oxidation by other tissue enzymes.

γ-Aminobutyrate Transaminase (EC 2.6.1.0) *(7,8)*

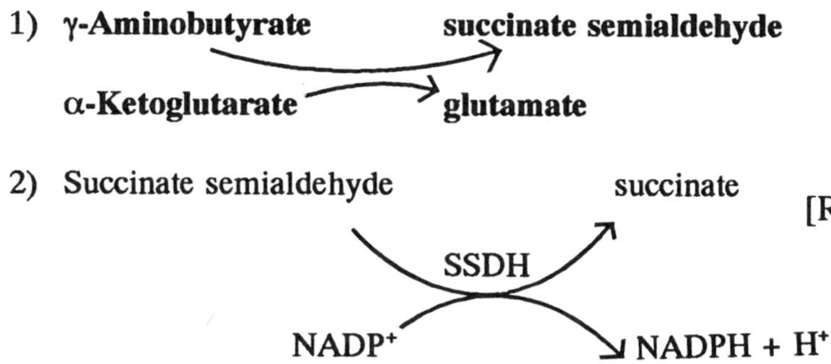

[Re. 7-5]

The auxiliary enzyme is succinate semialdehyde dehydrogenase (SSDH). The transaminase reaction is allowed to proceed in a first step; the α-ketoglutarate remaining is destroyed with H_2O_2. It is then possible to measure the succinate semialdehyde formed in the reaction with commercial GABAse, in spite of the fact that it contains both GABA transaminase and succinate semialdehyde dehydrogenase. The pH for the measurement of succinate semialdehyde is shifted to 7.0 because it is more favorable for the dehydrogenase reaction. At pH 7.0, the K_m for succinate semialdehyde is about 2 μ*M*, which minimizes the amount of enzyme required.

Sample tissue activities (mol/kg dry wt/h, 38°): retina (neuronal layers), rabbit 0.02–0.16; monkey, 0.05–0.33; and mouse brain, 0.5.

Spectrophotometer (0.5–2.5 nmol/min, 30–150 nmol Product)

Reagent: Tris-HCl buffer, pH 8.7 (85 m*M* Tris base, 15 m*M* Tris-HCl); GABA, 10 m*M*; α-ketoglutarate, 5 m*M*; pyridoxal phosphate, 100 μ*M*; mercaptoethanol, 1 m*M*; and bovine serum albumin, 0.02%. (The addition of α-ketoglutarate shifts the pH to 8.5.)

Reagent 2: Imidazole buffer, pH 6.5 (20 m*M* imidazole base, 80 m*M* imidazole-HCl); $NADP^+$, 500 μ*M*; GABAse, 0.025 U/mL (25 μg/mL); bovine liver catalase, 65 U/mL (1 μg/mL); and bovine serum albumin, 0.02%.

Procedure

Step 1: Samples are added in a vol of 5–10 μL (equivalent to 0.3–1.5 mg of brain, wet wt) to 100 μL of reagent 1 in a 10 × 75 mm tube; incubate at 38° for 60 min.

Step 2: Add 100 μL of 20 m*M* H_2O_2 in 0.05*M* NaOH; heat 30 min at 38°. This step destroys the α-ketoglutarate and tissue enzymes.

Step 3: Add 1 mL of reagent 2, and transfer to spectrophotometer cuvet to read when reaction is complete, approx 20 min.

Fluorometer Indirect Assay (0.02–0.2 nmol/min, 1–10 nmol of Product)

Reagent 1: The reagent is the same as for Spectrophotometer.

Reagent 2: The reagent is the same as for Spectrophotometer, except the $NADP^+$ is reduced to 100 μ*M*; and the GABAse is omitted.

Procedure

Step 1: The samples (equivalent to 10–100 μg of brain tissue, wet wt) are added in a vol of 5–10 μL to 50 μL of reagent in a 10 × 75 mm tube; incubate 60 min at 38°.

Step 2: Add 50 μL of 20 m*M* H_2O_2 in 0.05*M* NaOH; heat 30 min at 38°.

Step 3: Add 1 mL of reagent 2, and read at fluorometer setting equivalent to 2–10 μ*M* NADPH at full scale deflection. Add 0.001 U/mL (1 μg/mL) GABAse. (The commercial preparation usually has sufficient succinate semialdehyde dehydrogenase to complete the reaction in 10 min.)

Cycling Assay (0.1–0.5 pmol/min, 5–25 pmol Product)

Reagent: The reagent is the same as in Spectrophotometer.

Reagent 2: The reagent is the same as in Spectrophotometer, except that $NADP^+$ is reduced to 10 μ*M*.

Procedure

Step 1: Freeze-dried samples (10 ng brain, 20 ng retina, dry wt) are added to 0.2 μL of reagent 1 under oil. Standards are 0.2 μL of 0.5–2 μ*M* succinate semialdehyde. Incubate 60 min at 38°.

Step 2: Add 0.2 μL of 20 m*M* H_2O_2 in 0.05*M* NaOH, 30 min at 38°.

Step 3: Add 1 μL of reagent 2; 20 min at room temperature.

Step 4: Add 1 μL of 0.15 NaOH, 20 min at 80°

Step 5: Transfer 1 µL into a 10–75 mm tube containing 100 µL of NADP cycling reagent with enzyme levels to provide 1000-fold amplification for brain or 2000-fold amplification for retina in 60 min at 38°.

Step 6: Complete the indicator step as described in Chapter 5.

Comment

Succinate semialdehyde recently has been made available commercially. The succinate semialdehyde dehydrogenase activity in the mixed GABAse preparation can therefore be tested for the content of the dehydrogenase activity in advance of the tissue measurement of transaminase activity.

Aspartate Transaminase (EC 2.6.1.1) *(1,4,9)*

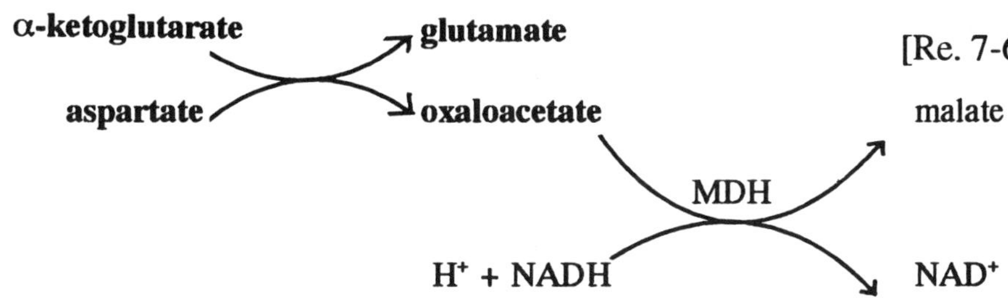

[Re. 7-6]

The auxiliary enzyme is malate dehydrogenase (MDH). The two enzyme steps can be combined because the activity is usually high enough so that nonspecific tissue oxidation of NADH would not be disturbing. (This contrasts with the alanine transaminase assay, *see* first paragraph of *Assay for Alanine Transaminase.*)

Sample tissue activities (mol/kg dry wt/h, 20°): rat kidney cortex, 25; mouse brain, 6; rabbit tibialis anterior muscle, 6; and rabbit soleus muscle, 8.

Spectrophotometer (1–5 nmol/min, 20–100 nmol of Product)

Reagent: Tris buffer, pH 8.4 (35 m*M* Tris base, 15 m*M* Tris-HCl); monosodium α-ketoglutarate, 8 m*M*; monosodium aspartate 40 m*M*; NADH, 100–150 µ*M*; bovine serum albumin, 0.02%; cytoplasmic malate dehydrogenase, (pig heart) 1.2 U/mL (1 µg/mL); and pyridoxal phosphate, 10 µ*M*.

Conduct of the Assay

To 1 mL of reagent in a cuvet is added 50 µL or less of homogenate containing an appropriate amount of tissue (e.g., 50 µg of kidney or 200 µg of skeletal muscle wet wt). Blanks are reagent without α-ketoglutarate or aspartate, with tissue added. Readings are made at 2-min intervals, and the rate calculated from the linear portion of the curve.

Fluorometer Direct Assay (0.05–1 nmol/min, 1–10 nmol of Product)

Reagent: The reagent is the same as for Spectrophotometer, except NADH, 5–10 µ*M*.

Conduct of the Assay

One milliliter of reagent is added to 10×75 mm tubes. Readings are made to give full-scale fluorescence. An appropriate amount of tissue is added (5–10 µg kidney; 20–50 µg skeletal muscle or brain, all wet wt) and readings are taken every 2 or 3 min. The activity is calculated from the linear portion of the curve.

Fluorometer Indirect Assay (2–20 pmol/min, 0.1–1 nmol of Product)

Reagent: The reagent is the same as for Spectrophotometer, except NADH, 5–10 µ*M*.

Procedure

Step 1: To 100 µL of reagent in a 10×75 mm tube is added the appropriate amount of tissue (0.5–1.5 µg of kidney, 0.1–0.3 µg of brain in a vol of 1–5 µL). Standards are 5–20 µ*M* NAD^+ in 100 µL. Incubate 60 min at 25°.

Step 2: Add 15 µL 1*M* HCl; room temperature 15 min.

Step 3: Add 1 mL 6*M* NaOH containing 10 m*M* imidazole; 60° for 20 min.

Step 4: Cool, dry, and read tubes on the fluorometer.

Cycling Assay (0.2–1 pmol/min, 10–50 pmol Product)

The assay was developed especially for human ova, in which a 0.05-µL sample from 1 µL of medium added to one ovum was used.

Reagent: The reagent was the same as for Spectrophotometer, except α-ketoglutarate, 2 m*M*; NADH, 80 µ*M*; and malate dehydrogenase, 0.6 U/mL (0.5 µg/mL), with the ammonium sulfate removed by centrifugation and replaced with reagent buffer.

Procedure

Step 1: Approximately 0.05 µL of tissue extract or, for example, 1 ng of freeze-dried brain is added to 2 µL of reagent under oil. Incubate 1 h at 20°. Standards are 0.05 µL of 0.8–2 m*M* NAD^+.
Step 2: Add 2 µL of 0.1*M* HCl; 10 min at room temperature.
Step 3: Transfer 1 µL into a 10 × 75 mm tube containing 100 µL of NAD cycling reagent with enzyme levels to provide 1000-fold amplification in 60 min at 38°.
Step 4: Complete the indicator step as described in Chapter 5.

Branched Chain Amino Acid *(1,10,11,12,13)* Amino Transferase (EC 2.6.1.42)

1) 

[Re. 7-7]

2) 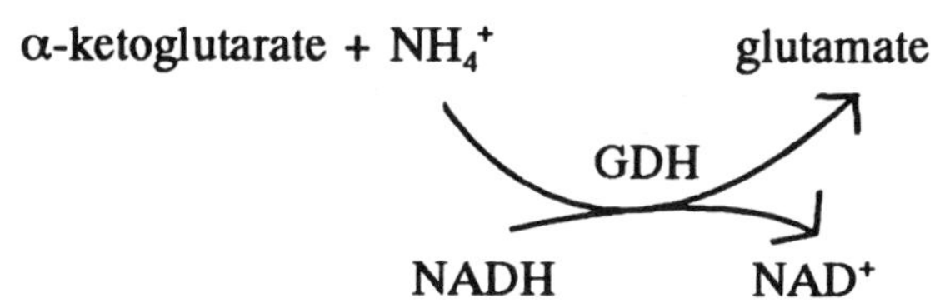

The auxiliary enzyme is glutamate dehydrogenase (GDH). The specific reaction proceeds in the reverse of the usual direction. The advantage is that the product measured (α-ketoglutarate) is much lower in tissue than is glutamate. The tissue blanks are thus lower, usually 1–30% of the transaminase product.

Sample tissue activities (mol/kg dry wt/h, 20°): rat kidney, 0.04; rabbit tibialis anterior muscle, 0.02; and rabbit soleus muscle, 0.03.

Fluorometer Indirect Assay (2–30 pmol/min, 0.1–1 nmol of Product)

Reagent: Imidazole buffer, pH 7.7 (85 m*M* imidazole base, 15 m*M* imidazole-HCl); monosodium glutamate, 25 m*M*; α-ketoisocaproate, 120 µ*M*; bovine serum albumin, 0.05%; and glycerol 20%.

Reagent 2: Imidazole buffer, pH 8.6 (220 m*M* imidazole base, 10 m*M* imidazole HCl); ammonium acetate, 50 m*M*; NADH, 40 μ*M*; ADP, 200 μ*M*; and beef liver glutamate dehydrogenase 1.2 U/mL (10 μg/mL).

Procedure

Step 1: To 100 μL of reagent 1 in 10 × 75 mm tubes are added 2–5 μL of homogenate (20–100 μg of muscle or kidney, wet wt); 60 min at 25°. Standards are 100 μL of 1–10 μ*M* α-ketoglutarate.
Step 2: Add 15 μL of 0.75*M* HCl; room temperature for 15 min.
Step 3: Add 100 μL reagent 2; room temperature 15 min.
Step 4: Add 20 μL 2*M* HCl; room temperature 15 min.
Step 5: Add 1 mL 6*M* NaOH containing 10 m*M* imidazole; 60° for 20 min. Cool, dry, and read tubes on the fluorometer.

Comment

It was found, under the analytical conditions used by Ichihara and Koyama *(12)* and by Taylor and Jenkins *(13)* (pyrophosphate buffer, pH 8.6), that over half of the enzyme activity of muscle homogenates was lost in 60 min. Addition of 40% glycerol gave complete protection, but activity was reduced by almost half. A compromise was to use 20% glycerol, which reduced initial activity <5% with a loss in 60 min of only 10%. The buffer was also changed to imidazole at pH 7.7, primarily to simplify the necessary acidification and neutralization steps. However, this also increased initial activity by ≈15%. The Michaelis constants under assay conditions were found to be 13 μ*M* for α-ketoisocaproate and 12 m*M* for glutamate. From this it may be calculated that observed activities are about 60% of the maximal rate of enzyme action (V_{max}).

Carnitine Acetyltransferase (EC 2.3.1.7) *(1)*

The acetyl-CoA generated is converted to citrate in the first of two enzyme sequences:

1)

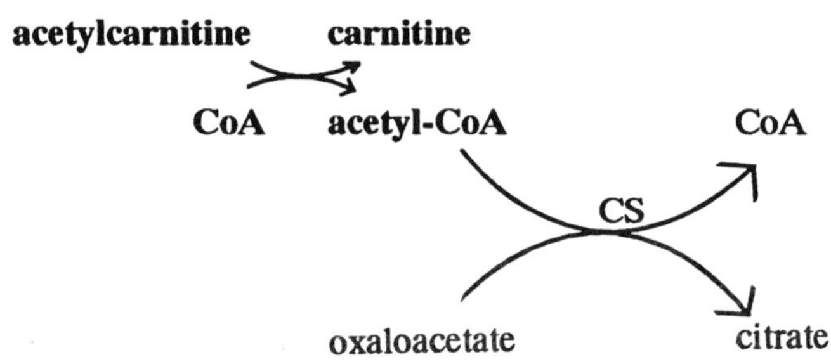

[Re. 7-8]

The auxiliary enzyme is citrate synthase (CS). Excess oxaloacetate is destroyed with alkali and heat, and the citrate is measured by the NAD^+ formed in the second enzyme sequence with citrate lyase (CL) and malate dehydrogenase (MDH):

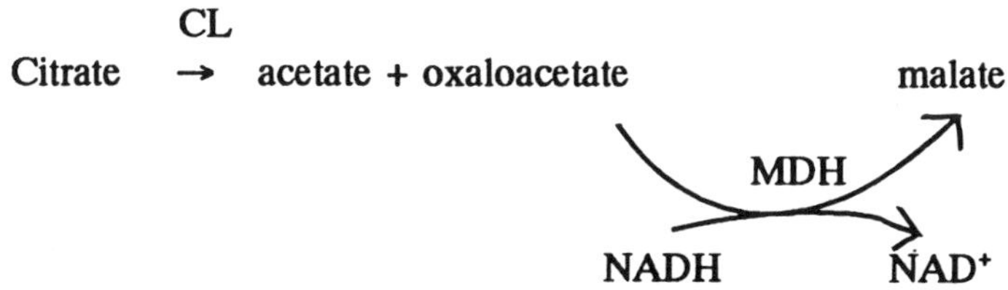

[Re. 7-9]

Sample tissue activities (mol/kg dry wt/h, 20°): rabbit tibialis anterior muscle, 0.30, and rabbit soleus muscle, 0.40.

Fluorometer Indirect Assay (10–150 pmol/min, 0.5–10 nmol of Product)

Reagent: Tris buffer, pH 8.1 (25 m*M* Tris base, 25 m*M* Tris HCl); acetylcarnitine, 5 m*M*; CoA, 2 m*M*; oxaloacetic acid, 0.5 m*M*; citrate-free, Norit-treated bovine serum albumin, 0.02%; and pig heart citrate synthase, 1.1 U/mL (10 µg/mL).

Reagent 2: Tris buffer 7.5 (20 m*M* Tris base, 80 m*M* Tris-HCl); NADH, 80 µ*M*; $ZnCl_2$, 100 µ*M*; bovine serum albumin, citrate-free, 0.01%; pig heart malate dehydrogenase, 1.2 U/mL (1 µg/mL); and *Aerobacter aerogenes* citrate lyase, 0.04 U/mL.

Procedure

Step 1: To 50 µL of reagent 1 in 10 × 75 mm tubes are added 1–2 µL of homogenate (e.g., 10–100 µg of muscle, wet wt), 25° for 60 min. Standards are 50 µL of 20–200 µ*M* citrate in reagent 1.

Step 2: Add 5 µL of 0.5*M* NaOH; 95° for 5 min.

Step 3: Add 0.5 mL of reagent 2; room temperature 20 min.

Step 4: Transfer 100 µL into 1 mL 6*M* NaOH with 10 m*M* imidazole; 60° for 20 min. Cool and read in the fluorometer.

Comment

The oxaloacetate in reagent 1 is destroyed in Step 2. The citrate lyase may contain citrate, which can be removed by washing the enzyme pellet with 2.5*M* $(NH_4)_2SO_4$. The assay is linear for 2 h. The volumes in Step 1 are minimized to reduce citrate contamination. The samples are diluted tenfold in Step 3 to avoid inactivation of the citrate lyase by citrate. The Michaelis constants of the rabbit muscle enzyme for CoA and acetylcarnitine are in the 0.05 and 0.2 m*M* range, respectively.

Citrate Synthase (EC 4.1.3.7) *(14,15,16)*

The assay is conducted in two separate enzymatic steps.

1) **Acetyl CoA + oxaloacetate → citrate + CoA**

2) Citrate $\xrightarrow{CL}$ acetate + oxaloacetate $\xrightarrow[H^+ + NADH \rightarrow NAD^+]{MDH}$ malate [Re. 7-10]

The auxiliary enzymes are citrate lyase (CL) and malate dehydrogenase (MDH). The substrate oxaloacetate of the first step is destroyed with hot alkali before enzyme step two. This indirect procedure allows the use of optimal substrate concentrations, and ends with a pyridine nucleotide step that permits amplification with cycling. Note the unusually high albumin level in the specific reagent. This is required for maximum activity.

The results may be erratic in the absence of preincubation, at least with frozen-dried muscle samples which, unlike most tissues, do not disintegrate in aqueous droplets. Other tissues should be tested to determine if preincubation is necessary or helpful. Cycling Assay provides an example of a preincubation procedure.

Sample tissue activities (mol/kg dry wt/h, 20°): rat kidney cortex, 4.6; rabbit tibialis anterior muscle, 2.7; rabbit soleus muscle, 33; mouse brain, 5.3; dog myocardium, 17; and human vastus muscle, 3.8.

Fluorometer Indirect Assay (5–100 pmol/min, 0.3–6 pmol of Product)

Specific reagent: Tris-HCl buffer, pH 8.0 (25 m*M* Tris base, 25 m*M* Tris-HCl); acetyl CoA, 0.4 m*M*; oxaloacetate, 0.5 m*M*; and citrate-free bovine serum albumin, 0.25%. The oxaloacetate is added just before use from a 1*M* solution stored in 0.5*M* HCl at –70°.

Citrate reagent: Tris-H buffer, pH 7.5 (10 m*M* Tris base, 40 m*M* Tris-HCl); $ZnCl_2$, 100 µ*M*; NADH, 30 µ*M*; bovine serum albumin, 0.01% (citrate-free); ascorbate, 4 m*M* (fresh); citrate lyase from *Aerobacter aerogenes*, 0.006 U/mL; and pig heart malate dehyrogenase, 0.6 U/mL (0.5 µg/mL) with most of the $(NH_4)_2SO_4$ removed by centrifugation.

Procedure

Step 1: Samples of 1–5 µL of homogenate are added to 50 µL of the specific reagent, 60 min at 25°. Standards are 1–5 nmol of citrate, equivalent to 1–5 µL of 1 m*M* solution.
Step 2: To all tubes add 5 µL of 0.5*M* NaOH; 95° for 5 min.
Step 3: Add 500 µL of citrate reagent at least 20 min at 25°.
Step 4: Add 30 µL of 1*M* HCl; 20 min at 25°.
Step 5: Transfer 100-µL aliquot to 1 mL of 6*M* NaOH containing 10 m*M* imidazole; 20 min at 60°. Cool and read in the fluorometer.

Cycling Assay (0.03–0.30 pmol/min, 2–20 pmol of Product)

Preincubation reagent for muscle: 2-Amino-2-methyl-1,3-propanediol, pH 8.8 (25 m*M* 2-amino-2-methyl-1,3- propanediol base, 25 m*M* 2-amino-2-methyl-1,3-propanediol-HCl); bovine serum albumin, 0.25% (citrate-free); and KCl, 0.6*M*. (The KCl is to dissolve the muscle myosin.)
Specific reagent: Tris-HCl buffer, pH 7.8 (17 m*M* Tris base, 33 m*M* Tris-HCl); otherwise the reagent is the same as Fluorometer Indirect Assay.
Citrate reagent: Tris-HCl buffer, pH 7.0 (20 m*M* Tris base, 180 m*M* Tris-HCl); NADH, 50 µ*M*; $ZnCl_2$, 80 µ*M*; bovine serum albumin, 0.02%; citrate lyase from *Aerobacter aerogenes*, 0.012 U/mL; and pig heart malate dehydrogenase, 1.2 U/mL (1 µg/mL).
Malate indicator reagent: *See* Chapter 5.

Procedure

Step 1: Each tissue sample is introduced into 1 µL of preincubation reagent in an oil well; 30–60 min at 25°. (This step is essential for muscle samples but may be omitted in the case of other tissues if tests show it to be unnecessary.)
Step 2: Add 3 µL of the specific reagent; 20° for 60 min. Standards are 3 µL of 25 and 50 µ*M* citrate in the specific reagent.
Step 3: Add 1 µL of 0.17*M* NaOH; 95° for 30 min.
Step 4: Cool and add 5 µL of citrate reagent; 20–25° for 20 min.
Step 5: Add 2 µL of 0.5*M* HCl; 10 min at 20–25°.
Step 6: Transfer 1 µL into 10 × 75 mm tubes containing 50 µL of NAD cycling reagent with enzyme levels to provide 1000-fold amplification in 1 h at 38°.
Step 7: Complete the indicator step as described in Chapter 5.

Creatine Phosphokinase (EC 2.7.3.2) *(3,14,17,18)*

The three enzyme reactions are combined:

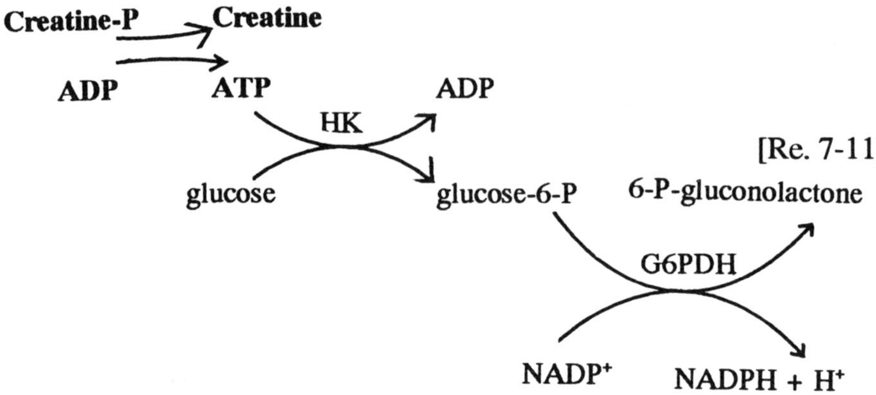

[Re. 7-11]

The auxiliary enzymes are hexokinase (HK) and glucose-6-P dehydrogenase (G6PDH).

Sample tissue activities (mol/kg dry wt/h, 20°): rat single muscle fibers, 80–525; human vastus muscle, 785; rat kidney, 0.18–18; rabbit tibialis anterior muscle, 1000; rabbit soleus muscle, 600; and mouse brain, 50.

Spectrophotometer (5–10 nmol/min 50–100 nmol of Product)

Reagent: Imidazole-acetate buffer, pH 7 (65 m*M* imidazole base, 35 m*M* imidazole-acetate); ADP, 1 m*M*; creatine-P, 25 m*M*; magnesium acetate, 10 m*M*; bovine serum albumin, 0.02%; dithiothreitol, 5 m*M*; AMP, 20 m*M* or diadenosine pentaphosphate, 10 n*M*; glucose, 2 m*M*; $NADP^+$, 1.5 m*M*; hexokinase, 2.8 U/mL (20 µg/mL); and glucose-6-P dehydrogenase, baker's yeast, 0.18 U/mL (0.5 µg/mL).

Conduct of the Assay

To 1 mL of reagent in a cuvet is added an appropriate dilution of the tissue to provide the amount of tissue to be analyzed (skeletal muscle, 2–4 µg wet wt; brain, 20–40 µg wet wt). Blanks are reagent without P-creatine, but with tissue added. Readings are taken every 2 min. The rate is calculated from the linear portion of the curve.

Fluorometer, Direct Method (0.1–1 nmol/min, 1–10 nmol of Product)

Reagent: The reagent is the same as in Spectrophotometer, except $NADP^+$, 100 µ*M*.

Conduct of the Assay

One milliliter of reagent is added to 10 × 75 mL test tubes. A reading is taken with the fluorometer sensitivity set at 10–20 divisions/μ*M*. Tissue samples are added (skeletal muscle, 0.05–0.1 μg; brain, 0.5–1 μg), mixed; and readings made every 2 min. The rate is calculated from the linear portion of the curve. Blanks are as for Spectrophotometer. Standards are 10 μL of 0.5 and 1 m*M* ATP.

Fluorometer Indirect Assay (0.02–0.2 nmol/min, 1–10 nmol Product)

Reagent: The reagent is the same as for Spectrophotometer, except that ADP is omitted.

Conduct of the Assay

Step 1: The reagent is placed in 10-μL volumes in oil wells. Standards are 10 μL of 0.1–10 m*M* ATP. Frozen-dried tissue samples are introduced into the reagent through the oil.

Step 2: The reaction is started by the addition of 1 μL of 10 m*M* ADP at measured time intervals.

Step 3: The reaction is stopped 60 min after introducing the first sample by the addition with the same time intervals of 2 μL of 100 m*M* EDTA that has been adjusted to pH 10 with NaOH.

Step 4: Aliquots of 10 μL are removed to 1 mL of 20 m*M* Tris-HCl, pH 8. The buffer fluorescence is read before and after the addition of the sample.

Fluorometer Indirect Assay (0.02–0.2 nmol/min, 1–10 nmol of Product)

This is a minor variation of the previous assay developed specifically for frozen-dried muscle sections that resist solubilization.

Reagent 1: Imidazole-acetate buffer, pH 7.0 (60 m*M* imidazole base, 40 m*M* imidazole-acetate); KCl 0.6*M*; bovine serum albumin, 0.05%; and dithiothreitol, 0.5 m*M*.

Reagent 2: Imidazole-acetate buffer as in reagent 1; P-creatine, 25 m*M*; ADP, 1 m*M*; $MgCl_2$, 10 m*M*; glucose, 3 m*M*; bovine serum albumin, 0.02%; dithiothreitol, 5 m*M*; AMP, 20 m*M* or diadenosine pentaphosphate, 10 n*M*; and yeast hexokinase 2.8 U/mL (20 μg/mL).

Reagent 3: Tris-HCl buffer, pH 8.0 (25 m*M* Tris base, 25 m*M* Tris-HCl); $NADP^+$, 100 μ*M*; EDTA, 1 m*M*; baker's yeast glucose-6-P dehydrogenase, 0.18 U/mL (0.5 μg/mL); and muscle P-glucoisomerase, 0.4 U/mL (1 μg/mL).

Conduct of the Assay

Step 1: Samples are introduced into 1 µL of reagent 1 in oil wells for preincubation; 60 min at 20°. Standards are 1 µL of 4–8 m*M* ATP.

Step 2: Reagent 2 is added in a vol of 10 µL; 60 min at 20°.

Step 3: The reaction is stopped by adding 2 µL of 0.5*M* NaOH; 15 min at room temperature.

Step 4: An aliquot of 10 µL is removed to 1 mL of reagent 3. A reading is made on the fluorometer when the reaction is complete, 10–20 min.

Comments

The AMP is added to inhibit the adenylokinase present in the tissue samples. Presumably the adenylokinase could be inhibited by the specific inhibitor, P^1,P^5–di(adenosine-5')pentaphosphate, which is effective at much lower concentrations (*see* GTP/GDP measurement). However, it has not been tested in this system. In the presence of AMP, the blank in samples incubated without P-creatine is equivalent to 1–2% of the creatine kinase activity of muscle. In some tissues, however, the activity in the absence of P-creatine is an appreciable fraction of creatine P-kinase activity and such blanks should always be tested. Preincubation was always included with frozen-dried samples to ensure activation by dithiothreitol and the solubilization of the dried muscle sections.

Fructose-1,6-Bisphosphatase (EC 3.1.3.11) *(3,18,19,20,21)*

The three enzyme reactions are combined:

[Re. 7-12]

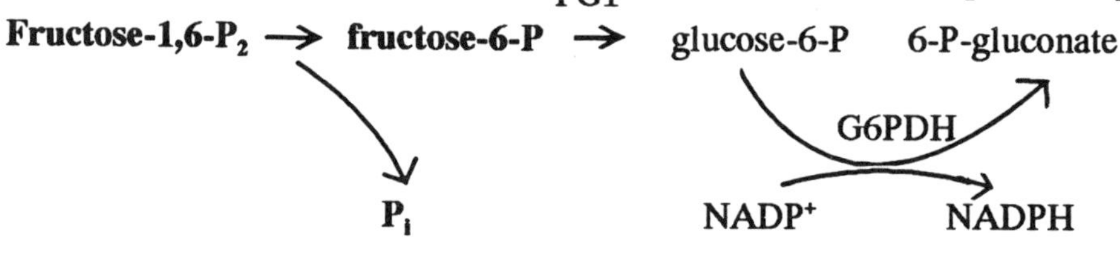

The auxiliary enzymes are P-glucoisomerase (PGI) and glucose-6-P dehydrogenase (G6PDH). The method described is one used for muscle and is based on the procedure of Bass et al. *(19)*. A somewhat modified reagent was used for kidney *(20)*.

Sample tissue activities (mol/kg dry wt/h 20°): rat kidney, 3; rabbit kidney, 1.25; rabbit tibialis anterior muscle, 0.40; rabbit soleus muscle, 0.04; rat single muscle fibers, 0.0003–0.40; and human vastus muscle, 0.38.

Fluorometer Direct Assay (0.1–0.5 nmol/min, 5–10 nmol of Product)

Reagent: Imidazole-HCl buffer, pH 7 (30 m*M* imidazole base, 20 m*M* imidazole-HCl); EDTA, 1 m*M*; bovine serum albumin, 0.05%; β-mercaptoethanol, 2 m*M*; $MgCl_2$, 2 m*M*; $NADP^+$, 100 μ*M*; fructose-1,6-P_2, 50 μ*M*; rabbit muscle P-glucoisomerase, 0.25 U/mL (0.5 μg/mL); and baker's yeast glucose-6-P dehydrogenase, 0.18 U/mL (0.5 μg/mL). Standards are 10 μL of 0.5 and 1.0 m*M* fructose-6-P.

Conduct of the Assay

A reading is taken at a fluorometer setting so that full-scale fluorescence is equivalent to 3–10 μ*M* NADPH. Samples are added in a small volume to give a rate of 2–5 divisions/min (0.05–0.5 μ*M*). Approximately 10–50 μg of kidney wet wt will give a satisfactory rate. Readings are taken and the activity calculated from the linear portion of the curve. Tissue blanks in reagent with fructose bisphosphate omitted are analyzed simultaneously with the samples.

Fluorometer Indirect Assay for Muscle (0.1–0.2 nmol/min, 5–10 nmol of Product)

Reagent: Imidazole-HCl buffer, pH 7.4 (45 m*M* imidazole base, 20 m*M* imidazole HCl); fructose bisphosphate, 50 μ*M*; $MgCl_2$, 40 m*M*; KCl, 0.3*M*; ATP, 10 μ*M*; EDTA, 1 m*M*; NAD^+, 50 μ*M*; bovine serum albumin, 0.02%; rabbit muscle P-glucoisomerase, 2.5 U/mL (5 μg/mL); and yeast glucose-6-P dehydrogenase, 0.5 U/mL (1.5 μg/mL).

Procedure

Step 1: To 1 mL of reagent in a 10 × 75 mm tube, add 10 μL of an appropriate dilution of the sample (40–100 μg of muscle). Standards are 10 μL of 0.5 and 1 m*M* fructose-6-P. Incubate 60 min at room temperature.

Step 2: Add 100 μL of 0.1*M* Na_2CO_3, mix and read on the fluorometer within 30 min (The shift in pH almost completely stops the reaction.)

Cycling Assay (0.1–0.25 pmol/min/sample, 5–15 pmol of Product)

Reagent 1: The reagent is the same as for Fluorometer Direct Assay (or Fluorometer Indirect Assay for Muscle).

Procedure

Step 1: Add freeze-dried sample to 5 μL of reagent 1 in oil wells (e.g., 10 ng kidney cortex, 25 ng fast-twitch muscle, dry wt). Standards are 5 μL of 1–3 μ*M* fructose-6-P (5–15 pmol). Incubate 1 h at 20°.

Step 2: Add 1 µL of 1*M* NaOH; heat at 95° for 10 min to destroy excess $NADP^+$.
Step 3: Transfer 2–4 µL into 10 × 75 mm tubes containing 100 µL of NADP cycling reagent with enzyme levels to provide 2000-fold amplification in 1 h at 38°.
Step 4: Complete the indicator step as described in Chapter 5.

Kinetics

The apparent K_m for fructose bisphosphate is very low, 2 µ*M* at pH 7 and 1 µ*M* at pH 8 (kidney). Higher levels are used, although they are somewhat inhibitory, to preserve linearity. $NADP^+$ causes some inhibition at levels higher than 100 µ*M*, perhaps owing to contamination with 5'AMP, which is known to be present in $NADP^+$ and to inhibit fructose bisphosphatase. The K_i for 5'AMP is approx 10 µ*M* at pH 7 and 30 µ*M* at pH 8. The addition of β-mercaptoethanol increased the activity twofold. The reaction was linear for at least an hour. The temperature coefficient is 8%/degree between 15 and 35°C.

Fumarase (EC 4.2.1.2) *(14,22)*

The analytical sequence is carried out in two steps to avoid oxidation of NADH by tissue constituents. The malate dehydrogenase (MDH) reaction is pulled to completion with aspartate transaminase (AT).

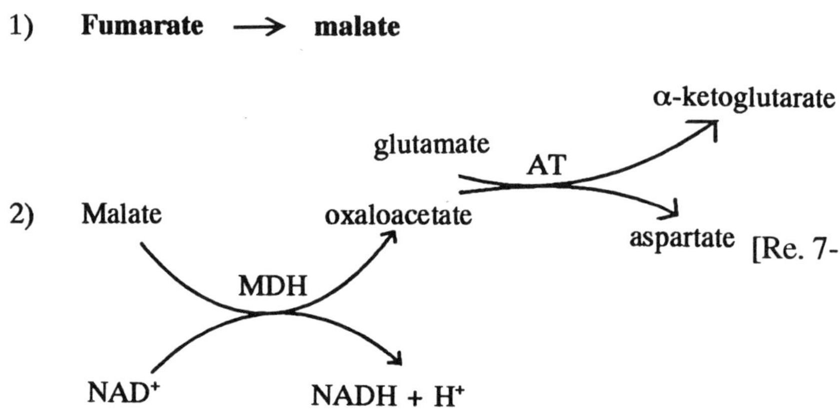

[Re. 7-13]

Sample tissue activities (mol/kg dry wt/h 20°): rat kidney cortex, 49; rat muscle, 5–22; human muscle, 1.9–4.8; and rabbit brain, 6.

Spectrophotometer (1–2.5 nmol/min, 60–150 nmol of Product)

Reagent: Potassium phosphate buffer, pH 7.0 (30 m*M* K_2HPO_4, 20 m*M* KH_2PO_4); disodium fumarate, 50 m*M*; and bovine plasma albumin, 0.05%.

Reagent 2: 2-Amino-2-methyl-1-propanol buffer, pH 9.6 (15 m*M* 2-amino-2-methyl-1-propanol base, 35 m*M* 2-amino-2 methyl-1-propanol HCl); NAD^+, 400 µ*M*; glutamate, 10 m*M*; malate dehydrogenase, 6 U/mL (5 µg/mL); and glutamate-oxaloacetate transaminase, 0.4 U/mL (2 µg/mL).

Procedure

Step 1: To 0.1 mL of reagent 1 in 10 × 75 mm tubes is added the appropriate amount of tissue (10 µg of kidney wet wt, 100 µg of skeletal muscle wet wt), incubate 1 h at 20°; heat 2 min at 100°.

Step 2: Add 1 mL of reagent 2 (without enzymes), and read at 340 nm. Add the enzymes in a vol of 5–10 µL. Read when reaction is complete, about 15 min.

Fluorometer Indirect Assay (0.05–0.15 nmol/min, 3–10 nmol of Product)

Reagent: The same as for Spectrophotometer.

Reagent 2: The same as for Spectrophotometer, except the buffer is reduced to 20 m*M*, NAD^+ to 50 µ*M*, and glutamate to 3 m*M*.

Procedure

Step 1: Add samples (e.g., 1 µg of kidney, wet wt) to 100 µL of reagent 1. Incubate at room temperature 60 min; heat 2 min at 100°. Standards are 5 µL of 0.8 and 2 m*M* malate added reagent 1.

Step 2: Add 1 mL of reagent 2 (with enzymes). Read when reaction is complete, about 15 min.

Cycling Assay (1–3 pmol/min, 50–150 pmol of Product)

Reagent: The reagent is the same as for Spectrophotometer.

Reagent 2: The reagent is the same as for Spectrophotometer (with enzymes), except NAD^+, 100 µ*M*.

Procedure

Step 1: To 1 µL of reagent in oil wells are added freeze-dried samples of appropriate size (20 ng muscle dry wt). Standards are 1 µL of 50 and 150 µ*M* malate in reagent 1. Incubate 60 min at 20°.

Step 2: Add 1 µL of 0.05*M* NaOH; heat 20 min at 80°.
Step 3: After cooling to room temperature, add 5 µL of reagent 2; 20 min at room temperature.
Step 4: Add 5 µL of 0.2*M* NaOH; 20 min at 80°.
Step 5: Transfer 2 µL into a 10 × 75 tube containing 100 µL of NAD cycling reagent with enzyme levels to provide 500-fold amplification in 1 h at 38°.
Step 6: Complete the indicator step as described in Chapter 5.

Glucose-6-Phosphatase (EC 3.1.3.9) *(20)*

The assay is conducted in two separate enzyme steps:

1) **Glucose-6-P → glucose + P_i**

2)

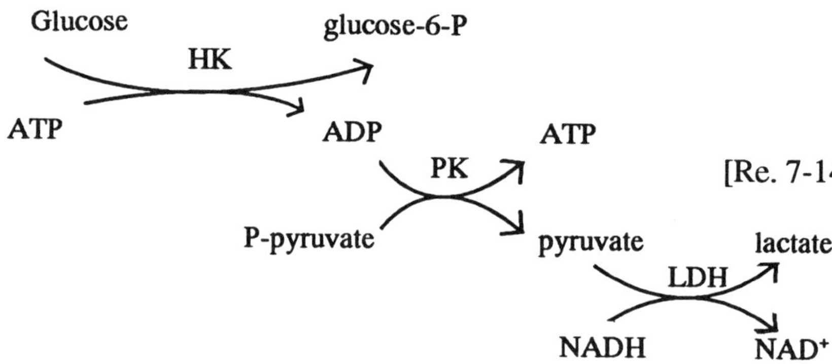

[Re. 7-14]

The auxiliary enzymes are hexokinase (HK), pyruvate kinase (PK), and lactate dehydrogenase (LDH). There are various ways described to measure glucose-6-phosphatase, but glucose release has proven to be the most satisfactory, particularly for small samples. The alternative of measuring P_i liberation gives higher and more erratic tissue blanks. The glucose is converted back to glucose-6-P with hexokinase and ATP, and the ADP formed is measured by the enzyme pathway given above.

Tissue Preparation for Larger Samples

Tissue is homogenized in 50 m*M* imidazole-HCl buffer, pH 7.1 (33 m*M* imidazole base, 17 m*M* imidazole HCl) containing 1 m*M* EDTA and 0.02% bovine serum albumin. Bovine serum albumin greatly increases the stability of the enzyme in the homogenate.

Sample tissue activity (mol/kg dry wt/h, 20°): rat kidney cortex, 2.0.

Spectrophotometer (0.3–2 nmol/min, 20–100 nmol of Product)

Reagent 1: Imidazole-HCl buffer, pH 6.8 (50 m*M* imidazole base, 50 m*M* imidazole-HCl); EDTA, 1 m*M*; bovine plasma albumin, 0.02%; and glucose-6-P 10 m*M*.

Reagent 2: Imidazole-HCl buffer, pH 7.2 (35 m*M* imidazole base, 15 m*M* imidazole-HCl); KCl, 75 m*M*; ATP, 100 μ*M*; P-pyruvate, 300 μ*M*; $MgCl_2$, 2 m*M*; β-mercaptoethanol, 2 m*M*; NADH, 100–150 μ*M*; rabbit muscle pyruvate kinase, 0.3 U/mL (2 μg/mL); beef heart lactate dehydrogenase, 6.25 U/mL (2.5 μg/mL); and yeast hexokinase, 0.3 U/mL (2 μg/mL) added after an initial reading.

Procedure

Step 1: To 100 μL of reagent 1 in 10 × 75 mm tubes are added 1–10 μL of tissue homogenate containing the appropriate amount of activity (e.g., 50–200 /μg wet wt of kidney), 60 min at 25°; 2 min at 95–100°. Blanks are reagent without glucose-6-P, but with tissue added.

Step 2: Add 1 mL of reagent 2 (without hexokinase) and mix. Transfer to cuvets, and read at 340 nm.

Step 3: Add 0.3 U of hexokinase in 2–10 μL, mix and read when reaction is over, 10–15 min.

Fluorometer Indirect Assay (0.02–0.2 nmol/min, 1–10 nmol of Product)

Reagent 1: The same as in Spectrophotometer.

Reagent 2: The same as in Spectrophotometer, except: ATP, 50 μ*M*; P-pyruvate, 30 μ*M*; NADH, 2–10 μ*M*; and pyruvate kinase, 0.6 U/mL (5 μg/mL).

Procedure

Step 1: To 100 of reagent 1 add 10 μL of tissue homogenate containing, for example, 2–20 μg wet wt of kidney. Incubate 60 min at 25°; heat 2 min at 95–100°. Blanks are as for Spectrophotometer. Standards contain 2–10 nmol glucose.

Step 2: Add 1 mL of reagent 2 without hexokinase, mix, and read at full-scale deflection on the fluorometer for the NADH level used.

Step 3: Add 0.3 U hexokinase in 2–10 μL, mix, and read when reaction is over, 10–15 min.

Cycling Assay (0.15–1.5 pmol/min, 10–100 pmol of Product)

Reagent 1: Same as for Spectrophotometer, except bovine serum albumin, 0.04%.

Reagent 2: The reagent contains double the concentration of all components in Fluorometer Indirect Assay, except: P-pyruvate, 40 μ*M*; NADH, 50 μ*M*; and β-mercaptoethanol, 2 m*M*. Before use, the enzymes are centrifuged and the supernatant ammonium sulfate fluid removed. All enzymes are included in the reagent.

Procedure

Step 1: Reagent in volumes of 1 μL are introduced under oil in the oil wells. Standards are 1 μL of 10–100 μ*M* glucose in complete reagent. Freeze-dried samples 5–50 ng of kidney (dry wt) are added to the reagent through the oil. Incubate 60 min 20°; and then heat 20 min at 60°.

Step 2: Add 1 μL of reagent 2; 30 min 22–25°.

Step 3: Add 5 μL of 50 m*M* HCl to stop the reaction and destroy excess NADH.

Step 4: Transfer 1 μL into a 10 × 75 mm tube containing 100 μL of NAD cycling reagent with enzyme levels to provide 500-fold amplification in 1 h at 38°.

Step 5: Complete the indicator step as described in Chapter 5.

General Comment

Reagent blanks can be troublesome. Many batches of glucose-6-P contain glucose. This can be removed with glucose oxidase, but it may be simpler to find a batch low in glucose. Acidification followed by heating after the first incubation increases the blank. Therefore, heat alone is used to stop the reaction. Other contributions to the blank are the result of slight contamination with

1. Pyruvate in the P-enolpyruvate (removable by heating for a few minutes at 100°C with a slight molar excess of H_2O_2);
2. ADP in the ATP (removable by conversion to ATP, *see* creatine assay, Chapter 6); and
3. NAD^+ in the NADH (removed by heating just before use in weak alkali [pH 10–12]). The last is only important for Cycling Assay.

Glucose-6-Phosphate Dehydrogenase (EC 1.1.1.49) *(16,23)*

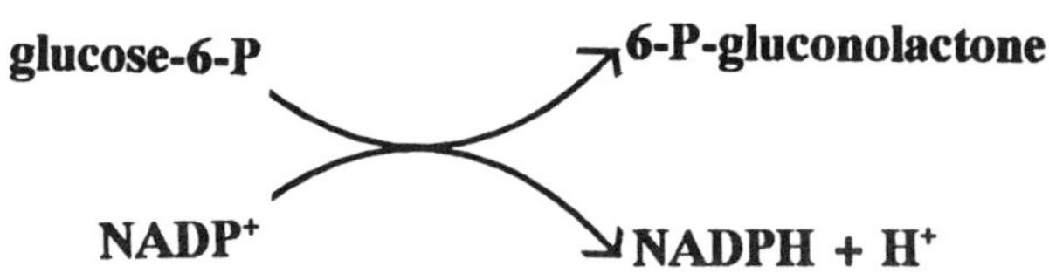

[Re. 7-15]

The activity is measured by the rate of formation of NADPH as shown. Modifications are described that permit the measurement of enzyme activity in 50 ng of dry tissue. (The spectrophotometric method is not recommended for unfractionated tissue in general, because the amount of tissue required in most cases would cause problems owing to turbidity and competing tissue enzymes.)

Sample tissue activities (mol/kg dry wt/h, 20°): human brain, 0.40; mouse brain, 0.60; rat liver, 0.60; and rat kidney, 0.50.

Spectrophotometer (2–10 nmol/min, 50–150 nmol Product)

Reagent: 2-Amino-2-methyl-1,3-propandiol-HCl buffer, pH 9.4 (70 m*M* 2-amino-2-methyl-1,3-propandiol base, 30 m*M* 2-amino-2-methyl-1,3-propandiol HCl); glucose-6-P, 2 m*M*; $NADP^+$, 0.5 m*M*; EDTA, 0.5 m*M*; and bovine serum albumin, 0.02%.

Conduct of the Assay

The sample is added to 1 mL of reagent in a cuvet (1–5 mg). Readings are taken every 2 min, and the rate calculated from the linear portion of the curve.

Fluorometer Direct Assay (0.02–1 nmol/min, 1–10 nmol of Product)

Reagent: The reagent is the same as for Spectrophotometer, except: $NADP^+$, 50 μ*M*, and glucose-6-P, 1 m*M*.

Conduct of the Assay

One-milliliter volumes of reagent are added to 10 × 75 mm tubes. The sensitivity of the fluorometer is set for full-scale fluorescence equivalent to 10 μ*M* NADPH. After an initial reading, an appropriate amount of tissue homogenate is added (15–500 μg, wet wt, of brain, muscle, or liver). Readings are made at 2-min intervals, and the rate calculated from the linear portion of the curve. Tubes containing reagent without glucose-6-P, but with added tissue serve as blanks. Standards are 1–10 nmol of glucose-6-P.

Cycling Assay (1.5–3 pmol/min, 5–25 pmol of Product)

Reagent: The reagent is the same as for Spectrophotometer, except: $NADP^+$, 100 μ*M*, and glucose-6-P, 1 m*M*.

Procedure

Step 1: To 5 µL of reagent in oil wells are added 10–50 ng (dry wt) samples of brain (for example). Standards are 5 µL of reagent containing 1–5 µ*M* NADPH. Incubate 60 min at 20°.

Step 2: Add 5 µL of 0.1*M* NaOH; 20 min at 80°.

Step 3: Transfer 2 µL into a 10 × 75 mm tube containing 100 µL of NADP cycling reagent with enzyme levels to provide 2000-fold amplification in 1 h at 38°.

Step 4: Complete the indicator step as described in Chapter 5.

Comment

The composition of the buffer is chosen to give near maximal velocity. Because the nonenzymatic hydrolysis of 6-P-gluconolactone occurs more rapidly at alkaline pH, the high pH serves to minimize the effect of the reaction nearing equilibrium.

Glutamate Decarboxylase (EC 4.1.1.15) *(8)*

1) Glutamate → γ-aminobutyrate (GABA) + CO_2

2)

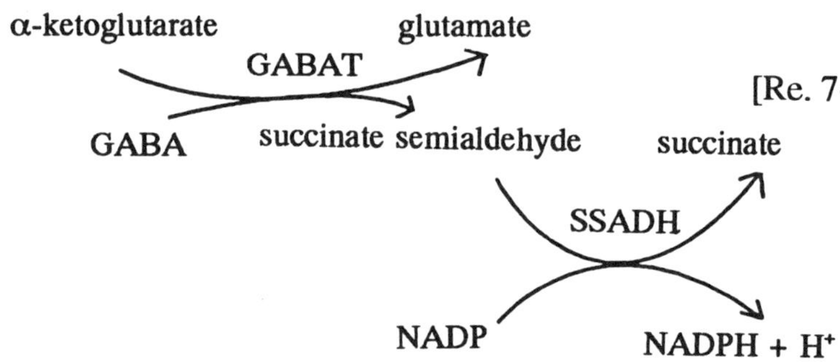

[Re. 7-16]

The auxiliary enzymes are GABA transaminase (GABAT) and succinate semialdehyde dehydrogenase (SSADH), which are both present in commercial "GABAse." The reaction is conducted in two steps, because the amount of auxiliary enzymes needed for a coupled assay would contribute disturbing blank values. (The spectrophotometric method is not recommended for unfractionated tissue in general, because the amount of tissue required in most cases would cause problems owing to turbidity and competing tissue enzymes.)

Sample tissue activities (mol/kg dry wt/h, 20°): rabbit retina, 0.05, and monkey retina, 0.072.

Spectrophotometer (0.5–2 nmol/min, 30–100 nmol of Product)

Reagent 1: Imidazole buffer, pH 7.0 (30 m*M* imidazole base, 20 m*M* imidazole HCl); pyridoxal phosphate, 160 µ*M*; glutamate 10 m*M*; and bovine plasma albumin, 0.05%.

Reagent 2: Potassium pyrophosphate buffer, pH 8.76 (33 m*M* $K_3HP_2O_7$, 67 m*M* $K_4P_2O_7$); α-ketoglutarate, 1 m*M*; GABAse, 0.4 U/mL (400 µg/mL); $NADP^+$, 1 m*M*; dithiothreitol, 1 m*M*; and bovine plasma albumin, 0.02%. (We are indebted to Diane Durham for the information that addition of dithiothreitol increases activity and decreases variability.)

Procedure

Step 1: Samples are added in 10 µL or less to 100 µL of reagent 1. Standards are 10 µL of 5–10 m*M* GABA in complete reagent. Incubate 1 h at 38°.

Step 2: Add 10 µL of 1*M* HCl; 10 min at 60°.

Step 3: Add 1 mL of reagent 2; incubate 60 min at room temperature, and read in cuvets in the spectrophotometer.

Fluorometer Indirect Assay (0.03–0.2 nmol/min, 1.8–10 nmol of Product)

Reagent 1: The reagent is the same as in Spectrophotometer.

Reagent 2: The reagent is the same as in Spectrophotometer, except: $NADP^+$, 50 µ*M*.

Procedure

Step 1: To 100 µL of reagent 1 in 10 × 75 mm tubes are added 10 µL of tissue (0.2–1 mg of brain, wet wt), or 10 µL of 0.5 and 1 m*M* GABA standard. Incubate 60 min at 37°.

Step 2: Add 10 µL of 1*M* HCl; 10 min at 60°.

Step 3: Add 1 mL of reagent 2; incubate 60 min at room temperature, and read on the fluorometer.

Cycling Assay (0.005–0.02 pmol/min, 0.3–1.2 pmol of Product)

Reagent 1: The reagent is the same as in Spectrophotometer.

Reagent 2: The reagent is the same as in Spectrophotometer, except: $NADP^+$, 25 µ*M*.

Procedure

Step 1: Samples (e.g., 5–20 ng freeze-dried brain) are added to 0.1 µL of reagent 1 in oil wells. Standards are 0.1 µL of 10 µ*M* GABA in reagent 1. Incubate 1 h at 38°.

Step 2: Add 0.1 µL of 75 m*M* HCl, and heat 20 min, 60°.
Step 3: Add 0.5 µL of reagent 2, and incubate 60 min at room temperature.
Step 4: Add 1 µL of 0.2*M* NaOH, and heat 20 min at 80°.
Step 5: Transfer 1 µL into a 10 × 75 mm tube containing 100 µL of NADP cycling reagent with enzyme levels to provide 15,000-fold amplification in 1 h at 38°.
Step 6: Complete the indicator step as described in Chapter 5.

Comment

Tissue blanks consist of tissue samples carried through the whole procedure with glutamate omitted from reagent 1. This is essential because preformed GABA, at least in brain, is present at a concentration that is a substantial fraction of the GABA generated.

Glutamate Dehydrogenase (EC 1.4.1.3) *(1,9,24,25,26)*

$$\begin{matrix} \mathbf{NH_4^+ + \alpha\text{-ketoglutarate}} & & \mathbf{glutamate} \\ & \rightleftharpoons & \\ \mathbf{H^+ + NADPH} & & \mathbf{NADP^+} \end{matrix} \qquad \text{[Re. 7-17]}$$

The activity of glutamate dehydrogenase can be measured with either NAD or NADP as coenzyme, and it is feasible to measure the activity either in the direction of oxidation or reduction of the coenzyme. The choice of pyridine nucleotide and direction of the assay may be influenced by the presence of disturbing side reactions in tissues. In brain and muscle tissue, at least, there is NADH-oxidizing activity, which would create high tissue blanks. For this reason, a method with NADPH as cofactor is described. In muscle, the activity with NADH was found to be 8% higher than with NADPH. (The spectrophotometric method is not recommended for unfractionated tissue in general, because the amount of tissue required in most cases would cause problems owing to turbidity and competing tissue enzymes.)

Sample tissue activities (mol/kg dry wt/h, 20°): rabbit tibialis anterior muscle, 0.2; rat soleus muscle, 0.40; and rabbit brain, 0.45.

Spectrophotometer (5–10 nmol/min, 25–150 nmol of Product)

Reagent 1: Imidazole buffer, pH 7.4 (35 m*M* imidazole base, 15 m*M* imidazole-HCl); α-ketoglutarate, monosodium salt, 2 m*M*; ammonium acetate, 25 m*M*; ADP, 100 μ*M*; bovine serum albumin, 0.05%; and NADPH, 200 μ*M*.

Conduct of the Assay

A reading is made on the spectrophotometer with 1 mL of reagent. Samples are added and readings taken every 1 or 2 min. Blanks contain samples added to cuvets containing reagent without α-ketoglutarate. Calculations are based on the linear portion of the assay.

Fluorometer Indirect Assay (15–50 pmol/min, 1–3 nmol of Product)

Reagent 1: The reagent is the same as for Spectrophotometer, except that NADPH is 100 μ*M*.
Reagent 2: NaOH, 6*M* containing 10 m*M* imidazole base.

Procedure

Step 1: To 100 μL of reagent 1 in 10 × 75 mm tubes, add 1–2 μL of muscle or brain diluted to contain 20–40 μg of tissue; 1 h at 20°. Standards are 1–3 nmol of $NADP^+$.
Step 2: Add 10 μL of 1*M* HCl; 10 min at room temperature.
Step 3: Add 1 mL of reagent 2; heat at 60° 20 min. Cool and read.

Cycling Assay (0.05–0.1 pmol/min, 3–6 pmol of Product)

The following assay was developed for analyses of enzyme activities in human ova.

Reagent: The reagent is the same as for Spectrophotometer, except that 50 μ*M* NADH is used (rather than NADPH).

Procedure

Step 1: To 2 μL of reagent in oil wells is added 0.1 μL of sample or standards of 25–50 μ*M* NAD^+.
Step 2: Add 2 μL of 0.1*M* HCl; 10 min at room temperature.
Step 3: Transfer 2 μL into a 10 × 75 mm tube containing 100 μL of NAD cycling reagent with enzyme levels to provide 5000-fold amplification in 1 h at 38°.
Step 4: Complete the indicator step as described in Chapter 5.

Glutaminase (EC 3.5.1.2) *(27,28)*

1) Glutamine + H_2O → glutamate + NH_4^+

2) 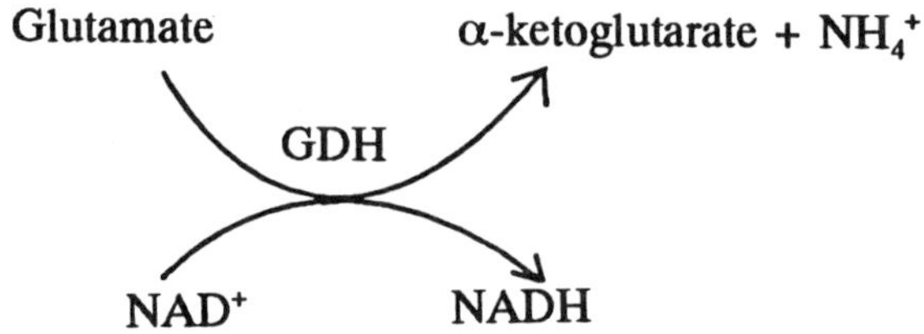

[Re. 7-18]

The auxiliary enzyme is glutamate dehyrogenase (GDH). The procedure consists of two steps, because owing to the low catalytic activity of glutamate dehydrogenase in the direction required, it is not practical to add sufficient enzyme to provide a coupled assay.

There are two distinct isozymes; one is P_i-dependent, and the other is independent of P_i, but activated 15-fold by maleate. Procedures for both forms are given so that the relative content of any tissue can be assessed. The P_i-independent isozyme has been found by Curthoys and Kuhlenschmidt *(28)* to be γ-glutamyltranspeptidase.

Sample tissue activities (mol/kg dry wt/h, 20°): rat kidney, 1.3; and mouse brain, 0.8.

Spectrophotometer (2–4 nmol/min, 25–100 nmol of Product)

Reagent 1: (Phosphate-dependent form) Tris-HCl buffer, pH 8.6 (40 m*M* Tris base, 10 m*M* Tris-HCl); glutamine 20 m*M*; phosphate pH 8 (140 m*M* K_2HPO_4, 10 m*M* KH_2PO_4); EDTA, 0.2 m*M*; and bovine serum albumin, 0.02%.

Reagent 1A: (Phosphate-independent, maleate-stimulated form) maleate buffer, pH 6.6 (50 m*M* disodium maleate, 10 m*M* monosodium maleate); glutamine, 10 m*M*; EDTA, 0.2 m*M*; and bovine serum albumin, 0.02%.

Reagent 2: Tris-HCl buffer, pH 8.4 (70 m*M* Tris base, 20 m*M* Tris-HCl); NAD^+, 1 m*M*; ADP, 0.3 m*M*; H_2O_2, 0.03%; and beef liver glutamate dehydrogenase in glycerol, 6 U/mL as measured in the opposite direction (50 µg/mL).

Procedure

Step 1: To 100 µL of reagent 1 or 1A in 10 × 75 mm tubes are added 10 µL of tissue homogenate containing, for example, 100 µg wet wt of kidney or brain. Standards are 10 µL of 5 and 15 m*M* glutamate. Incubate 60 min at 38°.

Step 2: Add 15 µL of 2*M* HCl; 10 min at room temperature.
Step 3: Add 1 mL of reagent 2, and read in the spectrophotometer after 20–30 min at room temperature.

Fluorometer Indirect Assay (0.03–0.3 nmol/min 1–10 nmol of Product)

Reagents 1 and 1A: The same as for Spectrophotometer.
Reagent 2: The same as for Spectrophotometer, except the buffer is 50 m*M* (35 m*M* Tris base, 15 m*M* Tris-HCl) and the NAD^+ is 300 µ*M*.

Procedure

Step 1: Reagent 1 or 1A is added in a vol of 20 µL to a 5 × 60 mm tube.
Step 2: Tissue homogenate samples are added in a vol of 1–3 µL containing, for example, 10 µg of kidney or brain. Standards are 3 µL of 1 and 2 m*M* glutamate. Incubate 60 min at 37°.
Step 3: Add 3 µL of 2*M* HCl; 10 min at room temperature.
Step 4: Transfer 20 µL to 1 mL of reagent 2. After 20 min at room temperature, the fluorescence is measured.

Comment

The peroxide in reagent 2 is included to destroy the α-ketoglutarate and ensure the completion of the dehydrogenase reaction. The ADP is included to stabilize the glutamate dehydrogenase.

Cycling Assay (0.4–4 pmol/min, 25–100 pmol of Product)

Reagents 1 and 1A: The same as for Spectrophotometer.
Reagent 2: The same as for Fluorometer Indirect Assay.

Conduct of the Assay

The assay is conducted in oil wells.

Step 1: Freeze-dried samples (20–100 ng) are introduced into 0.25 µL of reagent 1 or 1A in oil wells. Standards are 0.25 µL of 0.2–0.4 m*M* glutamate in complete reagent. Incubate 1 h at 20°.
Step 2: Add 0.25 µL of 0.3*M* HCl; 10 min at room temperature.
Step 3: Add 5 µL of reagent 2; 60 min at 20°.
Step 4: Add 2 µL of 0.25*M* NaOH; 20 min at 80°.
Step 5: Transfer 1 µL into 10 × 75 mm tubes containing 100 µL of NAD cycling reagent with enzyme levels to provide 500-fold amplification in 1 h at 38°.
Step 6: Complete the assay as described in Chapter 5.

Glycero-3-Phosphate Dehydrogenase (EC 1.1.1.8) *(5,6,14,16)*

$$\text{Dihydroxyacetone-P} + H^+ + \text{NADH} \rightarrow \text{glycero-P} + \text{NAD}^+ \qquad \text{[Re. 7-19]}$$

The reaction is followed by measuring the rate of disappearance of NADH or by measuring the strong alkali-induced fluorescence of the NAD^+ formed after the acid destruction of NADH. Further sensitivity is achieved by cycling the NAD^+ product.

Sample tissue activities (mol/kg dry wt/h, 20°): rat kidney, 5; rat liver, 15; mouse brain, 0.4; human brain, 0.27; rat tibialis anterior muscle, 6; and soleus muscle, 0.5.

Spectrophotometer (2–10 nmol/min, 20–80 nmol of Product)

Reagent: Imidazole buffer, pH 7.0 (30 m*M* imidazole base, 20 m*M* imidazole-HCl); dihydroxyacetone-P, 1 m*M*; NADH 100 µ*M*; bovine serum albumin, 0.02%; and sodium amytal, 2 m*M*.

Conduct of the Assay

One-milliliter volumes of reagent are added to cuvets and initial readings made. Tissue homogenate samples are added (50 µL or less containing in the order of 200 µg fast-twitch muscle or kidney, or 100 µg of liver, all wet wts). Readings are made at 2-min intervals. Blanks consist of tissue added to reagent without dihydroxyacetone-P. Calculations are made from the linear portion of the curve.

Comment

The sodium amytal is added to minimize the oxidation of NADH in the absence of dihydroxyacetone-P. Fresh homogenates have more amytal-sensitive, NADH-oxidizing activity than frozen or freeze-dried tissues. In the case of brain at least, the freeze-drying process eliminates the troublesome activity almost completely.

Fluorometer (0.1–0.5 nmol/min, 2–8 nmol of Product)

Reagent: The reagent is the same as Spectrophotometer, except NADH, 10 µ*M*.

Conduct of the Assay

One-milliliter volumes of reagent are pipeted into 10×75 mm tubes. Initial readings are made so that the NADH gives full-scale deflection on the fluorometer. Tissue is added in a 5–10-µL vol containing 200 µg slow-twitch or 20–50 µg fast-twitch muscle, 25 µg kidney, 200 µg brain, or 5 µg of liver, all wet wts. Readings are taken at 1–5 min intervals. Tissue blanks consist of tissue added to reagent without dihydroxyacetone-P (not needed for high-activity tissues).

Comment

The K_m for NADH is 0.3 µ*M*. Therefore, the average rate over the total interval should not fall below 97% of the initial rate until NADH has fallen from 10 to 2 µM, even though the rate from 3 to 2 µ*M* should be only 88% of the initial rate.

Fluorometer Indirect Assay (2–50 pmol/min, 0.1–3 nmol of Product)

Reagent: The reagent is the same as for Spectrophotometer, except NADH, 10 µ*M*; ascorbic acid, 2 m*M*; nicotinamide, 20 m*M*; and bovine serum albumin, 0.02%.

Procedure

Step 1: Place 0.1 mL of reagent in 10 × 75 mm tubes. In a timed sequence, add 5 or 10 µL vol of homogenate containing, for example, 1 µg kidney, 0.5 µg liver, and 5 µg brain, wet wts. Standards are 10 µL of 50–300 µ*M* NAD^+. Incubate 60 min at 20°.

Step 2: Add 10 µL of 1*M* HCl at the same rate as the samples were added; hold 15 min or more at room temperature.

Step 3: Add 1 mL 6*M* NaOH containing 10 m*M* imidazole; heat 20 min at 60°. Cool and dry tubes, and read in the fluorometer.

Cycling Assay (0.0.2–2 pmol/min, 10–100 pmol of Product)

Reagent: The reagent is the same as for Cycling Assay, except: NADH, 1 m*M*; nicotinamide, 10 m*M*; and bovine serum albumin, 0.05%.

Procedure

Step 1: To 0.5 µL of reagent under oil are added samples (25–200 ng dry wt of brain or slow-twitch muscle), 60 min at 20°.

Step 2: To each sample is added 1 µL of 0.05*M* HCl; 10 min at room temperature.

Step 3: Transfer 1 µL into a 10 × 75 mm tube containing 100 µL of NAD cycling reagent with enzyme levels to provide 500-fold amplification in 1 h at 38°.

Step 4: Complete the assay as described in Chapter 5.

Comment

The nicotinamide is added in Fluorometer Indirect Assay and Cycling Assay to prevent the enzymatic destruction of the product, NAD^+. The ascorbic acid is present to prevent the conversion of NADH to NAD^+ during incubation and acidification in very small volumes.

Modifications in Cycling Procedure for Frozen-Dried Muscle Sections (1–5 pmol/min, 100 pmol of Product)

Reagent: The reagent is the same as for Spectrophotometer, except dihydroxy-acetone-P, 200 µ*M*; 100 µ*M* NADH; and bovine serum albumin, 0.05%.

Preincubation reagent: Imidazole buffer, pH 7.0 (30 m*M* imidazole base, 20 m*M* imidazole-HCl); KCl, 0.6*M*; and bovine serum albumin, 0.05%.

Procedure

Step 1: To 1 µL of preincubation reagent in oil wells are added muscle samples (15 ng dry wt); 30 min at 20°C.

Step 2: Add 4 µL of reagent 1; 60 min at 20°. Standards are 4 µL of 25 and 50 µ*M* NAD^+.

Step 3: To each sample is added 5 µL of 0.1*M* HCl; 10 min at room temperature.

Step 4: Transfer 1 µL into a 10 × 75 mm tube containing 100 µL of NAD cycling reagent with enzyme levels to provide 1000-fold amplification in 1 h at 38°.

Step 5: Complete the indicator step as described in Chapter 5.

Glycogen Phosphorylase (EC 2.4.1.1) *(3,5,14,16,29,30,31)*

[Re. 7-20]

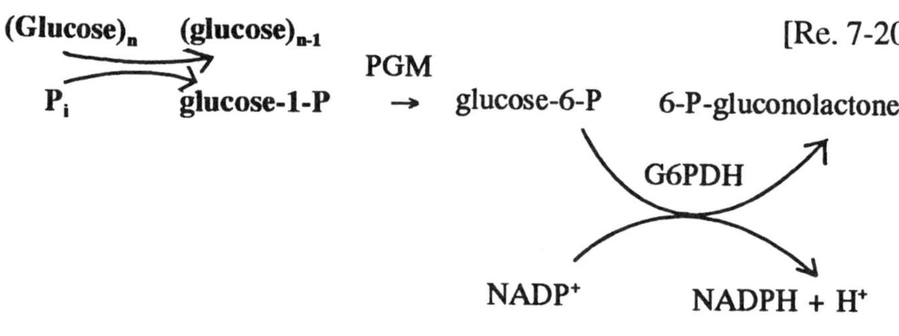

The auxiliary enzymes are P-glucomutase (PGM) and glucose-6-P dehydrogenase (G6PDH). The glucose-1-P generated may be converted to 6-P-gluconate as shown, combining all three reactions in a single step. Alternatively, the auxiliary sequence is carried out in a second step, which avoids some of the dangers from other tissue enzymes. Phosphorylase exists in two forms; The *b* form requires 5'AMP for activity, and the phosphorylated form *a* is active in the absence of the nucleotide. The kinetics of the *a* form are complex, and 5'AMP can alter the reaction rate at nonsaturating substrate concentrations *(29)*. Therefore, a very small amount of 5'AMP is included in the methods for phosphorylase *a*.

Sample tissue activities (mol/kg dry wt/h, 20°): dog myocardium, 2; human brain, 0.50; rabbit soleus muscle, 0.8; and rabbit soleus muscle, 25.

Tissue Homogenates

Tissue homogenates are prepared in 10–20 volumes of imidazole buffer, pH 7.0, containing 0.5 m*M* dithiothreitol, 5 m*M* EDTA, and 20 m*M* KF. The components are chosen to keep the *a* and *b* forms in the state in which they are homogenized. If the tissue is quick-frozen, the ratio of the *a* to *b* forms will be close to that of the in vivo conditions. If the tissues are removed without precautions to preserve the in vivo state, the state of the enzymes will be a function of the delay before homogenization. There is a rapid conversion from the *b* to *a* form during anoxia, followed by a slower transformation back to the *b* form. The reducing agent is essential for phosphorylase activity; the EDTA prevents the action of the phosphorylase *b* kinase and, thus, the conversion of phosphorylase *b* to the *a* form. The KF inhibits the action of the phosphatase that converts the *a* to the *b* form. (The spectrophotometric method is not recommended for unfractionated tissue in general because the amount of tissue required in most cases would cause problems owing to turbidity and competing tissue enzymes.)

Spectrophotometer (1–5 nmol/min, 15–75 nmol of Product)

Reagent: Imidazole-acetate buffer, pH 7.0 (35 m*M* imidazole base, 15 m*M* imidazole acetate); glycogen (glucosyl units), 20 m*M*; K_2HPO_4 20 m*M*; $MgCl_2$, 0.5 m*M*; EDTA, 1.0 m*M*; $NADP^+$, 1 m*M*; glucose-1,6-bisphosphate, 2 μ*M*; dithiothreitol, 0.5 m*M*; bovine serum albumin, 0.025%; muscle P-glucomutase, 0.6 U/mL (3 μg/mL); and yeast or *Leuconostoc* glucose-6-phosphate dehydrogenase, 1 U/mL. For the measurement of phosphorylase *a*, add 2 μ*M* 5'AMP; for phosphosphorylase *a* plus *b*, add 1 m*M* 5'AMP.

Conduct of the Assay

Add the sample to 1 mL of reagent in a spectrophotometer cuvet. Read at convenient intervals, 1–5 min. The change in OD should be linear with time after the first minute or two.

Fluorometer Direct Assay (0.02–1 nmol/min, 1–10 nmol of Product)

Reagent: The reagent is the same as for Spectrophotometer, except the $NADP^+$ is decreased to 100 μ*M*.

Conduct of the Assay

To 1 mL of reagent in 10 × 75 mm tubes, add 1–10 μL of homogenate equivalent to 25–100 μg of brain, wet wt. Read at convenient intervals, 1–5 min, depending on the rate of reaction.

Fluorometer Indirect Assay (15–75 pmol/min, 1–5 nmol of Product)

Reagent 1: The reagent is the same as for Spectrophotometer, except the glucose-6-P dehydrogenase and $NADP^+$ are omitted.

Reagent 2: Tris-HCl buffer, pH 8 (25 m*M* Tris base, 25 m*M* Tris HCl); $NADP^+$, 100 μ*M*; glucose-6-P dehydrogenase 0.2 U/mL; and rabbit muscle glucose-6-phosphate isomerase, 0.5 U/mL (1 μg/mL).

Procedure

Step 1: Homogenate samples (1–5 μL) containing, for example, 1 μg of fast-twitch muscle or 50 μg of brain, wet wt, are added to 100 μL of reagent 1 in 10 × 75 mm tubes at measured time intervals, and incubated for 60 min at room temperature. Standards are 5 μL of 0.1–1 m*M* glucose-1-P.

Step 2: Add 1 mL of reagent 2 in the same time sequence as Step 1. The tubes are read when the reaction is complete, 10–20 min.

Cycling Assay (0.2–1 pmol/min/sample, 10–50 pmol of Product)

Reaction vessels: Oil wells.

Reagent: The reagent is the same as in Fluorometer Direct Assay.

Procedure

Step 1: Freeze-dried samples (e.g., 25 ng of myocardium, 100 ng of brain, dry wt) are introduced through the oil into 1 µL of the reagent and incubated 60 min at room temperature. Standards are 1 µL of 10–50 µ*M* glucose-1-P in the reagent.
Step 2: Add 2 µL of 0.1*M* NaOH; 20 min at 80°.
Step 3: Transfer 1 µL into a 10 × 75 mm tubes containing 100 µL of NADP cycling reagent with enzyme levels to provide about 3000-fold amplification in 1 h at 38°.
Step 4: Complete the indicator step as described in Chapter 5.

Comments

The excess of EDTA over Mg^{2+} in the reagent is to prevent phosphorylase *b* kinase activity, but permit the action of P-glucomutase, which requires only a low level of Mg^{2+}.

The procedures given above have been used in the measurement of phosphorylase in brain tissue. If muscle phosphorylase activity is to be measured in frozen-dried sections, the tissue samples must be preincubated as described below.

Modification of Cycling Assay for Freeze-Dried Muscle (15 ng Dry Wt, 2–5 Pmol/min, 100–300 pmol of Product)

Preincubation reagent: Tris-HCl buffer, pH 8.0 (25 m*M* Tris base, 25 m*M* Tris HCl); 0.6*M* KCl, 0.2% bovine serum albumin, 1 m*M* 5'AMP, 10 m*M* K_2HPO_4, and 0.5 m*M* dithiothreitol.
Reagent: The same as for Spectrophotometer.

Procedure

Step 1: The freeze-dried samples are introduced through oil into 1-µL droplets of preincubation reagent; 30–60 min at 20°.
Step 2: Add 4 µL of reagent to samples. Standards are the same volume of 25 and 50 µ*M* glucose-1-P. Incubate 60 min at 20°.
Step 3: Add 5 µL of 0.1*M* NaOH;20 min at 80°.
Step 4: Transfer 1 µL into a 10 × 75 mm tube containing 100 µL of NADP cycling reagent with enzyme levels to provide 500-fold amplification in 1 h at 38°.
Step 5: Complete the indicator step as described in Chapter 5.

Glycogen Synthase (EC 2.4.1.11) *(9,16,21,31,32)*

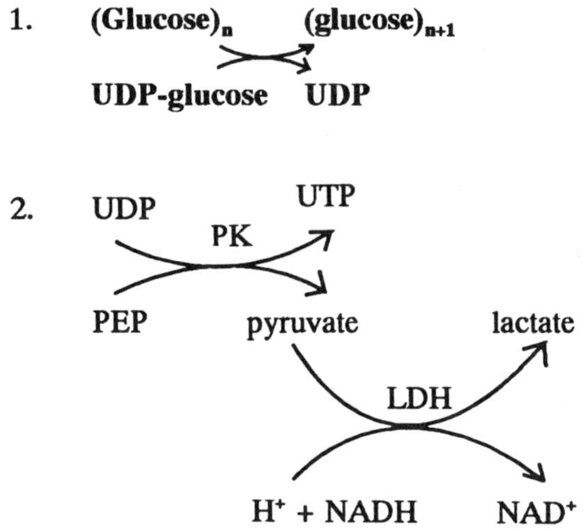

[Re. 7-21]

The auxiliary enzymes are pyruvate kinase (PK) and lactate dehydrogenase (LDH). PEP is phosphopyruvate. The activity is assessed by measurement of one of the products, UDP, by means of the enzyme sequence shown. The synthase exists in an unphosphorylated form, synthase *i*, which is fully active without the presence of glucose-6-P, and a series of phosphorylated forms that are variably dependent on glucose-6-P for full activity. Therefore, when total synthase activity is measured, glucose-6-P is incorporated in the reagent. (The spectrophotometric method is not recommended for unfractionated tissue in general, because the amount of tissue required in most cases would cause problems resulting from turbidity and competing tissue enzymes.)

Tissue Preparation

Crude homogenates are prepared by adding appropriate volumes of 50 m*M* Tris buffer, pH 7.5, containing 5 m*M* EDTA, 20 m*M* KF, and 5 m*M* dithiothreitol to the tissue and homogenizing in a glass homogenizer. The tissue preparations can be stored at –70°. The EDTA prevents the activity of the protein kinase, which phosphorylates synthase *i* in the presence of ATP. The KF prevents the removal of the phosphate from the phosphorylated enzyme by phosphatase activity present in the homogenate. The reducing agent dithiothreitol ensures reduction of the sulfhydryl groups of the enzyme, which are necessary for optimal activity.

Sample tissue activities (mol/kg dry wt/h, 20°): human brain, 0.03; dog myocardium, 0.17; rabbit tibialis anterior muscle, 0.4; rabbit soleus muscle, 0.28; and rat liver, 0.26.

Fluorometer Indirect Assay (0.05–0.5 nmol/min, 1–8 nmol of Product)

Reagent 1: Imidazole-HCl, pH 7.5 (35 m*M* imidazole base, 15 m*M* imidazole-HCl); uridine diphosphoglucose (UDPG) 2 m*M*; EDTA, 1 m*M*; NaF, 10 m*M*; dithiothreitol, 0.5 m*M*; and 0.02% bovine plasma albumin. When the sum of *i* and *d* forms is to be measured, 5 m*M* glucose-6-P is added.

Reagent 2: Imidazole acetate buffer, pH 6.7 (30 m*M* imidazole base, 70 m*M* imidazole acetate); KCl, 50 m*M*; $MgCl_2$, 2 m*M*; P-pyruvate, 100 µ*M*; NADH, 2–10 µ*M*; muscle pyruvate kinase, 3 U/mL (15 µg/mL); beef heart lactate dehydrogenase, 0.25 U/mL (1 µg/mL); beef liver catalase, 65 U/mL (1 µg/mL); and 0.02% bovine plasma albumin.

Comment

The tissue content of nucleoside diphosphates and pyruvate will contribute a substantial tissue blank. To correct for these components, replicates are carried through the procedure with reagent from which UDPG has been omitted. The alkaline H_2O_2 treatment is to destroy pyruvate and other keto acids.

Procedure

Step 1: Add the tissue (e.g., 20–100 µg of liver, skeletal muscle, or myocardium, wet wt) in 1–5 µL vol to 50 µL of reagent 1 in 10 × 75 mm tubes. Standards are 1–5 µL of 2–20 m*M* UDP. Incubate 60 min at 38°. Stop the reaction with 10 µL of 10 m*M* H_2O_2 in 0.5*M* NaOH followed by incubating 30 min at room temperature.

Step 2: Add 1 mL of reagent 2. After 20 min, read the fluorescence.

The assay can be conducted in one step, if desired, using the Tris buffer and all other components of reagents 1 and 2, and increasing the pyruvate kinase to 20 U/mL (100 µg/mL). This procedure is not recommended for most tissues, since the amounts of crude tissue homogenate required would add interfering turbidity, and there is oxidation of the NADH by the tissue in the absence of substrate. However, if a rapid assessment of the enzyme activity in tissue is desired, the addition of 10 m*M* sodium amobarbital will decrease the rate of oxidation of NADH by crude homogenates. The tubes can be read in the fluorometer before and after the addition of pyruvate kinase to increase precision.

Cycling Assay for Freeze-Dried Tissue Sections (0.1–1 pmol/min, 10–100 pmol of Product)

Reagents 1 and 2: The reagents are the same as for Spectrophotometer.

Procedure

Step 1: The samples (e.g., 1 µg of liver or muscle or 10 µg of brain, all dry wts) are introduced into 0.2 µL of the specific reagent in oil wells. Standards are 0.1 µL of 25–200 µ*M* UDP. Incubate 2 h at 20–22°.
Step 2: Add 0.5 µL of 3 m*M* H_2O_2 in 0.15*M* NaOH; 30 min at 20–25°.
Step 3: Add 1 µL of reagent 2; 30 min at 20–25°.
Step 4: Add 5 µL of 0.1*M* HCl; 20 min at 20–25°.
Step 5: Transfer 1 µL into a 10 × 75 mm tube containing 100 µL of NAD cycling reagent with enzyme levels to provide 500-fold amplification in 1 h at 38°.
Step 6: Complete the indicator step as described in Chapter 5.

Guanylate Kinase (EC 2.7.4.8) *(33,34)*

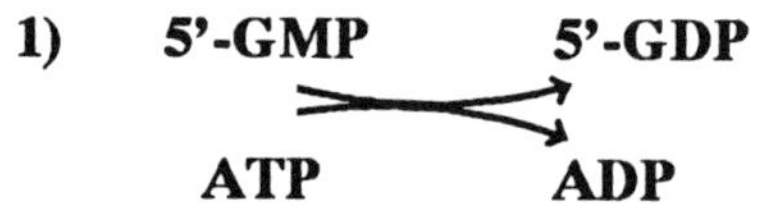

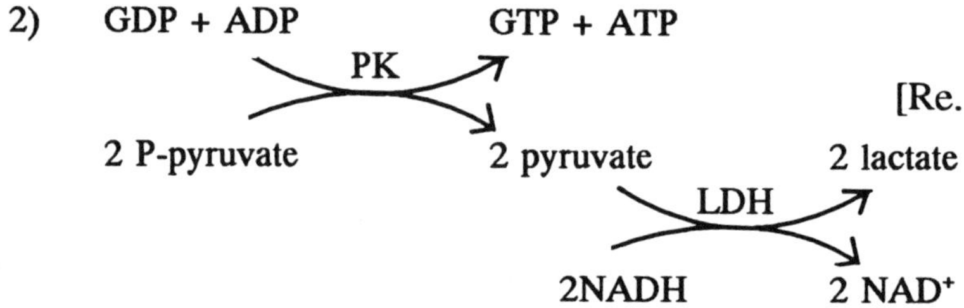

[Re. 7-22]

The auxiliary enzymes are pyruvate kinase (PK) and lactate dehydrogenase (LDH). Note, 2 mol of NAD^+ are generated for each mol of GMP phosphorylated, because both nucleotide products of reaction 1, GDP and ADP, are substrates for pyruvate kinase (although the reaction is about fivefold faster with ADP than GDP).

Sample tissue activities (mol/kg dry wt/h, 20°): monkey retina, peak level 7; rat kidney, 0.24–0.45; and rabbit brain, 2.5.

Spectrophotometer (1–3 nmol/min, 25–100 nmol of Product)

Reagent: Imidazole buffer, pH 7.0 (30 m*M* imidazole base, 20 m*M* imidazole-HCl); ATP, 1 m*M*; 5'GMP, 500 µ*M*; $MgCl_2$, 2 m*M*; and bovine serum albumin, 0.05%.

Reagent 2: Imidazole buffer, pH 7.0 (30 m*M* imidazole base, 20 m*M* imidazole HCl); $MgCl_2$, 2 m*M*; KCl, 75 m*M*; NADH, 100–150 μ*M*; P-pyruvate, 1 m*M*; rabbit muscle pyruvate kinase, 3 U/mL (20 μg/mL); and beef heart lactate dehydrogenase, 0.4 U/mL (2 μg/mL).

Procedure

Step 1: To 50 μL of reagent 1 in 10 × 75 mm tubes are added 5–10 μL of homogenate containing, for example, 25–100 μg of brain wet wt. Standards are 10 μL of 2 and 5 m*M* 5'GDP. Incubate at room temperature 60 min.
Step 2: Add 25 μL of 0.1*M* NaOH, and heat 20 min at 60°.
Step 3: To each tube add 1 mL of reagent 2, and read when reaction is over, about 30 min.

Fluorometer Indirect Assay (0.1–0.25 nmol/min, 2.5–5 nmol of Product)

Reagent 1: The reagent is the same as for Spectrophotometer, except ATP, 100 μ*M*; and 5'GMP, 100 μ*M*.
Reagent 2: The reagent is the same as for Spectrophotometer, except P-pyruvate, 20 μ*M*; and NADH, 5–10 μ*M*.

Procedure

Step 1: To 50 μL of reagent 1 in 10 × 75 mm tubes are added 5–10 μL of tissue samples (e.g., 5 μg of brain, wet wt); 60 min at room temperature. Standards are 10 μL of 0.1 and 0.2 m*M* GMP.
Step 2: Add 25 μL of 0.1*M* NaOH; heat 20 min at 60°.
Step 3: To each tube add 1 mL of reagent 2, except for pyruvate kinase. Read in the fluorometer with NADH giving full-scale fluorescence.
Step 4: Add pyruvate kinase, and read after 30 min.

Cycling Assay (5–30 fmol/min, 0.3–2 pmol of Product)

Reagent 1: The reagent is the same as for Spectrophotometer, except: ATP, 500 μ*M*; and 5'GMP, 75 μ*M*.
Reagent 2: Imidazole buffer, pH 6.8 (20 m*M* imidazole base, 30 m*M* imidazole-HCl); KCl, 75 m*M*; $MgCl_2$, 2 m*M*; P-pyruvate, 10 μ*M*; NADH, 1 μ*M*; bovine plasma albumin, 0.025%; pyruvate kinase, 3 U/mL (20 μg/mL); and beef heart lactate dehydrogenase, 0.25 U/mL (1 μg/mL).

Procedure

Step 1: Samples (e.g., 5 ng of freeze-dried kidney) are added to 1 μL of reagent 1 in oil wells. Standards are 1 μL of 0.5 and 0.1 μ*M* GDP; 60 min at 20°.
Step 2: Add 1 μL of 50 m*M* NaOH; 30 min or more at room temperature.

Step 3: Add 5 µL of reagent 2; 60 min at 25°.

Step 4: Add 2 µL of 0.2*M* HCl; 10 min at room temperature.

Step 5: Transfer 5 µL into a 10 × 75 mm tube containing 100 µL of NAD cycling reagent with enzyme levels to provide 10,000-fold amplification in 1 h at 38°.

Step 6: Complete the indicator step as described in Chapter 5.

Hexokinase (EC 2.7.1.1) *(1,3,16,23,35)*

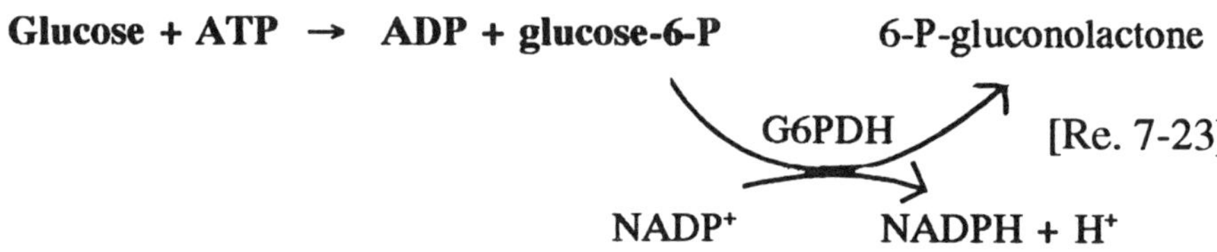

[Re. 7-23]

The auxiliary enzyme is glucose-6-phosphate dehydrogenase (G6PDH). Two basic methods are offered; the first combines the specific and auxiliary steps as shown above. This is suitable for tissues with high hexokinase activity and therefore relatively free of disturbing side reactions. The second method separates the specific and auxiliary reactions by an enzyme-destroying step, thereby eliminating most of the danger from side reactions. This also eliminates the problem arising from the fact that tissue 6-P-gluconate dehydrogenase is usually present with an activity at least as high as that of glucose-6-P dehydrogenase, and so will oxidize a portion of the 6-P-gluconate yielding an uncertain extra amount of NADPH. This is less disturbing for fluorometric assays, because the average 6-P-gluconate level during the assay can be kept to a small fraction of the K_m. In the two-step procedure, there is sure to be some conversion of the glucose-6-P formed to fructose-6-P owing to tissue P-glucoisomerase. Therefore, P-glucoisomerase is included in the second reagent to ensure that all the hexose phosphate is converted to 6-P-gluconate.

Sample tissue activities (mol/kg dry wt/h, 20°): mouse brain, 2; human brain, 3.4; myocardium, 1; rabbit tibialis anterior muscle, 0.2; rabbit soleus muscle, 0.60; and liver, 0.02 (The spectrophotometric method is not recommended for unfractionated tissue in general, because the amount of tissue required in most cases would cause problems resulting from turbidity and competing tissue enzymes.)

Spectrophotometer Assay (Method 1) (2–10 nmol/min, 50–150 nmol Product)

Reagent: Tris-HCl buffer, pH 8.0 (50 m*M* Tris base, 50 m*M* Tris-HCl); $MgCl_2$, 5 m*M*; glucose, 2 m*M*; ATP, 5 m*M*; NADP, 0.5 m*M*; bovine serum albumin 0.05%; Triton X-100, 0.5%; and glucose-6-P dehydrogenase (yeast or *Leuconostoc mesenteroides*), 0.3 U/mL.

Conduct of the Assay

Samples are added to 1 mL of reagent in a cuvet, and readings made every 1 or 2 min. The rate is calculated from the linear part of the curve.

Fluorometer Direct Assay (Method I) (0.05–1 nmol/min, 1.5–10 nmol of Product)

Reagent: The reagent is the same as for Spectrophotometer Assay, except $NADP^+$ is reduced to 100 μ*M*, and glucose to 1 m*M*.

Conduct of the Assay

One-milliliter volumes are pipeted into 10 × 75 mm fluorometer tubes. The fluorometer sensitivity is set so that full scale is equivalent to 10 μ*M* NADPH. After an initial reading, the sample is added and readings are made at 1–5-min intervals. The activity is calculated from the linear part of the curve. Reagent without glucose serves as a blank. Suitable amounts of tissue are 25 μg of brain, and 100 μg of slow-twitch muscle and heart muscle, wet wt.

Fluorometric Indirect Assay (15–150 pmol/min, 1–10 nmol of Product)

Reagent 1: The same as for Spectrophotometer Assay, except $NADP^+$ and G6PDH are omitted, and glucose is reduced to 1 m*M*.

Reagent 2: Tris-HCl buffer, pH 8.0 (25 m*M* Tris base, 25 m*M* Tris-HCl); $NADP^+$, 100 μ*M*; EDTA, 1 m*M*; glucose-6-P dehydrogenase (yeast or *Leuconostoc mesenteroides*), 0.3 U/mL; and P-glucoisomerase from rabbit muscle, 0.2 U/mL (1 μg/mL).

Procedure

Step 1: To 100 μL of reagent 1, add 10 μL of tissue homogenate containing, for example, 5 μg brain or 50 μg of muscle, wet wt, or 10 μL of 1 m*M* glucose-6-P standard. Incubate at 25° for 60 min.

Step 2: Add 10 μL 1*M* HCl; 65°, 10 min.

Step 3: Add 1 mL of reagent 2; read after reaction is complete, 15–30 min.

Cycling Assay (1–10 pmol Product)

Reagents: Steps 1 and 2 are the same as for Fluorometric Indirect Assay, except that $NADP^+$ in reagent 2 is reduced to 10 µ*M*.

Procedure

Step 1: The freeze-dried samples (e.g., 10 ng brain, 50 ng liver, dry wt) are introduced through the oil in oil wells into 1 µL of reagent 1. Standards are 1 µL of 1–10 µ*M* glucose-6-P in reagent 1. Incubate 60 min at 20°.

Step 2: Add 1 µL of 0.1*M* HCl; 10 min at 80°.

Step 3: Add 5 µL of reagent 2; 20 min at room temperature.

Step 4: Add 2 µL of 0.25*M* Na_3PO_4; 20 min at 80°. If NaOH is used at this heating step to destroy the $NADP^+$, there may be serious loss of NADPH owing to the presence of glucose.

Step 5: Transfer 2 µL into a 10 × 75 mm tube containing 100 µL of NADP cycling reagent with enzyme levels to provide 5000-fold amplification in 1 h at 38°.

Step 6: Complete the assay as described in Chapter 5.

β-Hydroxy-acyl Coenzyme A Dehydrogenase (EC 1.1.1.35) *(3,14,16,22)*

$$\text{Acetoacetyl CoA} + H^+ + \text{NADH} \rightarrow \text{ß-hydroxyacetyl CoA} + NAD^+ \qquad \text{[Re. 7-24]}$$

The activity is measured by the rate of disappearance of NADH in direct assays or the appearance of NAD^+ in indirect procedures.

Sample tissue activities (mol/kg dry wt/h, 20°): rat kidney cortex, 12; human vastus muscle, 7.5; dog myocardium, 18; mouse brain, 2.5; human brain, 2.0; rabbit tibialis anterior muscle, 1.6; and rabbit soleus muscle, 43.

Spectrophotometer (2–10 nmol/min, 20–150 nmol of Product)

Reagent: Imidazole buffer, pH 6.1 (15 m*M* imidazole base, 135 m*M* imidazole HCl); NADH, 100 µ*M*; acetoacetyl coenzyme A, 200 µ*M*; EDTA, 1 m*M*; and bovine plasma albumin, 0.05%.

Conduct of the Assay

To 1 mL of reagent in a cuvet are added 5–10 µL of tissue homogenate (350 µg brain or muscle, 50 µg kidney cortex, wet wts). Readings are taken every 2 min. Blanks are tissue added to reagent without acetoacetyl coenzyme A. Calculations are made from the linear portion of the curve.

Fluorometric Indirect Assay (2–50 pmol/min, 0.1–3 nmol of Product)

Reagent: The reagent is the same as for Spectrophotometer.

Procedure

Step 1: To 100 µL of reagent in 10 × 75 mm tubes are added 5–10 µL of tissue homogenate containing an appropriate amount of tissue (0.5 µg of muscle, 2 µg of brain, 0.5 µg of kidney, wet wts). Standards are 10 µL of 50–300 µM NAD^+. Incubate 60 min at 25°.

Step 2: To each sample are added 10 µL of 0.5*M* HCl; room temperature 10 min.

Step 3: To each sample are added 1 mL of 6*M* NaOH containing 10 m*M* imidazole, and heat 20 min at 60°. The tubes are cooled, dried, and read in a fluorometer.

Cycling Assay (0.3–2 pmol/min, 20–100 pmol of Product)

Reagent: The reagent is the same as Spectrophotometer, except: NADH, 50 µ*M*; and acetoacetyl coenzyme A, 50 µ*M*.

Procedure

Step 1: Each freeze-dried sample is added to 10 µL of the reagent in oil wells. For tissue other than muscle, the samples (10–40 ng of brain) are introduced into the reagent droplet directly through the oil. Standards are 10 µL of 3–10 µ*M* NAD^+. Incubate 60 min at 20°.

Step 2: To each sample add 2 µL of 0.3*M* HCl; 10 min at room temperature.

Step 3: Transfer 2 µL into a 10 × 75 mm tube containing 100 µL of NAD cycling reagent with enzyme levels to provide 500-fold amplification in 1 h at 38°.

Step 4: Complete the indicator step as described in Chapter 5.

Special note: If the samples are freeze-dried muscle, the tissue (10–25 ng) is preincubated in oil wells in 1 µL of 50 m*M* imidazole buffer, pH 7.0,

0.6*M* KCl; and 0.05% bovine serum albumin, 60 min at 20°. This is followed by Step 1, above.

Hypoxanthine (Guanine) Phosphoribosyl Transferase (EC 2.4.2.8)

1) Guanine + P-ribosylpyrophosphate → 5'-GMP + PPi

2) 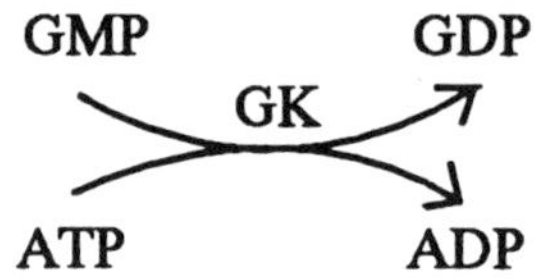

3) 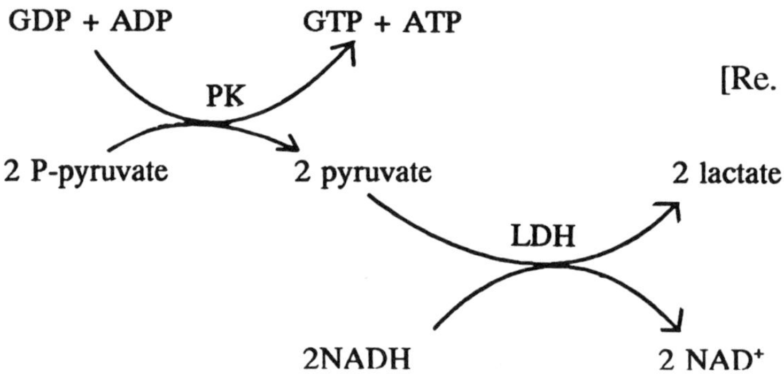

[Re. 7-25]

Auxiliary enzymes are guanylate kinase (GK), pyruvate kinase (PK), and lactate dehydrogenase (LDH). Two moles of NAD^+ are formed in the auxiliary enzyme sequence for each mol of GMP formed in the specific transferase reaction. NADH disappearance offers the most direct assay, but NAD^+ formation would be better suited for the most sensitive assays. In direct reading assays, as presented here, PK and LDH are allowed to react with preformed ADP, GDP, and pyruvate before adding GK.

Sample tissue activity (mol/kg dry wt/h, 20°): rabbit brain, 0.1.

Fluorometer Indirect Assay (0.015–0.15 nmol/min, 0.5–5 nmol of Product)

Reagent 1: Tris-HCl buffer, pH 7.8 (17 m*M* Tris base, 33 m*M* Tris-HCl); guanine HCl, 1 m*M*; phosphoribosylpylophosphate (PRPP), 0.25 m*M*; KCl, 50 m*M*; $MgCl_2$, 2 mm; EDTA, 0.1 m*M*; and dithiothreitol, 0.5 m*M*.

Reagent 2: Imidazole-HCl buffer, pH 7.0 (30 m*M* imidazole base, 20 m*M* imidazole-HCl); KCl, 20 m*M*; $MgCl_2$, 5 m*M*; P-pyruvate, 0.1 m*M*; ATP, 20 μ*M*; rabbit muscle pyruvate kinase, 0.8 U/mL (5 μg/mL); beef heart lactate dehydrogenase, 0.4 U/mL (2 μg/mL); and pork brain guanylate kinase, 0.05 U/mL (added after an initial reading on the fluorometer).

Procedure

Step 1: To 50 μL of reagent 1 in 10 × 75 mm tubes, add 10 μL of sample. For brain, samples should contain 25–250 μg of tissue. Standards are 10 μL of 0.05–0.5 m*M* 5'GMP. Incubate at 25° for 60 min; heat at 90–95° for 2 min.

Step 2: To each tube, add 1 mL of reagent 2 containing all the enzymes except guanylate kinase. Allow 10 min for PK and LDH to react, and then read on the fluorometer adjusted so that the high standard reading is full scale.

Step 3: Add guanylate kinase in a vol of 10 μL. The tubes are read again when the reaction is complete, about 10 min.

Isocitrate Dehydrogenase (NAD) (EC 1.1.1.41) *(36)*

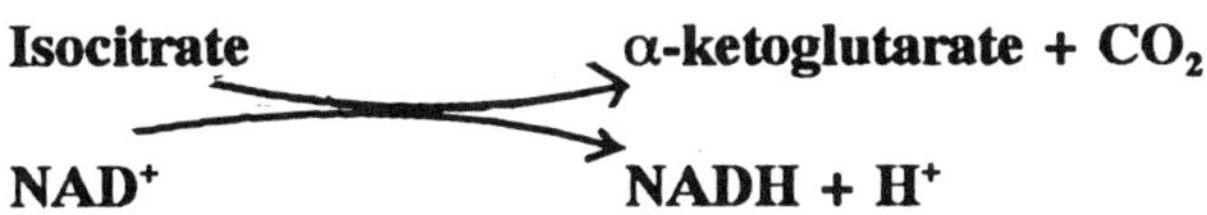

[Re. 7-26]

The activity is monitored by the rate of appearance of reduced NADH.

Sample tissue activities (mol/kg dry wt/h, 20°): kidney, 0.5; heart 3; and brain, 0.6 (David B. McDougal, Jr., personal communication).

Fluorometer (0.1–0.6 nmol/min, 1–10 nmol of Product)

Reagent: Phosphate buffer, pH 7.0 (30 m*M* K_2HPO_4, 20 m*M* KH_2PO_4); isocitrate, 2.5 m*M*; citrate, 18 m*M*; $MgCl_2$, 18 m*M*; ADP, 4.5 m*M*; NAD^+, 2 m*M*; bovine serum albumin, 0.05%; and 0.18% Triton X-100.

Conduct of the Assay

To one milliliter of reagent in 10 × 75 mm tubes are added tissue samples (e.g., 20 μg myocardium, 100 μg brain, wet wt). Readings are made every 1–5 min. Calculations are made from the linear portion of the curve.

Comment

NAD⁺-linked isocitrate dehydrogenase is a mitochondrial enzyme; the Triton X-100 is used to solubilize it. Citrate is added to ensure that the citrate-isocitrate ratio is near equilibrium. ADP is a positive effector of the enzyme and is added to yield maximal activity.

Isocitrate Dehydrogenase (NADP) (EC 1.1.1.42) *(23,37)*

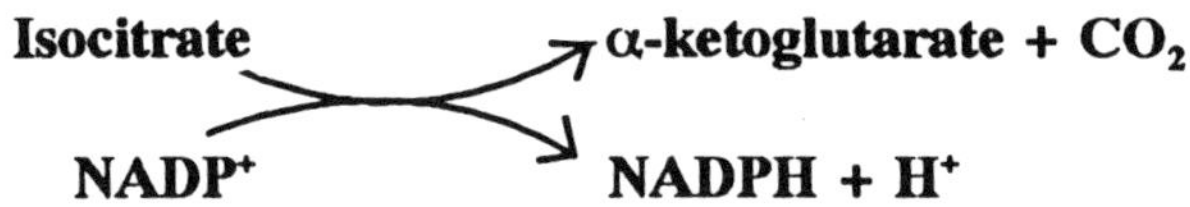

[Re. 7-27]

The activity is assessed by the rate of appearance of NADPH.

Sample tissue activities (mol/kg dry wt/h, 20°): mouse brain, 0.7; kidney, 12; and rat cerebellum, 0.6.

Fluorometer Indirect Assay (0.1–0.5 nmol/min, 5–20 nmol of Product)

Reagent: Tris-HCl buffer, pH 8.2 (60 m*M* Tris base, 40 m*M* Tris-HCl); NADP⁺, 1 m*M*; $MnCl_2$, 0.2 m*M*; isocitrate, 1 m*M*; and bovine plasma albumin, 0.05%.

Procedure

Step 1: To 50 µL of the reagent in 5 × 60 mm tubes are added appropriate tissue samples (4–10 µg of kidney, wet wt) in a vol of 2 µL. Standards are 2 µL of 1 and 5 µ*M* NADPH added to the reagent. Incubate 60 min at 38°. Chill to 0°.

Step 2: Transfer 20-µL aliquots to 1 mL of 0.05*M* K_2HPO_4 containing 1 m*M* EDTA to stop the reaction. Read on the fluorometer adjusted so that the high standard gives full-scale fluorescence.

Comment

Readings should be made within 30 min to be sure that no significant further reaction takes place.

Cycling Assay (0.03–0.15 pmol/min, 2–10 pmol Total Product)

Reagent: This is the same as for Fluorometer Indirect Assay, except NADP⁺ is 100 µ*M*.

Procedure

Step 1: Add freeze-dried samples (e.g., 5–20 ng of brain dry wt) at regular intervals to 2 µL of reagent in oil wells, and incubate 60 min at 20°.

Step 2: Add 5 µL of 0.5*M* NaOH at same time intervals as for the sample additions. Heat the oil well rack 20 min at 80°.

Step 3: Transfer 2 µL into a 10 × 75 mm tube containing 100 µL of NADP cycling reagent with enzyme levels to provide about 3000-fold amplification in 1 h at 38°.

Step 4: Complete the indicator step as described in Chapter 5.

3-Ketoacid CoA Transferase (EC 2.8.3.5) *(1,21,22)*

1) [Re. 7-28]

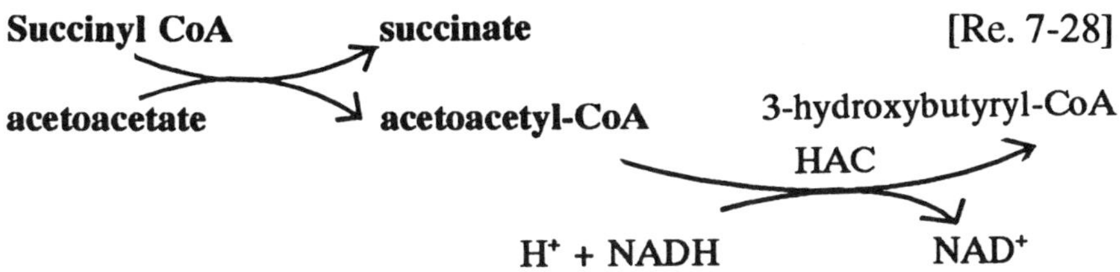

2)

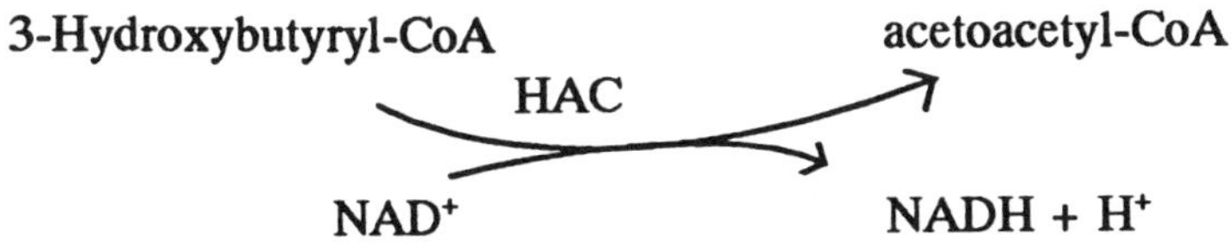

The auxiliary enzyme is 3-hydroxyacyl CoA dehydrogenase (HAC). The reaction is carried out in the direction of acetoacetyl-CoA formation, even though the rate is slower than in the opposite direction by a factor of about 5, in order to introduce a pyridine nucleotide step. The removal of the acetoacetyl-CoA avoids product inhibition. After excess NADH and potentially disturbing tissue enzymes have been destroyed with acid, the dehydrogenase reaction is reversed and the pH shifted to make the reaction more favorable. (Direct measurement of NAD after Step 1 would run the risk of including NAD^+ formed by nonspecific NADH oxidation by other tissue enzymes.) One of the final products is NADH, which offers the possibility of increasing sensitivity by amplification with enzymatic cycling.

Sample tissue activities (mol/kg dry wt/h, 20°): rat kidney cortex, 2; rabbit tibialis anterior muscle, 0.15; and rabbit soleus muscle, 0.45.

Fluorometer Indirect Assay (0.03–0.08 nmol/min, 2–10 nmol of Product)

Reagent: Tris-HCl buffer, pH 8.1 (12.5 m*M* Tris base, 12.5 m*M* Tris-HCl); succinyl-CoA, 0.5 m*M*; acetoacetate, 2 m*M*; NADH, 150 µ*M*; bovine serum albumin, 0.05%; and pig heart β-hydroxyacyl-CoA dehydrogenase, 1.5 U/mL (10 µg/mL). If the dehydrogenase is supplied as a suspension in $(NH_4)_2SO_4$, most of the salt is removed by centrifugation, the supernatant solution is removed, and the enzyme pellet dissolved in the original volume of reagent buffer.

Reagent 2: 2-Amino-2-methyl-1,3-propanediol buffer, pH 9.4 (40 m*M* propanediol base, 10 m*M* propanediol-HCl); NAD^+, 200 µ*M*; and β-hydroxyacyl-CoA dehydrogenase, 0.75 U/mL (5 µg/mL).

Procedure

Step 1: To 200 µL of reagent 1 in 10 × 75 mm tubes add tissue samples of an appropriate dilution to form 2–5 nmol of product in an hour (20–40 µg muscle, 5–25 µg kidney, wet wt), 60 min at 25°. Standards are 200 µL of 10–50 µ*M* acetoacetyl CoA.

Step 2: Add 15 µL of 1*M* HCl; 5 min at 95°.

Step 3: Add 1 mL of reagent 2. The tubes are read in the fluorometer when the reaction is complete, about 10 min at room temperature.

Cycling Assay (0.03–0.4 pmol/min, 2–25 pmol of Product)

Reagents 1 and 2: The same as for Fluorometer Indirect Assay.

Procedure

Step 1: Freeze-dried samples (e.g., 10–40 ng muscle, 10 ng kidney cortex, dry wt) are introduced through oil into oil wells containing 1 µL of reagent 1. Incubate 60 min at 20–25°.

Step 2: Add 1 µL 50 m*M* HCl; 80°, 20 min.

Step 3: Add 5 µL of reagent 2; 20 min at room temperature.

Step 4: Add 1 µL of 0.4*M* NaOH; 80°, 20 min.

Step 5: Transfer 2 µL into a 10 × 75 mm tube containing 100 µL of NAD cycling reagent with enzyme levels to provide 1500-fold amplification in 1 h at 38°.

Step 6: Complete the indicator step as described in Chapter 5.

α-Ketoglutarate Dehydrogenase (EC 1.2.4.2) *(36,38)*

[Re. 7-29]

Measurement of the α-ketoglutarate dehydrogenase portion of the multienzyme α-ketoglutarate dehydrogenase complex is made by the appearance of NADH.

Sample tissue activity: mouse brain, 0.4 mol/kg dry wt/h, 20°.

Fluorometer Direct Assay (0.02–1 nmol/min, 1.5–10 nmol Product)

Reagent: Phosphate buffer, pH 7.0 (30 m*M* K_2HPO_4, 20 m*M* NaH_2PO_4); α-ketoglutarate, 1 m*M*; coenzyme A, 50 μ*M*; pig heart diaphorase, 0.04 U/mL (10 μg/mL); EDTA, 1 m*M*; Triton X-100, 0.025% (v/v); β-mercaptoethanol, 2 m*M*; $MgCl_2$, 5 m*M*; NAD^+, 1 m*M*; and sodium amytal, 4 m*M*.

Conduct of the Assay

One milliliter of reagent is added to 10 × 75 mm tubes. A reading is taken with the fluorometer set so that full-scale deflection is equivalent to 5–10 μ*M*. Tissue samples (100–200 μg of brain, or 20 μg of liver mitochondrial protein) are added, and readings are made every 2 min. Calculations are made from the linear portion of the curve.

Comment

Diaphorase stimulated the α-ketoglutarate dehydrogenase activity approx 30%. Amytal is added to inhibit tissue NADH oxidase.

Lactate Dehydrogenase (EC 1.1.1.27) *(3,14,16,21)* (Method I)

[Re. 7-30]

Sample tissue activities (mol/kg dry wt/h, 20°): mouse brain, 30; human brain, 11, dog myocardium, 30; rabbit tibialis anterior muscle, 350; rabbit soleus muscle, 40; human vastus muscle, 42; and rat liver, 20.

Spectrophotometer (0.3–1.5 nmol/min, 20–100 nmol of Product)

Reagent: Imidazole buffer, pH 7 (50 m*M* imidazole base, 50 m*M* imidazole-HCl); 1 m*M* sodium pyruvate; 150 μ*M* NADH; and 0.05% bovine serum albumin.

Conduct of the Assay

To 1 mL of reagent in a cuvet are added 5 or 10 μL of tissue homogenate containing, for example, 50 μg of brain or liver, 20 μg of fast-twitch muscle (wet wts). Readings are taken at 2– 5-min intervals.

Fluorometric Indirect Assay
(2–80 pmol/min, 0.1–5 nmol of Product)

Step 1: To 100 µL of reagent in 10 × 75 mm tubes are added 5 or 10 µL of tissue homogenate containing, for example, 0.5 µg of muscle, and 1 µg of liver or brain. Standards are 10 µL of 10–500 µ*M* NADH. Incubate 60 min at 25°.

Step 2: To each sample, add 10 µL of 1*M* HCl; 10 min or more at room temperature.

Step 3: Add 1 mL of 6*M* NaOH containing 10 m*M* imidazole, and heat 20 min at 60°. The tubes are cooled exactly to room temperature and read in the fluorometer.

Comment

Lactate dehydrogenase, at least that present in muscle, is unstable at high dilution, especially in dilute buffer. It is recommended that the original homogenates be not more diluted than 1:100, and that any further dilution be made in 200 m*M* imidazole buffer containing 0.1% bovine plasma albumin at 0° and used within an hour.

Cycling Assay
(1–10 pmol/min, 50–500 pmol of Product)

Reagent: The reagent is the same as for Spectrophotometer.

Conduct of the Assay

Step 1: Each freeze-dried sample (e.g., 10 ng of brain, liver or kidney) is added to 5 µL of reagent in an oil well. Standards are 5 µL of reagent containing 20–100 µ*M* NAD^+. Incubate 60 min at 20°.

Step 2: To each sample, add 5 µL of 0.1*M* HCl; 10 min or more at room temperature.

Step 3: Transfer 1 µL into a 10 × 75 mm tube containing 100 µL of NAD cycling reagent with enzyme levels to give 200-fold amplification in 1 h at 38°.

Step 4: Complete the indicator step as described in Chapter 5.

Lactate Dehydrogenase (EC 1.1.1.27) *(3,16,18,22)* (Method II)

1) 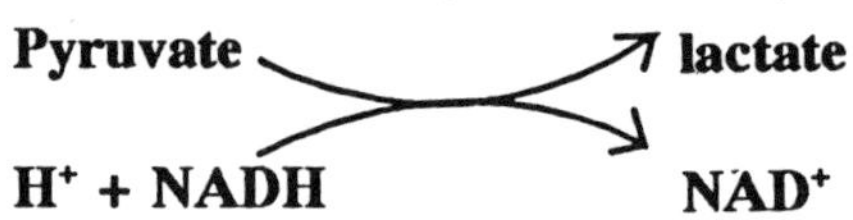

[Re. 7-31]

2) 

The auxiliary enzyme is glucose-6-P dehydrogenase (G6PDH) from *Leuconostoc mesenteroides*, which can use either NAD^+ or $NADP^+$. Step 2 is carried out after destroying NADH from Step 1 with HCl. The advantages of this two-step process are twofold. There is a large tolerance in the size of the NADH excess in Step 1, and the final fluorescence or absorption reading of the NADH increases with activity rather than decreases, as would be the case if Step 1 were followed directly in the fluorometer or spectrophotometer.

Sample tissue activities (mol/kg dry wt/h, 20°): mouse brain, 30; human brain, 11; rat brain barrels, 10; dog myocardium, 30; rabbit tibialis anterior muscle, 350; rabbit soleus muscle, 40; human vastus muscle, 42; and rat liver, 20.

Fluorometer Indirect Assay (0.05–0.2 nmol/min, 2–10 nmol of Product)

Reagent 1: Imidazole-HCl buffer, pH 7.0 (12 m*M* imidazole base, 8 m*M* imidazole HCl); pyruvate, 100 µ*M*; NADH, 50 µ*M*; and bovine serum albumin, 0.02%.

Procedure

Step 1: To 1 mL of reagent are added appropriate amounts of tissue samples (30–60 µg of muscle, 50–100 µg of brain, 10–20 µg of liver, wet wts); 60 min at 25°. Standards are 10 µL of 0.5 and 1 m*M* NAD^+.
Step 2: Add 15 µL of 2*M* HCl; 15 min at room temperature.
Step 3: Add 15 µL of 2*M* NaOH. Take a reading.
Step 4: Add 10 µL imidazole buffer, pH 7, containing 10 m*M* glucose-6-P and 25 U/mL (50 µg/mL) glucose-6-P dehydrogenase from *Leuconostoc mesenteroides*. Read when reaction is over, 5 min or less.

Malate Dehydrogenase (EC 1.1.1.37) (Method I)

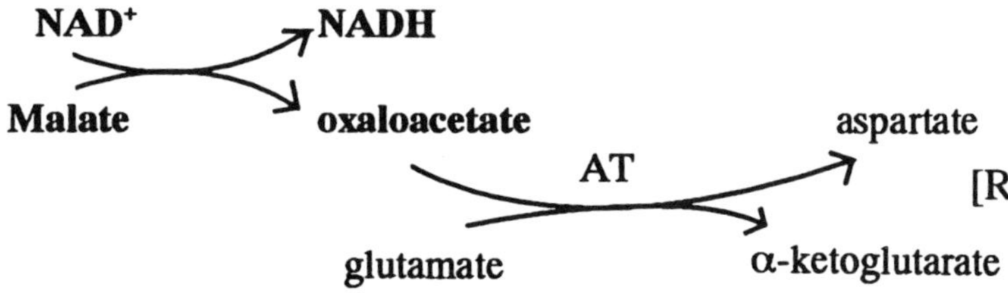

[Re. 7-32]

The auxiliary enzyme is aspartate aminotransferase (AT). Because of the instability of oxaloacetate, the reaction is measured in the direction of malate oxidation, even though the rate is slower than in the reverse direction by a factor of 5 or 6. At neutral pH, oxaloacetate breaks down to pyruvate at a rate of 5%/h, thus creating the likelihood that some portion of lactate dehydrogenase activity will be included in the assay. Nevertheless, a fluorometric assay is also included (Method II) using oxaloacetate and NADH as substrates, since the measurement can be accomplished very rapidly in the fluorometer, and the sensitivity is very high. Because of the unfavorable equilibrium in the direction of malate oxidation, the oxaloacetate formed in the other methods is removed with aspartate transaminase.

Sample tissue activities (mol/kg dry wt/h, 20°): (malate oxidation)—rat kidney cortex, 30; rabbit tibialis anterior muscle, 6; rabbit soleus muscle, 11; dog myocardium, 36; and rabbit brain, 10.

Spectrophotometer (1–5 nmol/min, 60–150 nmol of Product)

Reagent: 2-Amino-2-methyl-1,3-propanediol-HCl buffer, pH 9.1, 50 m*M* (B:A, 35:15); sodium malate, 10 m*M*; sodium glutamate, 10 m*M*; NAD^+, 1 m*M*; aspartate aminotransferase, 1 U/mL (5 µg/mL); and bovine serum albumin, 0.05%.

Conduct of the Assay

Step 1: To 1 mL of reagent in a cuvet add appropriate dilutions of tissue homogenate (25–75 µg of kidney, 100–250 µg of brain, wet wts). Readings are taken every 2 min. Calculations are made from the linear portion of the curve. Blanks are comprised of tissue added to reagent without malate.

Fluorometer (0.1–1 nmol/min, 2–10 nmol of Product)

Reagent: The reagent is the same as Spectrophotometer, except NAD^+, 200 µ*M*.

Conduct of the Assay

To 1 mL of reagent is added the appropriate amount of tissue sample (1–5 µg of kidney, 3–15 µg of brain, 2–10 µg of slow-twitch muscle, wet wts). Blanks are tissue added to reagent without malate. Standards are 5 or 10 µ*M* NADH added to the reagent. Readings are made every 1–3 min at a fluorometer setting to give full-scale deflection with the chosen standard. Calculations are made from the linear portion of the curve.

Fluorometer Indirect Assay (3–30 pmol/min, 0.2–2 nmol of Product)

Reagent: The reagent is the same as for Spectrophotometer.

Procedure

Step 1: To 10 µL of reagent in oil wells are added freeze-dried samples (20–200 ng dry wt of brain). Standards are 10 µL of 10–100 µ*M* NADH. Incubate 1 h at 20°.

Step 2: To each sample are added 3 µL of 0.25*M* Na_3PO_4, 20 min at 80°.

Step 3: Transfer 10 µL to 1 mL 6*M* NaOH containing 10 m*M* imidazole and 3 m*M* H_2O_2; 15 min at 60°. Cool, dry, and read in the fluorometer.

Cycling Assay (1–5 pmol/min, 50–250 pmol of Product)

Reagent: The reagent is the same as Spectrophotometer, except: NAD^+, 300 µ*M*.

Preincubation reagent (for freeze-dried muscle): The same buffer as the reagent for spectrophotometer; KCl, 0.6*M*; and bovine plasma albumin, 0.05%.

Procedure

Step 1: Freeze-dried samples (10–20 ng, dry wt) are added through the oil into 1 µL of preincubation reagent in oil wells; 30 min at 20°.

Step 2: Add 10 µL of reagent to each sample; 60 min at 20°. Standards are 10 µL of 5–10 µ*M* NADH.

Step 3: Add 2 µL of 0.2*M* NaOH to each sample; 20 min at 80°.

Step 4: Transfer 1 µL into a 10 × 75 mm tube containing 100 µL of *NADP* cycling reagent with 0.7 U/mL (20 µg/mL) glutamate dehydrogenase and 0.6 U/mL (2 µg/mL) glucose-6-P dehydrogenase from *Leuconostoc mesenteroides* to give about 500-fold amplification in 1 h at 38°C.

Step 5: Complete the assay with the 6-P-gluconate dehydrogenase indicator step according to Chapter 5.

Comment

The standard NAD cycle of Chapter 5 depends on final measurement of malate in the indicator step; the malate in the specific reagent would consequently give a high blank. Therefore, the NADH generated here is measured with the standard *NADP* cycle, which as explained in Chapter 5, because it contains *Leuconostoc* glucose-6-P dehydrogenase and reacts with the NAD at a slower, but still sufficiently rapid rate for the present purpose.

Malate Dehydrogenase (EC 1.1.1.37) (Method II)

1)
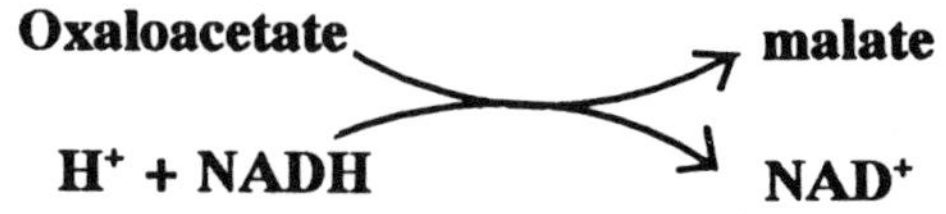

2)

[Re. 7-33]

The auxiliary enzyme is glucose-6-P dehydrogenase (G6PDH) from *Leuconostoc mesenteroides,* which can use either NAD^+ or $NADP^+$. Step 2 is carried out after destroying NADH from Step 1 with HCl. The advantages of this two-step process are twofold: There is a large tolerance in the size of the NADH excess in Step 1, and the final fluorescence or absorption reading of the NADH increases with activity rather than decreases as would be the case if Step 1 were followed directly in the fluorometer or spectrophotometer.

Sample tissue activities (mol/kg dry wt/h, 20°): (oxaloacetate reduction)—rat kidney cortex, 150; rabbit soleus muscle, 40; dog myocardium, 150; and brain, 40.

Fluorometer Indirect Assay (0.05–0.2 mmol/min, 2–10 nmol of Product)

Reagent: Tris buffer, pH 8.1 (50 m*M* Tris-base, 50 m*M* Tris-HCl); NADH, 50 μ*M*; oxaloacetate, 50 μ*M* (prepared immediately before use from 1*M* stock solution in 0.5*M* HCl); and bovine serum albumin, 0.05%.

Procedure

Step 1: To 1 mL of reagent in a 10 × 75 mm tube, add 10–20 μL of tissue homogenate samples equivalent (for example) to 1 μg of brain or slow-twitch muscle, or 4 μg of kidney cortex or myocardium. Incubate 60 min at 25°.

Step 2: Add 15 μL 2*M* HCl; 15 min at room temperature.

Step 3: Add 15 μL 2*M* NaOH; take a reading.

Step 4: Add 10 μL of pH 8 buffer containing 10 m*M* glucose-6-P and 25 U/

mL (50 μg/mL) glucose-6-P dehydrogenase from *Leuconostoc mesenteroides*. Read when reaction is over, 5 min or less.

Malate Dehydrogenase *(22)* (Decarboxylating) (EC 1.1.1.40)

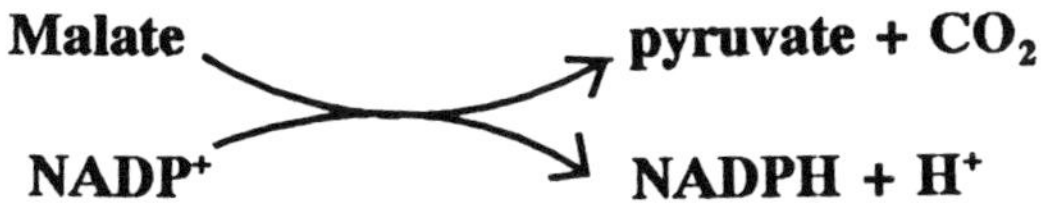

[Re. 7-34]

The activity is measured by the rate of formation of NADPH.

Sample tissue activities (mol/kg dry wt/h, 20°): rat kidney cortex, 0.25; mouse brain, 0.35; and rat muscle, 0.1.

Fluorometer Direct Assay (0.1–0.5 mmol/min, 5–10 nmol of Product)

Reagent: Tris buffer, pH 7.8 (20 m*M* Tris base, 40 m*M* Tris-HCl); malate, 5 m*M*; $MnCl_2$, 50 μ*M*; $NADP^+$, 50 μ*M*; dithiothreitol, 1 m*M*; and bovine plasma albumin, 0.02%.

Conduct of the Assay

To 1 mL of reagent in 10 × 75 mm tubes is added the appropriate amount of tissue (0.1–0.5 mg of kidney, brain, or muscle, wet wt). Readings are made at 2-min intervals. Blanks are comprised of tissue added to reagent without malate. Standards are 5–10 μ*M* NADPH added to reagent.

Cycling Assay (30 fmol/min, 2 pmol of Product)

Reagent: The same as for Fluorometer Direct Assay.

Procedure

Step 1: The freeze-dried sample (10–20 ng of kidney, brain, or muscle, dry wt) is added to 5 μL of reagent in oil wells; 60 min at 20°. Standards are 5 μL of 0.5 μ*M* NADPH.

Step 2: To each sample are added 5 μL of 0.2*M* NaOH at the same rate at which the tissue was added. Heat 20 min at 80°.

Step 3: Transfer 5 μL into a 10 × 75 mm tube containing 10 μL of NADP cycling reagent with enzyme levels to provide 10,000-fold amplification in 1 h at 38°.

Step 4: Complete the indicator step as described in Chapter 5.

Nucleoside Diphosphokinase (EC 2.7.4.6) *(33,34,39,40)*

1) **GDP + ATP → GTP + ADP**

[Re. 7-35]

2)

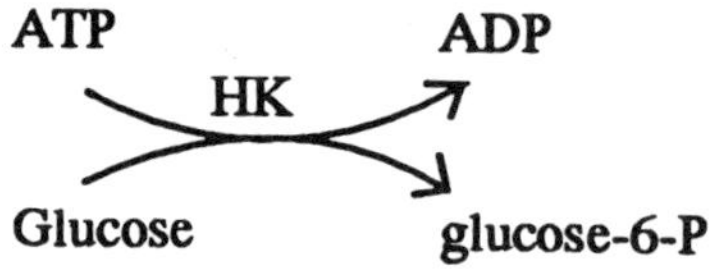

3)

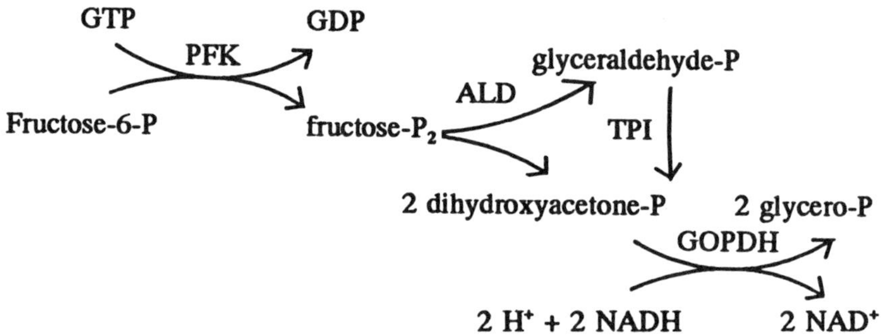

After the specific reaction, excess ATP is removed with hexokinase (HK), and GTP is measured by the NAD^+ produced in the above enzyme sequence involving phosphofructokinase (PFK), aldolase (ALD), triose-P isomerase (TPI), and glycero-P dehydrogenase (GOPDH). Note the yield of 2 mol of NAD^+/mol of GTP.

Blanks and GTP standards (prepared in reagent 1) are carried through the entire procedure. The hexokinase in reagent 2 is kept to a minimum, because hexokinase also reacts with GTP (at about 0.25% of the rate with ATP). For the same reason, the incubation time with reagent 2 should be approximately the same for all samples and standards. The glycero-P dehydrogenase in reagent 3 is kept to a minimum, because commercial preparations have been contaminated with adenylokinase, which produces a progressive blank from the ADP formed from ATP by hexokinase.

It would have been a simpler assay if the kinase could have been measured in the opposite direction with determination of the ATP produced with hexokinase, glucose-6-P dehydrogenase, and $NADP^+$. Attempts to do this failed, because the necessary high ratio of GTP (substrate) to ATP (product) gave an intolerable blank owing to the nonspecificity of hexokinase mentioned above.

Nucleoside diphosphokinase has unusual kinetics in that it is inhibited by one substrate, GDP, and this is overcome by the other, ATP *(39,40)*. The reagent composition chosen is a compromise. Activity would have been somewhat higher with lower GDP and higher ATP, but the first would restrict the linear range of assay, and the second would increase the hexokinase requirement *(see above)* and the danger of a higher blank from adenylokinase contamination. (This blank increases as the square of the ADP concentration.)

Sample tissue activities (mol/kg dry wt/h, 20°): rabbit brain, 4, and rat kidney cortex, 35.

Fluorometric Indirect Assay (2–200 pmol/min/Sample, 0.1–10 nmol of Product)

Reagent 1: Imidazole-HCl buffer, pH 7.4 (40 m*M* imidazole base, 10 m*M* imidazole-HCl); GDP, 200 μ*M*; ATP, 1 m*M*; $MgCl_2$, 4 m*M*, bovine serum albumin, 0.05%; and mercaptoethanol, 5 m*M*.

Reagent 2: Imidazole-HCl buffer, pH 7.4 (40 m*M* imidazole base, 10 m*M* imidazole-HCl); glucose, 1 m*M*; $MgCl_2$, 1 m*M*; yeast hexokinase, 0.14 U/mL (1 μg/mL); and 0.02% bovine serum albumin.

Reagent 3: Imidazole-HCl, pH 7.4 (80 m*M* imidazole base, 20 m*M* imidazole-HCl); fructose-6-P, 100 μ*M*; $MgCl_2$, 4 m*M*; EDTA, 0.6 m*M*; K_2HPO_4, 20 m*M*; NADH, 50 μ*M*; P-fructokinase, 0.6 U/mL (10 μg/mL); aldolase, 0.018 U/mL (2 μg/mL); triose-P isomerase, 20 U/mL (4 μg/mL); and glycero-P dehydrogenase, 0.07 U/mL (0.4 μg/mL) (all four enzymes from rabbit muscle).

Procedure

Step 1: Add freeze-dried samples (0.05–2 μg brain, 10–200 ng kidney cortex, dry wt) through oil to 5 μL of reagent 1 in oil wells; 60 min at 20°.

Step 2: Add 2 μL of 0.1*M* NaOH to each sample; 10 min at 25°.

Step 3: Add 5 μL of the sample to 50 μL of reagent 2 in a 10 × 75 mm tube; 30 min at 25°.

Step 4: To each sample add 5 μL of reagent 3; 30 min at 25°.

Step 5: Add 10 μL of 1*M* HCl to destroy excess NADH.

Step 6: Add 1 mL of 6*M* NaOH containing 10 m*M* imidazole, with immediate mixing. Heat 10 min at 60°, cool to room temperature, and read on fluorometer.

Phosphoenolpyruvate Carboxykinase (EC 4.1.1.32) *(20)*

1) Oxaloacetate + ITP → P-pyruvate + IDP + CO_2

[Re. 7-36]

2) ADP + P-pyruvate $\xrightarrow{PK}$ ATP + pyruvate

pyruvate + H^+ + NADH $\xrightarrow{LDH}$ lactate + NAD^+

The auxiliary enzymes are pyruvate kinase (PK) and lactate dehydrogenase (LDH). The rate of disappearance of NADH or appearance of NAD^+ is the measure of activity.

Sample tissue activity (mol/kg dry wt/h, 20°): kidney, 1.1.

Fluorometer Indirect Assay (0.2–0.4 nmol/min, 2–8 nmol of Product)

Reagent 1: Imidazole sulfate buffer, pH 7.2 (35 m*M* imidazole base, 15 m*M* imidazole sulfate); $MgSO_4$, 10 m*M*; $MnSO_4$, 0.1 m*M*; EGTA, 0.1 m*M*; β-mercaptoethanol, 1 m*M*; ITP, 5 m*M*; oxaloacetate, 2.5 m*M*; and bovine plasma albumin, 0.05%.

Reagent 2: Imidazole buffer, pH 7.2 (35 m*M* imidazole base, 15 m*M* imidazole HCl); ADP, 0.1 m*M*; KCl, 50 m*M*; EDTA, 1 m*M*; NADH, 10 μ*M*; beef heart lactate dehydrogenase, 0.5 U/mL (2 μg/mL); rabbit muscle pyruvate kinase, 0.2 U/mL (1 μg/mL); and 0.02% bovine plasma albumin.

Procedure

Step 1: To 50 μL of reagent 1 in 10 × 75 mm tubes is added an appropriate amount of tissue sample (1–5 μg of kidney cortex, wet wt). Incubate 60 min at 25°; 2 min at 100°. Blanks are tissue added to reagent without substrate (ITP or oxaloacetate omitted).

Step 2: To each tube is added 1 μL of 15% H_2O_2 (to destroy keto acids); 2 min at 100°.

Step 3: Add 2 μL of beef liver catalase, 1500 U/mL (25 μg/mL); 10 min 25°.

Step 4: Add 1 mL of reagent 2 (without the lactate dehydrogenase); and read in the fluorometer with setting to give full scale.

Step 5: Add the lactate dehydrogenase in 10 μL or less, and read in 5 or 10 min.

Comments

Imidazole sulfate buffer is used to avoid inhibition by the chloride ion. Peroxide is used to destroy keto acids that react with the lactate dehydrogenase, causing troublesome blanks. Low levels of Mn^{2+} and EGTA increase activity twofold.

Cycling Method (0.1–1 pmol/min, 5–50 pmol of Product)

Reagent 1: The reagent is the same as for Fluorometer Indirect Assay.

Reagent 2: Imidazole HCl buffer, pH 6.8 (100 m*M* imidazole base, 100 m*M* imidazole HCl); KCl, 150 m*M*; ADP, 0.3 m*M*; $MgCl_2$, 2 m*M*; EGTA, 1 m*M*; ascorbic acid, 4 m*M*; bovine plasma albumin, 0.02%; and beef liver catalase 120 U/mL (2 µg/mL).

Reagent 3: Imidazole buffer, pH 7 (12 m*M* imidazole base, 8 m*M* imidazole HCl); beef heart lactate dehydrogenase, 50 U/mL; pyruvate kinase, 3 U/mL; and NADH, 60 µ*M*. The enzymes are centrifuged and the supernatant ammonium sulfate replaced with buffer before use.

Conduct of the Assay

Step 1: Freeze-dried samples are introduced through oil into oil wells containing 5 µL of reagent 1. Standards are 5 µL of 1–5 µ*M* P-pyruvate. Incubate 1 h at 20°.

Step 2: Add 1 µL of 0.25*M* NaOH containing 0.1% H_2O_2, 30 min at 80°. Neutralize with 1 µL of 0.25*M* HCl.

Step 3: Add 5 µL of reagent 2; 30 min at room temperature.

Step 4: Add 1 µL of reagent 3; 20 min at room temperature.

Step 5: Add 1 µL of 1*M* HCl; 10 min at room temperature.

Step 6: Transfer 1–5 µL (depending on expected activity) into a 10 × 75 mm tube containing NAD cycling reagent with enzyme levels to provide 2000-fold amplification in 1 h at 38°.

Step 7: Complete the indicator step as described in Chapter 5.

Comment

The ascorbic acid minimizes the NADH oxidation to NAD^+, which can occur during acidification. Tissue sample size may need to be varied according to activity, or conversely, the cycling rate may be adjusted. The temperature coefficient is unusually high, about 13%/degree C, for the kidney enzyme.

Phosphofructokinase (EC 2.7.1.11) *(3,5,14,16,41)*

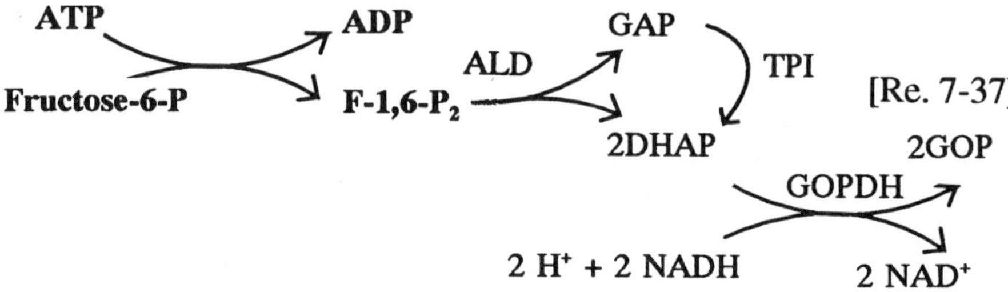

[Re. 7-37]

Abbreviations are F-1,6-P_2, fructose-1,6-bisphosphate; GAP, glyceraldehyde-P; DHAP, dihydroxyacetone-P; GOP, glycero-P; ALD, aldolase; TPI, triose-P isomerase; and GOPDH, glycero-P dehydrogenase.

Sample tissue activities (mol/kg dry wt/h, 20°): rabbit kidney cortex, 0.11; rat muscle, 0.8–4; human vastus muscle, 11; dog myocardium, 5.5; human brain, 2.8; mouse brain, 2.2; rabbit tibialis anterior muscle, 25; rabbit soleus muscle, 2.5; and rat liver, 0.6, all dry wts.

Spectrophotometer (1–5 nmol/min, 60–200 nmol Product)

Reagent: Tris-HCl buffer, pH 8.1 (25 m*M* Tris base, 25 m*M* Tris-HCl); ATP, 1 m*M*; fructose-6-P, 1 m*M*; K_2HPO_4, 10 m*M*; $MgCl_2$, 2 m*M*; 5'AMP, 1 m*M*; β-mercaptoethanol, 1 m*M*; NADH, 0.1–0.15 m*M*; bovine serum albumin, 0.05%; rabbit muscle aldolase, 0.10 U/mL (12 µg/mL); rabbit muscle triose P-isomerase, 15 U/mL (3 µg/mL); and rabbit muscle α-glycero-P dehydrogenase, 1.7 U/mL (10 µg/mL).

Conduct of Assay

The reagent is added to cuvets in a vol of 1 mL, and a reading made at 340 nm. Tissue blanks are run in reagent without substrate (fructose-6-P). The appropriate amount of tissue is added (e.g., 200–500 µg brain, 20–100 µg skeletal muscle, wet wts), and readings are taken every 1–3 min. The rate is calculated from the linear portion of the curve. If significant nonspecific oxidation of NADH is observed, the addition of 2 m*M* sodium amytal will reduce this activity to a negligible level. High levels of ATP are inhibitory to P-fructokinase; the additions of K_2HPO_4 and 5'AMP overcome the inhibition.

Fluorometer Direct Assay (0.05–1 nmol/min, 1–10 nmol of Product)

Reagent: The reagent is the same as for Spectrophotometer, except that NADH is reduced to 5–10 µM.

Conduct of the Assay

One milliliter of reagent is added to 10 × 75 mm tubes. Tissue samples are added (20–50 µg of brain, 5–10 µg of skeletal muscle, 100–250 µg of liver, wet wt), mixed, and readings taken every 2 min. Blanks are as for Spectrophotometer. Standards are 10 µL of 0.25 and 0.5 m*M* fructose-1,6-bisphosphate. The rate is calculated from the linear portion of the curve.

Fluorometer Indirect Assay (2–100 pmol/min, 0.1–5 nmol of Product)

Reagent: The reagent is the same as Spectrophotometer, except: NADH, 200 µ*M*.

Procedure

Step 1: Reagent, 100 µL are added to 10 × 75 mm tubes at 0°.
Step 2: Tissue homogenate is added in a vol of 2 µL containing, for example, 1 µg of muscle, 2 µg of brain, and 10 µg of liver, wet wts. Standards are 2 µL of 10 and 25 µ*M* fructose-1,6-bisphosphate.
Step 3: Incubate at 25°, 60 min.
Step 4: Add 10 µL of 0.75*M* HCl; room temperature for 10 min.
Step 5: Add 1 mL of 6*M* NaOH containing 10 m*M* imidazole; 20 min at 60°.
Step 6: Cool and wipe tubes; and read in fluorometer.

Phosphoglucoisomerase (EC 5.3.1.9) *(5,25,42,43)*

[Re. 7-38]

The auxiliary enzyme is glucose-6-P dehydrogenase (G6PDH). Although the reaction is in the reverse (nonphysiological) direction, the V_{max} under the analytical conditions is about the same as in the forward direction.

Sample tissue activities (mol/kg dry wt/h, 20°): mouse brain, 8; rabbit muscle, 50; and rat liver, 15.

Spectrophotometer (2–10 nmol/min, 25–150 nmol of Product)

Reagent: Tris-HCl buffer, pH 8 (25 m*M* Tris base, 25 m*M* Tris-HCl); fructose-6-P, 0.5 m*M*; $NADP^+$, 1 m*M*; and yeast or *Leuconostoc* glucose-6-P dehydrogenase, 0.1 U/mL.

Procedure

Reagent is added in a vol of 1 mL to cuvets, and a reading made. Appropriate amounts of tissue are added (e.g., 100 µg brain, 20 µg muscle, wet wt). Blanks consist of tissue added to reagent without fructose-6-P. Readings are taken at 2-min intervals, and calculations made from the linear portion of the curve.

Fluorometer Direct Assay (0.2–1 nmol/min, 2–10 nmol of Product)

Reagent: The reagent is the same as for Spectrophotometer, except $NADP^+$, 100 µ*M*.

Conduct of the Assay

Reagent is added in a vol of 1 mL to 10 × 75 mm tubes. A reading is taken with full-scale deflection equivalent to 5–10 µ*M*. Standards are 5 and 10 µL of 1 m*M* glucose-6-P.

Tissue homogenate is added equivalent to 5 µg brain or liver wet wt, and readings are made at 1–5-min intervals. Calculations are made from the linear portion of the curve.

Fluorometer Indirect Assay (2–10 pmol/min, 100–500 pmol of Product)

Reagent: The same as for Spectrophotometer.

Reagent 2: 6*M* NaOH containing 10 m*M* imidazole and 3 m*M* H_2O_2.

Step 1: Samples of freeze-dried tissues (5–25 ng dry wt) are introduced into 5 µL of reagent 1 in oil wells; 60 min at 20°.

Step 2: To each sample is added 1 µL of 0.125*M* Na_3PO_4; 20 min at 80°.

Step 3: Transfer 5 µL into 1 mL of reagent 2; 15 min at 60°, cool, and read.

Phosphoglucomutase (EC 2.7.5.1) *(1,30,31)*

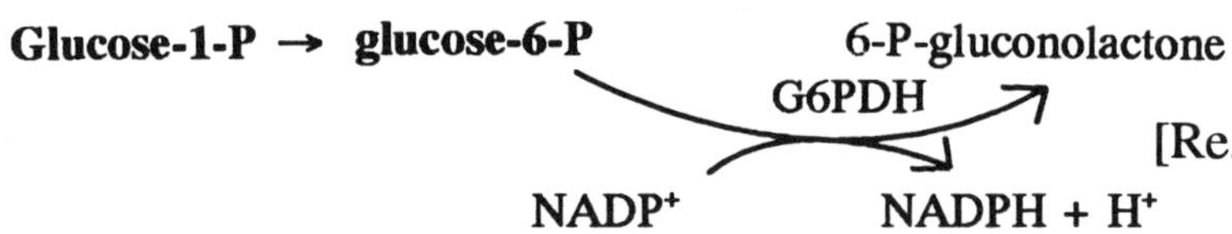

[Re. 7-39]

The auxiliary enzyme is glucose-6-P dehydrogenase (G6PDH). The rate of appearance of NADPH can be measured directly, enhanced by strong alkali or amplified by cycling (not illustrated).

Sample tissue activities (mol/kg dry wt/h, 20°): dog myocardium, 6; rabbit tibialis anterior muscle, 70; rabbit soleus muscle, 10; and mouse brain, 5.

Spectrophotometer (3–10 nmol/min, 60–150 nmol of Product)

Reagent: Imidazole buffer, pH 6.9 (25 m*M* imidazole, 25 m*M* imidazole-HCl); glucose-1-P, 1 m*M*; glucose-1,6-P_2, 2 μ*M*; $MgCl_2$, 1 m*M*; $NADP^+$, 0.5 m*M*; dithiothreitol, 0.5 m*M*; EGTA, 0.5 m*M*; bovine plasma albumin, 0.02%; and glucose-6-P dehydrogenase from *Leuconostoc mesenteroides,* 1 U/mL (2 μg/mL).

Conduct of the Assay

Samples containing, for example, 20–100 μg of skeletal muscle or 100–300 μg cardiac muscle, wet wt, are added to cuvets with 1 mL of reagent. Tissue samples added to reagent without glucose-1-P serve as blanks. Readings are made at 2-min intervals, and calculations are based on the linear portion of the curve.

Fluorometer Direct Assay (0.1-1 nmol/min, 2–10 nmol of Product)

Reagent: The reagent is the same as for Spectrophotometer, except: $NADP^+$, 50 μ*M*.

Conduct of the Assay

One milliliter of reagent is added to 10 × 75 mm tubes. After an initial reading, an appropriate amount of tissue is added (1–15 μg of skeletal muscle, 5–30 μg of cardiac muscle or brain, wet wt). Blanks are tissue added to reagent without glucose-1-P. Standards are 5 and 10 μL of 1 m*M* glucose-6-P. Readings are made every 2 min at a fluorometer setting so that 10 μ*M* NADPH give full-scale reading. The activity is calculated from the linear portion of the curve.

Fluorometer Indirect Assay (0.05–0.15 nmol/min, 3–10 nmol of Product)

Reagent 1: The reagent is the same as Spectrophotometer.

Reagent 2: 2-Amino-2-methyl-1,3-propanediol buffer, pH 9.1 (18 m*M* 2-amino-2-methyl-1,3-propanediol base, 7 m*M* 2-amino-2-methyl-1,3-propanediol HCl), and EDTA, 2 m*M*.

Conduct of the Assay

Step 1: To 100 µL of reagent 1 in 10 × 75 mm tubes are added 1–3 µL of homogenate containing 2–10 µg of heart or brain, wet wt. Standards are 100 µL of 30–100 µ*M* glucose-6-P in reagent 1.
Step 2: After incubation for 60 min at 25°, to each tube is added 1 mL of reagent 2. The fluorescence is read 5–60 min after this addition.

Comment

Step 2 almost completely stops the reaction by the tenfold dilution of substrate and tissue; the pH shifts away from the optimum and the high affinity of EDTA for Mg^{2+}. Nevertheless, readings should not be unduly delayed, because there is a slight upward drift of fluorescence (about 0.5%/h).

Fluorometer Indirect Assay (3–30 pmol/min, 0.2–2 nmol of Product)

Reagent: The reagent is the same as for Spectrophotometer, except glucose-1-P, 0.5 m*M*; $NADP^+$, 100 µ*M*; and bovine plasma albumin, 0.05%.
Preincubation reagent: Imidazole buffer 6.9 (25 m*M* imidazole base, 25 m*M* imidazole HCl); KCl, 0.6*M*; and bovine plasma albumin, 0.05%.

Conduct of the Assay

The preincubation step is necessary for frozen-dried muscle samples only.

Step 1: The freeze-dried sample (25–250 ng cardiac muscle, dry wt) is added to 1 µL of preincubation reagent under oil in an oil well; 30 min at 20°.
Step 2: To each sample is added 10 µL of reagent. Standards are 10 µL of 100–300 µ*M* glucose-6-P in the reagent.
Step 3: To each sample are added 2 µL of 0.25*M* NaOH; 80° for 30 min.
Step 4: A 10-µL aliquot is transferred into 1 mL of 6*M* NaOH containing 10 m*M* imidazole and 3 m*M* H_2O_2; 20 min at 60°. The tubes are wiped dry, cooled to room temperature, and read on the fluorometer.

6-Phosphogluconate Dehydrogenase (EC 1.1.1.43) *(5,16,37)*

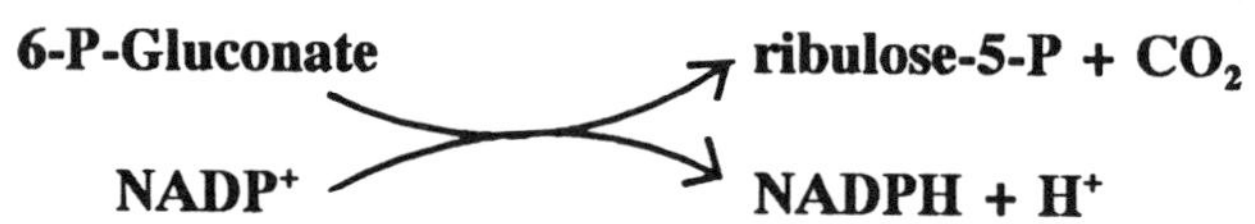

[R. 7-40]

The activity is measured by the rate of appearance of NADPH.

Sample tissue activities (mol/kg dry wt/h, 20°): human brain, 0.27; mouse brain, 0.25; and liver, 1.3.

Spectrophotometer (2–10 nmol/min, 50–150 nmol of Product)

Reagent: Tris-acetate buffer, pH 8.1 (50 m*M* Tris base, 50 m*M* tris acetate); $MgCl_2$, 5 m*M*; EDTA, 0.1 m*M*; $NADP^+$ 500 μ*M*; ammonium acetate, 30 m*M*; dithiothreitol, 5 m*M*; and bovine plasma albumin, 0.05%.

Conduct of the Assay

To 1 mL of reagent in a cuvet is added the appropriate amount of tissue (3 mg brain, 1 mg of liver, wet wt). Blanks are reagent without substrate (6-P-gluconate), but with tissue added. Readings are made at 2-min intervals. Calculations are based on the linear portion of the curve.

Fluorometer Direct Assay (0.05–1 nmol/min, 1–10 nmol of Product)

Reagents: The reagent is the same as for Spectrophotometer, except: $NADP^+$, 50 μ*M*.

Conduct of the Assay

To 1 mL of reagent in a 10 × 75 mm tube is added the appropriate amount of tissue (0.1–0.5 mg brain, 10–100 μg of liver, wet wt). Blanks are tissue added to reagent without substrate. Readings are made at 2-min intervals. Calculations are made from the linear portion of the curve.

Cycling Assay (0.2–1 pmol/min, 10–50 pmol of Product)

Reagent: The reagent is the same as for Fluorometer Direct Assay.

Procedure

Step 1: To 5 μL of reagent under oil is added each freeze-dried sample (e.g., 50–200 ng brain, 10–50 ng liver, dry wt), with incubation for 60 min at 20°. Standards are 5 μL of 2–10 μ*M* NADPH in complete reagent.

Step 2: Add 5 μL of 0.1*M* NaOH; 20 min at 80°.

Step 3: Transfer 5 μL into a 10 × 75 mm tube containing 100 μL of NADP cycling reagent with enzyme levels to provide 500-fold amplification in 1 h at 38°.

Step 4: Complete the indicator step as described in Chapter 5.

3-Phosphoglycerate Kinase (EC 2.7.2.3) *(44,45)*

The enzyme is measured in the direction of 1,3-diphosphoglycerate (1,3-DPGA) *dephosphorylation.* This very unstable substrate is generated in the reagent as shown:

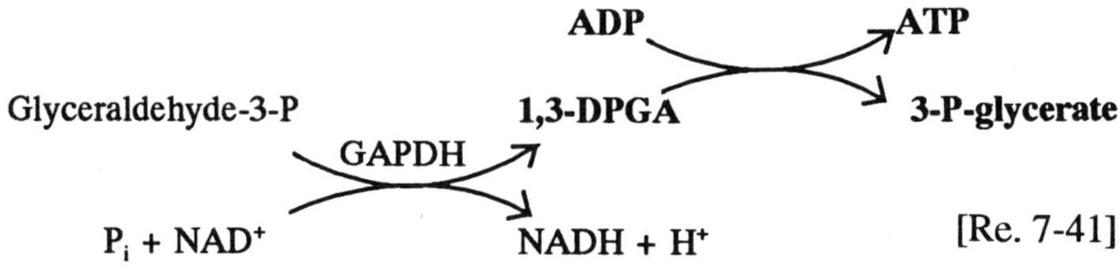

[Re. 7-41]

The auxiliary enzyme is glyceraldehyde-P dehydrogenase (GAPDH).

Because of the position of the GAPDH equilibrium, only a low level of NADH is formed until the 1,3-DPGA is removed by the kinase. Consequently, the rate of NADH formation is a measure of the kinase activity. In indirect assays, the low initial NADH level is included in the reagent blank.

Sample tissue activity (mol/kg dry wt/h, 20°): mouse brain, 20.

Spectrophotometer Method (5–10 nmol/min/Sample, 100–200 nmol of Product)

Reagent: Imidazole buffer, pH 7.2 (35 m*M* imidazole base, 15 m*M* imidazole-HCl); $MgCl_2$, 2 m*M*; K_2HPO_4, 5 m*M*; bovine plasma albumin, 0.02%; mercaptoethanol, 5 m*M*; NAD^+, 0.5 m*M*; glyceraldehyde-3-P, 0.5 m*M*; ADP, 1 m*M*; and yeast glyceraldehyde-3-P dehydrogenase, 0.2 U/mL (2 µg/mL).

Conduct of the Assay

One milliliter of reagent without enzyme is added to cuvets and a reading taken. Glyceraldehyde-3-P dehydrogenase is added, and another reading is made to determine the level of NADH and, thus, of the 1,3-diphosphoglycerate formed. Tissue samples (e.g., 5–10 µg of brain) are added, and readings taken at 2-min intervals. The activity is calculated from the linear portion of the curve. Blanks are tissue samples added to reagent without ADP.

Fluorometer (0.05–1 nmol/min/Sample, 5–10 nmol of Product)

Reagent: The reagent is the same as for Spectrophotometer, except: NAD^+, 35 µ*M*.

Conduct of the Assay

One milliliter of reagent without glyceraldehyde-3-P dehydrogenase is added to 10 × 75 mm tubes, and a reading taken in the fluorometer with full-scale deflection equivalent to 5–10 μ*M*. The glyceraldehyde-3-P dehydrogenase is added, and a reading taken to measure the NADH level and, therefore, that of the 1,3-diphosphoglycerate. An appropriate amount of tissue is added (0.1–1 μg of brain), and readings taken at 2–3-min intervals. Calculations are made from the linear portion of the curve. Blanks are tissue added to reagent without ADP.

Comment

The amount of 1,3-diphosphoglycerate formed can be predicted by using the equilibrium constant *(44)*. At low ionic strength, as shown here, the K_{eq} is about 0.2 at pH 7. The K_m for 1,3-diphosphoglycerate is about 10 μ*M* in brain. Therefore, the maximum velocity would be difficult to measure in the fluorometer, since the fluorescence of NADH departs from linearity at that concentration.

Pyruvate Kinase (EC 2.7.1.40) *(14,16)*

[Re. 7-42]

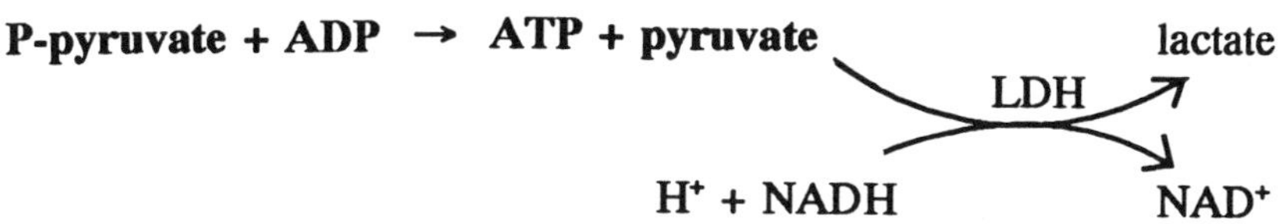

The auxiliary enzyme is lactate dehydrogenase (LDH). The specific reaction and auxiliary reaction are combined into a single step.

Sample tissue activities (mol/kg dry wt/h, 20°): mouse brain, 28; human brain, 23; rat kidney cortex, 4; kidney medulla, 14; and rat muscle fibers, 2–40.

Spectrophotometer (3–10 nmol/min, 50–150 nmol of Product)

Reagent: Imidazole-HCl buffer, pH 7.0 (50 m*M* imidazole-HCl, 50 m*M* imidazole base); KCl, 100 m*M* $MgCl_2$, 3 m*M*; P-pyruvate, 1.5 m*M*; ADP, 1 m*M*; NADH, 150 μ*M*; beef heart lactate dehydrogenase, 0.6 U/mL (2.5 μg/mL); and bovine plasma albumin, 0.05%.

Conduct of the Assay

To 1 mL of reagent in a cuvet is added an appropriate amount of tissue (50–100 µg of brain, fast-twitch muscle, or kidney medulla, wet wts). Blanks are tissue added to reagent without ADP. Readings are taken at 2-min intervals, and the activity calculated on the linear portion of the curve.

Fluorometer Direct Assay (0.05–1 nmol/min, 1–5 nmol of Product)

Reagent: The same as Spectrophotometer, except NADH 10 µ*M*.

Conduct of the Assay

To 1 mL of reagent in 10 × 75 mm tubes is added the appropriate amount of tissue (2 µg of brain or muscle, or 10 µg of kidney cortex, wet wt). Blanks are as for Spectrophotometer. Standards are 10 µL of 0.1–0.8 m*M* pyruvate added to the reagent. Readings are taken at 1–5-min intervals, and calculations are based on the linear portion of the curve.

Fluorometer Indirect Assay (2–100 pmol/min, 0.1–5 nmol of Product)

Reagent 1: The reagent is the same as for Spectrophotometer: except NADH and lactate dehydrogenase are omitted.

Reagent 2: Imidazole-HCl buffer, pH 7 (25 m*M* imidazole base, 25 m*M* imidazole-HCl), beef heart lactate dehydrogenase, 0.6 U/mL, NADH, 0.5 m*M*; and bovine serum albumin, 0.05%.

Procedure

Step 1: To 5 µL of reagent 1 in oil wells are added freeze-dried samples (e.g., 5–200 ng of brain, dry wt); 60 min at 20°. Standards are 5 µL of 2–20 µ*M* pyruvate.

Step 2: Stop the reaction with 1 µL of 0.4*M* NaOH; 10 min at 80°; cool.

Step 3: Add 3 µL of reagent 2; 15 min at 20°.

Step 4: Add 1 µL of 1*M* HCl; 10 min at 20°.

Step 5: Transfer 5 µL to 1 mL of 6*M* NaOH containing 10 m*M* imidazole; 20 min at 60°. Cool, dry, and read in the fluorometer.

Comment

Pyruvate kinase from various tissue sources shows differing activation by fructose-1,6-bisphosphate at low levels of P-pyruvate. To test the response of the kidney enzyme, for example, phosphopyruvate was lowered to 0.25 m*M*, and samples incubated with and without 0.5

m*M* fructose-1,6-bisphosphate. The activity in the kidney cortex was increased by fructose-1,6-bisphosphate; that in the kidney medulla was not.

Succinate Semialdehyde Dehydrogenase (EC 1.2.1.24) *(7,8,46)*

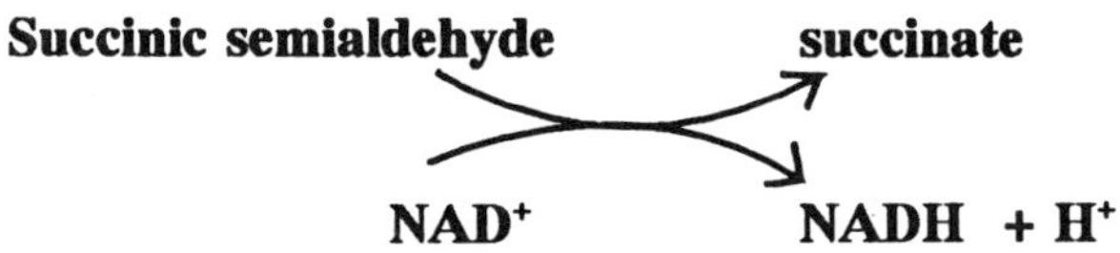

[Re. 7-43]

Sample tissue activity (mol/kg dry wt/h, 20°): rat brain, 0.08. (The activity is ninefold higher at 45° *[45]*).

Fluorometer Direct Assay (0.05–0.5 nmol/min, 1–10 nmol of Product)

Reagent: Tris-HCl buffer, pH 8.8 (85 m*M* Tris base, 15 m*M* Tris-HCl); KCl, 50 m*M*; dithiothreitol, 1 m*M*; NAD^+, 50 μ*M*; succinate semialdehyde, 300 μ*M*; and bovine serum albumin, 0.02%.

Conduct of the Assay

To 1 mL of reagent in 10 × 75 mm tubes is added an appropriate amount of tissue (50–250 μg brain, wet wt), and readings are made every 2 min. Blanks are tissue added to reagent without succinate semialdehyde. The activity is calculated from the linear portion of the curve.

Cycling Assay (0.2–2 pmol/min, 10–100 pmol of Product)

Reagent: 2-Amino-2-methyl-1,3-propanediol-HCl buffer, pH 8.9 (50 m*M* 2-amino-2-methyl-1,3-propanediol base, 50 m*M* 2-amino-2-methyl-1,3-propanediol HCl); KCl, 50 m*M*; succinate semialdehyde, 1.2 m*M*; NAD^+, 2.5 m*M*; dithiothreitol, 1 m*M*; and bovine serum albumin, 0.02%.

Procedure

Step 1: Freeze-dried samples (15–150 ng of brain, dry wt) are introduced through oil into 1 μL of reagent; 60 min at 45°. Standards are 1 μL of 10 μ*M* NADH.

Step 2: To each sample are added 10 μL of 0.025*M* NaOH; 20 min at 80°.

Step 3: Transfer 2 μL into a 10 × 75 mm tube containing 100 μL of NAD cycling reagent with enzyme levels to provide 500-fold amplification in 1 h at 38°.

Step 4: Complete the indicator step as described in Chapter 5.

Thiolase (EC 2.3.1.9) *(1,9,47)*

[Re. 7-44]

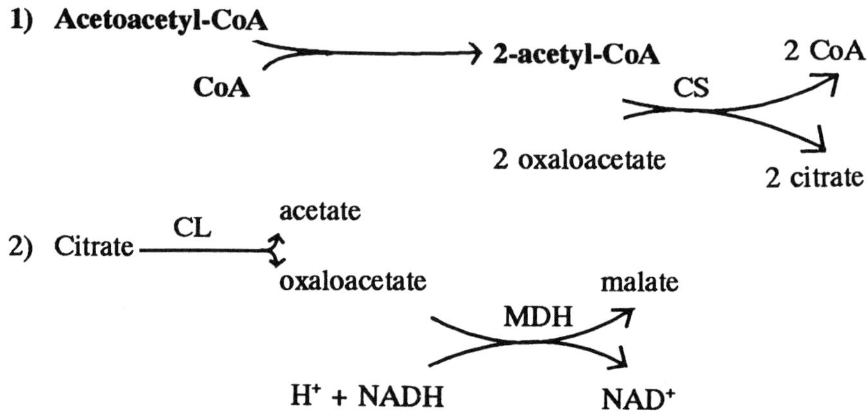

The auxiliary enzymes are citrate synthase (CS), citrate pro-3S lyase (CL), and malate dehydrogenase (MDH). The major mitochondrial and cytoplasmic thiolases are distinguished by the differences in the effects of the two substrates on their activities. The mitochondrial enzyme (M) is strongly inhibited by acetoacetyl-CoA, which is counteracted by high levels of the other substrate, CoA. The cytoplasmic thiolase (C) is not markedly inhibited by either substrate and has a very low K_m for CoA *(46)*.

Sample tissue activities (mol/kg dry wt/h, 20°): rabbit tibialis anterior (M) 0.8; soleus muscle (M) 2.0; Human ova (C) 232 pmol/ovum/h, (M) 576 pmol/ovum/h; and mouse ova (C) 6.8 pmol/ovum/h; (M) 62 pmol/ovum/h (all at 20°).

Fluorometer Indirect Assay (30–75 pmol/min, 2–5 nmol of Product)

Reagent 1C: (Cytoplasmic enzyme) Tris-HCl buffer, pH 8 (25 m*M* Tris base, 25 m*M* Tris-HCl); acetoacetyl CoA, 1 m*M*; CoA 0.25 m*M*; oxaloacetate, 0.5 m*M*; and pig heart citrate synthase 1.1 U/mL (1 µg/mL); and bovine plasma albumin (citrate-free), 0.02%.

Reagent 1M: (Mitochondrial enzyme) The reagent is the same as reagent 1C, except: acetoacetyl CoA, 0.25 m*M*; and CoA, 10 m*M*.

Reagent 2: Triethanolamine-HCl buffer, pH 7.9 (50 m*M* triethanolamine, 50 m*M* triethanolamine-HCl); $ZnCl_2$, 500 μ*M*; K_2HPO_4, 1 m*M*; Na_2SO_4, 25 m*M*; NADH, 50 μ*M*; bovine serum albumin (citrate-free), 0.01%; citrate lyase from *Aerobacter aerogenes*, 0.04 U/mL (5 μg/mL); and pig heart malate dehydrogenase, 1.2 U/mL (1 μg/mL). The high Zn^{2+} level is needed to compensate for the high CoA level in the mitochondrial reagent (CoA inhibition is probably owing to reducing the concentration of free Zn^{2+}). The sulfate helps to protect the lyase from inactivation.

Procedure

Step 1: To 50 μL of reagent 1C or 1*M* in 10 × 75 mm tubes is added an appropriate amount of sample (5–25 μg of muscle, wet wt). Standards are 50 μL of 50–100 μ*M* citrate. Incubate 60 min at 25°.

Step 2: Add 5 μL of 0.5*M* NaOH; 95° for 5 min.

Step 3: Add 500 μL of reagent 2; 20 min at room temperature.

Step 4: Add 30 μL of 1*M* HCl; 10 min at room temperature.

Step 5: Transfer 100 μL of sample to 1 mL of 6*M* NaOH containing 10 m*M* imidazole; 20 min at 60°. Cool and dry tubes, and read on a fluorometer.

Cycling Assay (0.25–1 pmol/min, 15–60 pmol of Product)

The reaction is carried out in oil wells.

Reagents 1C and 1M: The reagents are the same as in Fluorometer Indirect Assay.

Reagent 2: The reagent is the same buffer as Fluorometer Indirect Assay with the following changes: NADH, 10 μ*M*; citrate lyase, 0.4 U/mL (40 μg/mL); and malate dehydrogenase, 3 U/mL (2.5 μg/mL).

Procedure

Step 1: Samples and standards are added in a vol of 0.05 μL to oil wells containing 0.5 μL of reagent 1C or 1M. Standards are 0.05 μL of 03–1.2 m*M* citrate; 60 min at 20°.

Step 2: Add 0.5 μL of 75 μ*M* NaOH; 30 min at 80°.

Step 3: Add 5 μL of reagent 2; 30 min at room temperature.

Step 4: Add 1 μL of 0.5*M* HCl; 10 min at room temperature.

Step 5: Transfer 2 μL into a 10 × 75 mm tube containing 100 μL of NAD cycling reagent with enzyme levels to provide 500-fold amplification in 1 h at 38°.

Step 6: Complete the indicator step as described in Chapter 5.

Uridinediphosphoglucose Pyrophosphorylase (EC 2.7.7.9) *(31,48)*

[Re. 7-45]

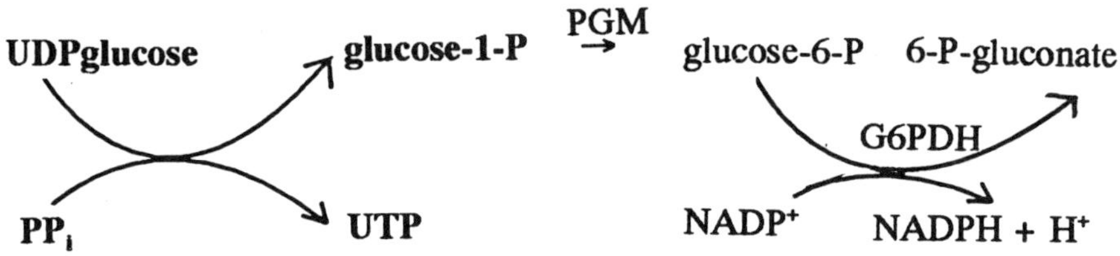

The auxiliary enzymes are P-glucomutase (PGM) and glucose-6-P dehydrogenase (G6PDH). The specific and auxiliary enzyme reactions are combined into a single step.

Sample tissue activities (mol/kg dry wt/h, 20°): rat kidney, 1.2; and dog myocardium, 1.5.

Fluorometer Direct Assay (0.2–0.5 nmol/min, 5–10 nmol of Product)

Reagent: Tris-HCl buffer, pH 8.2 (60 m*M* Tris base, 40 m*M* Tris-HCl); imidazole, 0.1*M*; UDPglucose, 1 m*M*; PP_i, 4 m*M*; $MgCl_2$, 5 m*M*; $NADP^+$, 0.4 m*M*; bovine serum albumin, 0.04%; baker's yeast glucose-6-P dehydrogenase, 0.18 U/mL (0.5 µg/mL); and P-glucomutase, 4 U/mL (20 µg/mL).

Conduct of the Assay

To 1 mL of reagent are added tissue samples of appropriate size (25–100 µg of kidney or heart, wet wt). Blanks consist of tissue added to reagent without UDPglucose. Standards are 10 µL of 0.1–0.5 m*M* glucose-1-P. The tubes are read at timed intervals with the rate calculated from the linear part of the reaction.

Comment

The imidazole is included because it completely inhibits the zinc-activated dinucleotide pyrophosphatase which cleaves UDPglucose to yield UDP and glucose-1-P and would cause erroneously high results. Imidazole has a K_i of 12 m*M* for the pyrophosphatase.

Cycling Assay (0.5–1.5 pmol/min, 15–50 pmol Product)

Reagent: The reagent is the same as in Fluorometer Direct Assay.

Step 1: To 1 µL of the reagent in oil wells are added freeze-dried tissue samples (10–25 ng heart or kidney, dry wt); 60 min, 20°. Standards are 1 µL of 15–50 µ*M* glucose-1-P.

Step 2: To all samples are added 3 µL of 0.6*M* phosphate buffer (0.3*M* Na_3PO_4, 0.3*M* K_2HPO_4); 80° for 30 min.

Step 3: Transfer 2 µL into a 10 × 75 mm tube containing 100 µL of NADP cycling reagent with enzyme levels to provide 500-fold amplification in 1 h at 38°.

Step 4: Complete the indicator step as described in Chapter 5.

Chapter 8

Improvement, Modification, Adaptation, Troubleshooting, and Development of New Methods

This chapter is intended to offer suggestions for simplifying or improving methods, for adapting methods to a different order of sensitivity, and for developing new methods. It also discusses some of the difficulties that may arise and ways to circumvent them.

Simplification and Improvement

It is a rare method that cannot be improved or simplified for the purposes at hand. Once methods have been published, they tend to crystallize, and most users hesitate to tamper with the procedure, even though it might be easy to make changes that would save time and increase precision.

Eliminating Steps

Often one of the easiest things to do is eliminate a step or combine several steps in a procedure. Merely asking if this can be done will usually suggest a way to do so. Elimination of even one step is worthwhile as a time saver when there are many assays to run and will often decrease the chances for error. If the reaction requires many components, these can frequently be combined ahead of time into a single reagent. It may even be possible to prepare a complete or partially complete reagent, and store it frozen for repeated use. The advent of commercial freezers that maintain temperatures of –50° or below has made this even more practicable. Reagent components that do not store well at –20° usually are quite stable at –50°.

Neutralization

Often a method calls for acidification or alkalinization followed by neutralization to a particular pH. To neutralize each sample by titration would be a tedious process, whereas it is almost always possible to accomplish the same result by adding a single calculated volume of buffer, or of acid or base containing buffer. For example, 1 mL of unbuffered solution is acidified with 100 µL of 0.5*M* HCl (50 µequiv). This can subsequently be brought to pH 8.1 by adding 100 µL of 1*M* Tris base (100 µequiv, $pK_a = 8.1$) or if less buffer is wanted, 100 µL of 0.4*M* NaOH (40 µequiv) containing 0.2*M* Tris base (20 µequiv).

A variation of this would be a situation calling for neutralization to pH 8.1 of a sample, such as the above, followed by addition of two other components unstable in alkali, for example, an enzyme and NAD. Instead of making three separate additions, all three components could be added together by neutralizing with a buffer composed of 1.1*M* Tris base: 0.1*M* Tris-HCl (pH 9.2).

The enzyme and NAD^+ would be added to the buffer shortly before use. Unless the enzyme were unusually sensitive to alkali, it should be able to withstand brief exposure to pH 9.2, and as a precaution, the buffer at this time could be chilled to 0°.

Simplification by Reducing the Scale

Decreasing the scale and increasing sensitivity for its own sake will be discussed in another section. The point to be made here is that many methods can be carried out faster on a reduced scale. Smaller volumes are quicker to pipet and mix. More small samples can go into a centrifuge than large ones. It is quicker to clean 1-mL tubes than 100-mL flasks.

Substituting Less Time-Consuming Steps

There are several analytical manipulations that can be accomplished in alternative ways: one slow, one fast. Making a quantitative transfer from one vessel to another in the classical manner involves pouring or pipeting out of the first vessel, rinsing the first vessel, and diluting to volume. It is much faster to transfer a large measured aliquot from vessel 1 to vessel 2. The fraction left behind can usually be calculated with precision. Another classic operation, filtration, can usually be replaced by centrifugation with a great saving in time and glassware, and with less risk of contamination.

If a sample is to be diluted, it is much more time-consuming to "dilute to volume" than it is to dilute by addition of a measured volume. For example, a 1-mL sample is to be diluted tenfold. The classic procedure would be to pipet 1 mL into a 10-mL volumetric flask, dilute to volume, and mix (hard to do in a small volumetric flask). Far quicker is to pipet 1 mL into a 25–mL vessel and add 10 mL with a pipet (with easy subsequent mixing). The dilution in the second case is 1:11 instead of 1:10, but calculators are quicker than the hand.

Improvement of an Enzyme Method

In addition to eliminating steps, a method may often be improved in other ways. It frequently pays to observe the effects of modest changes in the concentrations of the reactants or in the pH to see if any of the conditions are critical or suboptimal. This is important in any assay, but is usually of particular importance in an enzyme assay.

Reagent pH

Ordinarily, it is desirable to measure an enzyme at its optimal pH. This is not just because it gives maximal sensitivity, but also because small deviations in pH will usually cause the least change in velocity in the neighborhood of the pH optimum. However, there is no fundamental reason for measuring an enzyme under optimal conditions; few if any enzymes operate under optimal conditions in vivo! The actual velocity observed in the assay is arbitrary in any event. The important thing is reproducibility to permit accurate comparisons or possibly calculation of the absolute amount of enzyme present. There may be good reasons to measure the enzyme at a pH different from the optimum. At some other pH, the enzyme or one of the other reactants may be more stable, the equilibrium may be more favorable, thereby making it easier to achieve linearity, or a disturbing side reaction may cause less interference.

Incubation Temperature

Similarly, there is no inherent reason why an enzyme has to be measured at some particular temperature. There is a tradition for the use of 38°, but other temperatures may be preferable. Occasionally, a rugged enzyme of low activity can be measured at much higher temperature. More frequently, there is an advantage in a temperature lower than 38°. Usually enzyme *stability* has a much larger temperature coefficient than enzyme *activity*. There are cases where it is actually advantageous to carry out the incubation at 0°. Some enzymes at 0° have 20% as much activity as at 38°, or even more.

Substrate Level

Usually in an enzyme assay, an effort is made to "saturate" the enzyme with substrate. The fact is often overlooked that to achieve 99% saturation for an enzyme with nornal kinetics requires a substrate level 100 times the K_m. It is very often undesirable to use so much substrate. There may be trace impurities in the substrate that can cause high blanks or even erroneous results, the substrate may itself inhibit, and so on. Again, there is nothing sacred about measuring the maximum velocity. It may actually be better in some cases to use much less than saturating levels of substrate and to calculate the V_{max} if that is the wish. An example of a case in which low substrate is an advantage is the measurement of malate dehydrogenase with oxaloacetate and NADH. Not only is oxaloacetate inhibitory at high levels, but it is unstable and breaks down partially to pyruvate. Since most tissues contain rather high levels of lactate dehydrogenase, there is clearly a danger of a substantial positive error. The more oxaloacetate used, the greater the amount of pyruvate that will be present and the greater this danger. Compare the case of two reagents that contain oxaloacetate, in one instance at a concentration equal to the K_m and in the other at a concentration 20 time the K_m. The second reagent will give 90% greater velocities for malate dehydrogenase, but 2000% greater errors owing to lactate dehydrogenase.

Improvement by Dilution

As pointed out for the P-glucomutase assay in Chapter 7, a method for measuring an enzyme in a crude tissue preparation can often be improved by increasing the tissue dilution. The activity of the enzyme being measured should not suffer by the dilution, if the concentrations of its substrates and cofactors remain the same. Other enzymes, however, that might interfere are often rendered less effective. An example of this is the measurement of hexokinase via glucose-6-P dehydrogenase. There are several tissue enzymes that could cause trouble: ATPase could deplete one of the substrates; P-glucoisomerase and P-fructokinase could sidetrack some of the glucose-6-P, converting it to fructose diphosphate; 6-P-gluconate dehydrogenase could partially convert 6-P-gluconate to ribulose-5-P, producing an uncertain extra amount of NADPH; and other enzymes could reoxidize part of the generated NADPH. Glucose-6-P (a potent inhibitor of hexokinase), if generated too fast, might accumulate in sufficient concentration to slow the reaction. At high dilution, all these difficulties would be reduced without any necessary decrease in hexokinase activity.

Improvement by Changing the Number of Steps

Often, a potential side reaction in an enzyme assay can be eliminated by splitting the assay into two steps. In the case of P-glucomutase, the first step (the conversion of glucose-1-P to glucose-6-P) is carried out in the absence of $NADP^+$ and glucose-6-P dehydrogenase. This procedure eliminates the potential error of further oxidation of glucose-6-P to 6-P-gluconate with a variable amount of NADPH generated. In other situations, combining steps may be advantageous. A good example is the assay just cited for hexokinase. Here it is highly desirable to combine the two enzyme steps. Otherwise, there is almost certain to be conversion of much of the glucose-6-P to fructose diphosphate and triose phosphates. Glucose-6-P dehydrogenase added in generous amount can successfully compete with P-glucoisomerase (and indirectly with P-fructokinase) for the glucose-6-P as it is generated.

There are in most tissues enzymes capable of cleaving NAD^+ and $NADP^+$ (NADase and organic pyrophosphatases). These become important for enzyme assays if the final product is NAD^+ or $NADP^+$, if the enzyme is of moderate to low activity, and if the assay is indirect. (Obviously, neither enzyme would be disturbing in a direct assay in which the decrease in absorption or fluorescence is to be measured.) NADase can usually be inhibited sufficiently by nicotinamide. The product of organic pyrophosphatase is nicotinamide monophosphate, which gives the same strong fluorescent product in 6*M* NaOH as NAD^+ or $NADP^+$. Therefore, this enzyme would not interfere if the nucleotide product were to be measured in strong alkali. It would, however, interfere if the product had to be amplified by enzymatic cycling, which, of course, requires the intact pyridine nucleotide. If, in an assay with two or more enzyme steps, tests show the $NAD(P)^+$ cannot be recovered satisfactorily, the assay can usually be divided into two parts, with destruction of tissue enzymes by acid, alkali, or heat after the first part.

There is in all tissues another group of enzymes that is potentially disturbing. This consists of enzymes that catalyze direct oxidation of NADH or NADPH by O_2. Direct oxidation of NADH is the most troublesome. The greater part of this can be blocked with amytal, as well as with several other electron transport inhibitors. Most of the activity has also been found to disappear in frozen-dried tissue. There is, however, always a sufficient residue of NADH oxidase activity to be highly disturbing for measurement of enzyme with low activity. Here, as with NADase, the solution in the case of a multistep assay may be to divide the assay into two steps, with destruction of tissue enzymes after the first step before the addition of the pyridine nucleotide.

Improvement of an Enzymatic Metabolite Assay

Conditions are usually less critical for a metabolite assay than for an enzyme assay, but it may be harder to achieve adequate sensitivity and specificity. Because enzymes are never completely pure or absolutely specific, it is best to use no more enzyme than necessary. The more favorable the conditions for the enzyme action, the less enzyme will be required. Conditions that give maximum enzyme velocity with saturating levels of substrate may be very different from most favorable conditions with the low levels of substrate encountered in an assay. A striking example is the case of α-ketoglutarate measurement given in Chapter 6. For metabolite assays, the important kinetic parameter is not V_{max} or the K_m, but the apparent first-order rate constant k, which is equal to V_{max}/K_m (*see* Chapter 2). To measure this constant, an elaborate kinetic study is not required. It is not even essential to know the V_{max} or the K_m (although it is well to know these under final assay conditions). This constant, or simply the half-time, can be determined directly with a low level of the substrate. Optimal conditions can be found by making this determination with a suitable variety of pHs, buffers, and so on.

In a multistep metabolite assay, as in a multistep enzyme assay, there may be an advantage in breaking the assay into two parts. This is frequently true when one of the enzymes has a disturbing contaminant or side action. For example, a method for measuring glycerol consists of converting it to glycero-P with glycerokinase and ATP, and then measuring the glycero-P with glycero-P dehydrogenase and NAD^+. If the glycerokinase were contaminated with lactate dehydrogenase, malate dehydrogenase, or NADH oxidase, there could be a serious error in a one-step assay, wherease the error could be eliminated if the assay were split into two steps.

Adaptation of Pyridine Nucleotide Methods to Increase Sensitivity

Only a few general remarks will be offered here. The spectrophotometer is well suited to measuring reduced pyridine nucleotides in the 10–150 μ*M* range or a total in 0.5 mL of 5–75 nmol. The fluorometer is suited to measuring reduced pyridine nucleotides directly or oxidized nucleotides indirectly in the 0.05–10 μ*M* range, or a total in 1 mL of 0.05–10 nmol. Through the use of enzymatic cycling, sensitivity can be increased another 100,000-fold or more.

Conversion of a spectrophotometric method to a fluorometric one of moderate sensitivity is usually simple and straightforward, as has been seen. Often it is possible and desirable to reduce the concentration of cofactors and auxiliary enzymes. Fewest changes are needed if the reaction is measured in the direction of NADH or NADPH formation, because it is then usually possible to retain near saturation levels of the coenzyme (NAD^+ or $NADP^+$). With reactions proceeding in the opposite direction, it is necessary, in the case of direct assays in the fluorometer, to reduce the NADH or NADPH levels, possibly to concentrations far below enzyme saturation. This means that the amount of auxiliary enzyme may have to be increased to compensate, or in the case of a metabolite assay that the reaction time must be prolonged. Prolonging the time may present no problem in a fluorometric assay. In the spectrophotometer, with a limited number of cuvets per rack, it is usually desirable to complete the reaction promptly. In the fluorometer, with an unlimited number of tubes, a great many samples can be followed if necessary for an hour or more without disadvantage.

In adapting a method to the fluorometer, the alternative to reducing NADH or NADPH levels is to measure the increase in NAD^+ or $NADP^+$ (*see* α-ketoglutarate method, Chapter 6). This introduces an extra step, but, because it is unnecessary to follow the reaction, it actually decreases analytical time with a large number of samples.

Troubleshooting

Surely the most important thing in quantitative analysis is to recognize trouble, and the next most important is to know how to cure it. It is too much to expect that a method can be applied to all biological materials without ultimately encountering difficulty. This is especially true of methods for substances present in small quantity, accompanied as they are in biological specimens, by innumerable other substances, many of them unknown.

Detecting Trouble in Metabolite Assays

Trouble can consist of nonspecificity (high results), incomplete reaction (low results), loss of product (low results), or high or/and erratic blank values (erratic results). In a metabolite assay, one of the simplest and most effective tests is to carry out simultaneous time curves with a standard, a blank, and a biological sample. In a valid assay the sample will follow the same time curve as the standard, and there will be no further change in either direction after the reaction is finished. The reagents used for making the extract may affect the enzyme activities; therefore, the best comparison is made with a standard that includes simulated extract equal to the sample. If the assay is indirect, it is still possible to obtain several points on a time curve with replicate samples.

The following examples illustrate the value of this kind of test:

1. When $HClO_4$ extracts of brain were analyzed for pyruvate (with lactate dehydrogenase and NADH), it was noted that after a rapid drop in NADH at the same rate as for a pyruvate standard, there occurred a long, slow fall that finally stopped when NADH consumption had nearly doubled. This second step could be accelerated by increasing the amount of enzyme. (This phenomenon was not observed with extracts of liver or kidney.) The unknown substance, presumably another keto acid, was not identified. In consequence of this test, brain pyruvate assays were made with carefully controlled amounts of lactate dehydrogenase just sufficient to allow complete reaction with pyruvate.
2. When liver extracts were analyzed for P_i with glycogen phosphorylase, P-glucomutase, and glucose-6-P dehydrogenase (Chapter 6), it was observed that NADPH rose at the expected rate, but fell thereafter. This fall did not occur with standards and was less noticeable in extracts of several other tissues. The difficulty was attributed to oxidized glutathione in the liver extract reacting with a glutathione reductase contaminant of the glucose-6-P dehydrogenase to oxidize the NADPH. It was cured by addition of dithiothreitol to reduce the glutathione.
3. In an enzymatic method for P-creatine (Chapter 6), the tissue extract, made with $HClO_4$, is assayed with a reagent containing hexokinase, glucose, glucose-6-P dehydrogenase, and $NADP^+$. After NADPH has been produced in stoichiometric yield, creatine kinase and extra ADP are added, resulting

in extra NADPH equivalent to the P-creatine present. It was observed, that with certain tissue samples, the reaction did not stop at the same time as the standards. The difficulty was traced to failure to remove the last traces of protein (inadequate centrifugation). This protein included some adenylokinase, known to be very acid-resistant, that generated extra ATP from the ADP present.

One might anticipate increased opportunities for trouble in a complex assay system with one or more sequential enzyme steps. Although it is true that more things could go wrong, such an assay system actually offers greater specificity and more opportunities for testing that specificity. Omission of an auxiliary enzyme provides an excellent null test. Intermediate standards can be introduced to test each step individually. For example, if there should be difficulty in measuring P_i in the example given above, the system could be quickly checked with standard solutions of glucose-1-P, glucose-6-P, and NADPH.

Erratic Results

There are many possible causes of erratic results. One of the most common is an erratic blank (*"cherchez le blanc"*). If the blanks are the trouble, they will vary as much as the standards in absolute terms. The most common cause is contamination, but its source may not be obvious (glassware, dust, fingers, one of the reagent components, even the distilled H_2O). The higher the blank, the greater the error it is likely to introduce; it is therefore well to know where the blank comes from and if possible to reduce it.

Experience has shown that one of the most likely causes of error is failure to mix properly. It is difficult to describe how much mixing is enough for all volumes and all sizes and shapes of containers, but until proven otherwise, faulty mixing should be suspected as the cause of erratic results. In some enzyme assays, mixing that is too violent can be almost as bad as too little mixing because of surface denaturation. It cannot be overemphasized that samples or reagents stored frozen become unmixed (the H_2O freezes out) and must be thoroughly remixed after complete thawing.

In chasing down the source of erratic results, it is helpful to know whether the faulty values are high or low, or both. If they are all high it is probably contamination; if all are low, it is probably something interfering with one of the enzymes or causing destruction of the product.

With reagents containing many components, it may sometimes be difficult to locate the source of a high reagent blank or the cause of a slow reaction rate. The trouble cannot always be located by omitting one component at a time, since the faulty component may be essential for the reaction to occur. Usually, however, the offender can be identified by increasing or decreasing each component one at a time by a factor of 2 or 3.

Detecting Trouble in Enzyme Assays

Troubles in an enzyme assay can include any of those listed for metabolites, but there is a special hazard of underestimation owing to inhibition, or lag in an auxiliary enzyme step or inactivation during the reaction.

Specificity

Specificity may be easier to check than in a metabolite assay. Often a good test is to run a "tissue" blank, that is, to incubate the tissue (or other) sample with reagent from which the specific substrate has been omitted. This may occasionally fail. For example, in assays for glucose-6-P dehydrogenase in liver and muscle, the addition of sample to a reagent without glucose-6-P may not provide a valid tissue blank. The reason is that breakdown of rich amounts of glycogen in these tissues can provide enough glucose-6-P to give substantial dehydrogenase rates. (In this case, a test can be made by running the reaction at very high tissue dilution, which, except in rare cases, will lower the endogenous substrate level to insignificance.)

Proportionality with Time and Amount

In any new situation or where trouble is suspected, a time curve should be made as well as a test of proportionality between product formation and amount of sample. Fall off with time could be the result of enzyme destruction, product inhibition, depletion or destruction of substrate, or destruction of the product that is being measured (e.g., reoxidation of NADH produced). Enzyme destruction can usually be detected by preincubating the sample with reagent from which the substrate(s) has been omitted and then determining whether this affects the subsequent rate when substrate is finally added. Product inhibition and product destruction are easily tested directly. Depletion or destruction of substrate can be detected by adding more substrate when the rate begins to fall off; however, in this test, if acceleration should occur it could also be owing to counteracting product inhibition.

Acceleration

Acceleration with time, rather than fall off, can occur. In a multienzyme reaction, acceleration is expected to at least a small degree, the amount depending on the activity of the auxiliary enzyme(s). Acceleration can also be the result of activation of the enzyme in the assay or further reaction of the product if this yields more of the substance being measured. (For example, in a glucose-6-P dehydrogenase assay, the product 6-P-gluconate can be further oxidized, yielding NADPH at an accelerating rate.)

It is usually very informative to measure velocity (and fall off or acceleration, if any) at several enzyme dilutions. (Big differences in dilution tell more than small differences.) At increased dilution, the effects of possible stimulatory, inhibitory, or destructive factors are all reduced, and velocity will be affected accordingly. If, however, velocity is proportional to concentration, this would be taken as evidence that none of these factors is operative.

Cure of Trouble in an Enzyme Assay

The location of the difficulty will often suggest the cure. As already said, perhaps *ad nauseum,* running the reaction at greater dilution may be the solution. If the trouble is enzyme stability, there is no guaranteed cure, but it is worth trying the addition of plasma albumin (total protein should not fall below 0.02% in any event), mercaptoethanol or dithiothreitol (0.2–5 m*M*), or/and EDTA (0.1–1 m*M*). If Mg^{2+} is required by the enzyme, EDTA can nevertheless be used at a lower level. Stability might also be better at another pH. It is worth sacrificing sensitivity for an increase in stability. The same is true of lowering the temperature, which as said elsewhere may greatly increase stability with comparatively small decrease in enzyme velocity.

Development of a New Method

The increasing commercial availability of crystalline enzymes makes it relatively easy to develop new specific sensitive methods. Most "new" methods are really adaptations of old methods or the application of well established reactions to analytical use. Once the basic reaction is selected and the necessary enzymes are available, a tentative procedure is set up and tested. The method is then worked out on the *convenient* sensitivity scale, which is often not the same as the scale ultimately required. If preliminary trials make the method appear feasible, the rest is simply a matter of improvement, troubleshooting, and finally adapting to the scale of sensitivity required *(see above)*.

A typical example is the method for P_i in Chapter 6. A method was required for measuring P_i in the 10 pmol range (for analyses of 0.1 µg histochemical samples). Existing colorimetric methods fell far short of the necessary sensitivity. A number of enzymatic possibilities were considered. The basic reaction selected (glycogen phosphorolysis) was discovered by the Coris in the 1930s. The availability of crystalline phosphorylase, P-glucomutase, and glucose-6-P dehydrogenase made it appear feasible to devise a practical enzymatic pyridine nucleotide method.

Most of the experimental work was carried out by direct observation of fluorescence change using 1-mL volumes of reagent and 5 nmole of P_i (i.e., 10,000 times as much P_i as would have to be finally measured in a much smaller volume by an indirect procedure). In this case, the kinetics of all three enzymes were well known; otherwise, a cursory kinetic study would have been conducted. Rates were merely confirmed under assay conditions and minor adjustments made. As described in Chapter 6, the pH and buffer were a compromise based on the different behavior of the three enzymes concerned. One of the major troubles encountered was contamination with P_i, which seemed to be everywhere, including the enzyme preparations. The presence in the phosphorylase of a phosphatase for 5'AMP was discovered during work with standard solutions. The presence of an ATPase in the phosphorylase and glutathione reductase in the glucose-6-P dehydrogenase was not discovered until trials were made with tissues. Cures for these trouble were not difficult once they were spotted. The final operation was to increase the sensitivity 10,000-fold. This was straightforward, and consisted essentially of decreasing all volumes and introducing an enzymatic cycling step. The chief problems had all been solved on the macro level.

PART 3

Quantitative Histochemistry

Chapter 9

Preparation of Tissue and Sections

Introduction

The following portion of this book represents a coherent system of quantitative analysis for histochemical purposes. It derives from the original system of Linderstrom-Lang and Holter *(1)*, but differs in several significant details. In the original methodology of the Carlsberg Laboratory, serial frozen sections were cut from a uniform tissue block. Alternate sections were either analyzed, or fixed and stained for histological control. This procedure has been further developed and exploited with great success by David Glick, and is fully described in book form *(2)*.

One essential change in the system to be described is that before analysis the sections are dried from the frozen state and then cleanly dissected into defined histological portions. This permits direct histological control of the actual sample analyzed and the study of very small structures that might not extend into adjacent sections. One of the virtues of this modification is that most biological materials are relatively stable in the frozen-dried state. Consequently, a given set of sections, perhaps from a tissue in an otherwise transient physiological state, can serve for years as a source of material for analysis.

A second essential modification from the Carlsberg procedure is that sample size is determined by dry weight rather than by the dimensions of the tissue section. This permits analysis of irregular pieces and of much smaller specimens.

Finally, sensitivity has been increased almost without limit by the introduction of enzymatic cycling and the oil well technique. These changes now permit single cells to be analyzed for enzymes and metabolites of all kinds, even when present at very low concentrations.

The complete analytical system is designed for study of solid tissues. It is, of course, possible to use the tools and methods for other purposes. For example, the system has been extensively used for analyzing short portions of individual freeze-dried muscle fibers and, in one instance at least, for measuring enzymes in several portions of single freeze-dried renal tubules *(3)*. It has also been applied to studies of individual freeze-dried mouse and human ova *(4,5)*, and to single stomata from freeze-dried leaves *(6)*.

Freezing

The procedure used in preparation of tissues for histochemical analysis will vary somewhat with the intended purpose. If the material is to be used for the analysis of biologically labile metabolites, the tissue is frozen *in situ* or as rapidly as possible after removal, to preserve the in vivo state *(see* Chapter 4). When the tissue is intended for the measurement of enzyme activity, rapid changes following removal of the tissue are less likely to interfere. In either case, the freezing process itself is made as rapid as possible, and an effort is made to keep ice crystals small and thereby maintain the structural integrity of the tissue.

Samples larger than a gram or two are probably best frozen directly in liquid N_2 (Chapter 4). If the sample is a gram or less, it is plunged with vigorous stirring into Freon-12 (CCl_2F_2) brought to its freezing point (–150°) with liquid N_2. With small samples, the freezing is faster than with liquid N_2 itself. The use of chilled propane or isopentane is avoided because of the fire hazard. This hazard is enhanced by the danger of condensation of O_2 from the air, even though the chilling is accomplished with liquid N_2.

For small samples, another means of freezing is available, that was originally suggested by Fernandez Moran. It is particularly convenient for situations such as may occur in the operating room, when samples are presented on short notice. This is to freeze by immersion in liquid nitrogen that has been chilled from its boiling point (–210°) by rapid evaporation. Ordinarily, the gas that is evolved at the surface of liquid N_2 at it boiling point delays heat transfer and is thus not satisfactory for rapid freezing. The use of N_2 chilled below its boiling point prevents the formation of gas. To chill the N_2, the Dewar flask is connected to a vacuum pump through a rubber hose and rubber stopper. A portion of the liquid N_2 is pumped off. Freezing of the liquid N_2 begins when about 10% of the volume has been evaporated. In the Dewar flask, the liquid N_2 will remain below its boiling point for an hour or two. If a substantial part of the liquid N_2 is frozen by the evaporation process, it will stay below its boiling point for even longer periods. This method is not suitable for large samples, because the heat capacity of liquid N_2 is low.

When identification is needed and time permits, tissue samples may be placed on a piece of labeled hardened filter paper before freezing.

Storage of Frozen Tissues

The frozen tissue samples can be stored before cutting sections, but the temperature must be kept very low to prevent ice crystal growth, enzyme action, or possible migration of constituents. As discussed in Chapter 4, labile metabolites are drastically changed in a few hours at –20° and a few minutes at –5°. Many enzymes are unstable at –20° during storage for a few weeks. On the other hand, after several months at –80°, tissues have been cut without noticeable crystal growth or loss of enzyme activity, and after a week at –35°, there is no discernible loss of ATP or P-creatine. In general, however, it is recommended that sections be cut and dried as soon after preparation as convenient, since the dangers of loss or migration of materials are greatly reduced after drying.

Mounting the Tissue for Cutting

Experience has shown that mounting the block of tissue for cutting is one of the most critical steps. It is important to mount the tissue without thawing, since this would result in rapid diffusion of soluble compounds and formation of large ice crystals on refreezing.

The frozen tissue is softened somewhat by allowing it to warm up to –15 to –20°. Blocks suitable for mounting are cut out with a razor blade. If labile metabolites are to be measured, the time at this warmer temperature level should be kept to a minimum (*see* Chapter 4). In this case, until ready to mount, the blocks should be kept at dry ice temperature. Wooden dowel rods, 5–6 cm long, 5–8 mm in diameter, and drilled at one end to form a shallow cup, are used for mounting (Fig. 1). The diameter of the dowel and depth of cup will vary with the tissue to be mounted. The tissue block is mounted with either a commercial paste (e.g., O.C.T. compound) or (what we prefer) a paste of brain tissue, prepared for this purpose, applied in generous amount to the dowel cup. (Adherence to the tissue may be improved if the paste is

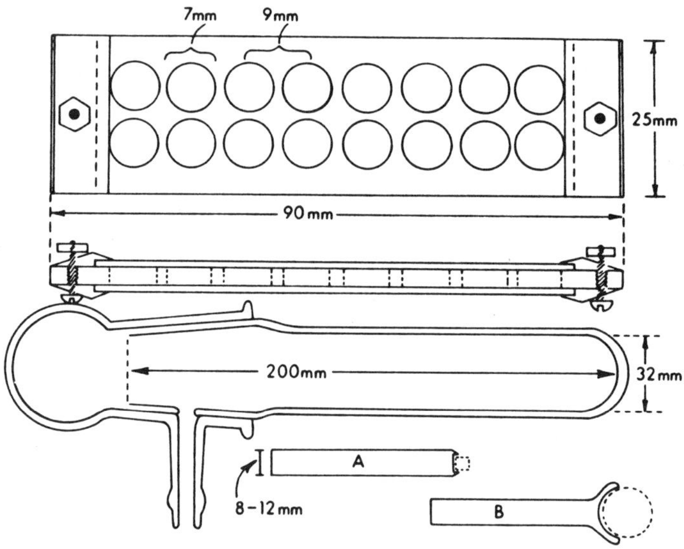

Fig. 1. Aluminum holder for tissue sections, glass vacuum tube for drying and storing sections (in aluminum holders), and wooden holders for tissue blocks. Wooden holder A is for routine use with tissue blocks that can be trimmed to fit the slight depression. Holder B is an example of a special holder for use with an eyeball for cutting retinal sections.

diluted with one-third volume of H_2O.) With the paste above freezing, the tissue block is picked up with insulated forceps and pushed into the paste with suitable orientation, and at once plunged into the freezing medium. This medium consists of hexane or heptane in a small beaker that is maintained at a low temperature by placing it inside a larger beaker containing dry ice. Liquid N_2 or Freon-12 chilled to –150°, is not suitable for this purpose, because the temperature is too cold and the tissue blocks may crack. To mount successfully, the block should be cold enough to freeze the paste, rather than itself be melted. To make sure this is the case, the tissue block, if not already at dry ice temperature, can be placed briefly on a small piece of dry ice.

As soon as safely frozen in place in the hexane or heptane bath, the block and holder is removed and excess solvent is blotted off. If difficulty is encountered in having the block crack loose during sectioning, it can be further anchored in place by painting on a little more of the paste after the solvent has been completely removed. The mounted tissue is kept at –50° or below until just before cutting, particularly if labile metabolites are to be measured.

Sectioning the Tissue

The holder and block of tissue are placed in a microtome in a cryostat. We strongly recommend the use of the closed type of cryostat because of the much better temperature control. The cryostat temperature is maintained at –18 to –30°, depending on the thickness of the sample desired *(see below)*. The angle of the blade of the microtome should be set so that the leading edge of the bevel is nearly parallel to the plane of motion of the tissue block.

The thicker the section, the greater the tendency for it to crack as it comes off the blade and, therefore, the higher the temperature must be. Sections of 5 μm can usually be cut at –25 to –30°, whereas at 25 μm, it may be necessary to raise the temperature to –18 to –20°. If the temperature is too warm, the section will be compressed. The sections are cut with a slow, steady motion and kept flat by holding a small camel's hair brush against the surface of the block as the section is cut. If the sections tend to fragment, in spite of a sharp clean knife and suitable temperature, this can usually be controlled by cutting very slowly and smoothly, and increasing the pressure from the brush.

The sections are transferred serially as they are cut to a suitable holder with a sharp pointed wooden or plastic stick. Metal instruments are not suitable because of heat conduction. Satisfactory tissue holders are made of drilled aluminum blocks sandwiched between glass slides (Fig. 1). If the cut sections are nearly as large as the hole, several sections may be placed in the holder without mixing the order. Moreover, the exact order is seldom important. It is usually not wise to crowd more than five or six sections in a hole. As soon as the holder is filled, the cover is replaced to prevent drying at the temperature of the cryostat. Great care should be exercised not to warm the holder with the gloved hand. Whenever possible the holder should be handled with rubber-tipped forceps. If the sections are to be used for the measurement of labile constituents, they should not remain longer than necessary at the comparatively warm temperature of cutting.

Drying of Samples

The section holders are placed in a large, round-bottomed tube fitted with a combined glass cap and stopcock (Fig. 1). This improved model was designed by Wenger *(7)*. The cap is greased with silicone stopcock grease with care not to get grease into the airway. The tube is transferred without warming to a constant temperature box at –35 to –40°. After the tube and contents are cooled to the lower temperature, the sections are dried by applying a vacuum equivalent to 0.01 mm Hg or less. A dry ice trap is used to receive the moisture. Most of the water can be removed in 1–6 h, depending on the number and thickness of the sections, but it is usually safer to continue evacuation overnight. Care is exercised not to get crystals of ice into the evacuation tube, since these, being solid, are slow to evaporate. A minute crystal of ice (1 mg) if left in the tube could ruin the samples. (Air saturated with moisture contains only 0.5 mg of H_2O/100 mL at 0°, and 2.5 mg at 24°.) As a precaution against possible residual moisture or vagrant ice crystals, the tube is taken from the cold box and evacuation is continued outside until the contents reach room temperature.

The evacuation temperature recommended, –35 to –40°, avoids dangers encountered at higher temperatures and difficulties in dessication at lower temperatures. At temperatures above –25°, there is some shrinkage on drying, and the possibility exists that salts present might result in a small fluid phase permitting diffusion or enzyme action. Below –40°, the vapor tension of ice falls to very low pressures. At –40, –60, and –80°, the vapor tension is 0.1, 0.008, and 0.0004 mm of Hg, respectively. Thus, it would take 250 times longer to dry tissue at –80° than at –40°.

At –40°, there is little advantage in using a vacuum of <0.01 mm Hg, since this gives a pressure gradient between the tissue and the trap of 0.09 mm of Hg out of a possible 0.1 mm. A pressure of 0.01 mm can be easily maintained with a mechanical pump (and a dry ice trap). The most important element in achieving this pressure is not the pump, but the prevention of air leaks. It pays to use a minimum of rubber tubing. For a given pressure at the pump, the rate of drying is governed by the diameter and length of the pathway from sections to trap. A narrow or partially plugged stopcock or small-bore connecting tubing will greatly delay drying.

Storage of Dry Sections

In contrast to the fresh frozen tissue, the dried sections can be stored for long periods in the evacuated tubes as –50° or below without loss of enzyme activities or breakdown of labile metabolites, such as ATP or P-creatine. Even if air leaks in, enzyme activities tested have not been noticeably affected by many years of storage. As a precaution against getting ice crystals into the stopcock orifice, it is plugged with a cork or covered with a rubber cap.

The time of exposure to room temperature and room air is more critical. A test in dry brain sections of three enzymes, hexokinase, 6-P-gluconate dehydrogenase, and glucose-6-P dehydrogenase showed that the first two lost 40% activity in 3 d at 25° in room air, but lost little or no activity under vacuum at 25° during the same period. Glucose-6-P dehydrogenase was more unstable and, under the same conditions, lost 65% of its activity in air and 45% under vacuum.

The stability of several metabolites in dry sections has also been tested at room temperature. Under vacuum at 20° glucose, glycogen, and ATP were not significantly changed in 10 d. In air at 20° with 50% humidity, glucose fell 25% in 10 d, and ATP fell 35% in 16 d. In 12 d, glycogen fell <10%. In some cases the humidity at room temperature is very critical. For example, oxidized and reduced pyridine nucleotides are stable for several hours at 25° if the relative humidity is 50% or less. However, if the humidity reaches 80%, NADH and NADPH are rapidly oxidized, and NAD^+ and $NADP^+$ are destroyed.

Preparation of Samples by Other Means than Sectioning

Muscle Fibers

Essén et al. *(8)* were the first to prepare muscle fibers for analysis by freeze-drying a portion of the muscle, and then dissecting out 2–5 mm lengths of single fibers. These were then weighed and analyzed individually. To freeze-dry safely at –35°, a bundle of fibers that is 1 mm in diameter requires 4–6 d. The dried fibers can be stored separately under vacuum in aluminum racks at low temperature, as described for tissue sections. Each fiber can provide many samples 25–100 µm long (10–50 ng dry wt) to permit assessment of a variety of enzymes and metabolites. *See,* for example Lowry et al. *(9)* and Hintz et al. *(10).*

Mammalian Ova

Individual ova are easily freeze-dried and stored under vacuum. The ova are freeze-dried on a microscope slide in a minimal amount of fluid. If there is no need to identify individual ova, a number of ova are dried together and are subsequently stored on the original slide, held in place by the dried medium *(4)*. When identity of the cell must be maintained, as is the case for human ova, the dried cell is stored in a small well drilled into a microscope slide and capped with a portion of another microscope slide tightly clipped in place. This prevents loss of the ovum when the vacuum is applied or released *(5)*.

Chapter 10

Dissection and Histological Control

Dissection and Balance Room

A room in which temperature and humidity can be controlled facilitates dissection. As noted above, humidity >50% increases the danger of changes in tissue constituents. It also seems wise to keep the temperature as cool as comfort permits (20° is suggested) to increase further the stability of sensitive components. Some tissues, specifically myelinated portions of the nervous system, become sticky and difficult to handle if the temperature rises much above 25°.

Since the tissue sections and dissected fragments are very light, they are easily blown about. It is almost essential to surround the dissection bench and balance area with lightweight curtains (preferably of fiberglass to avoid a fire hazard). A most satisfactory arrangement is to place a metal screen 3 or 4 feet above the bench top with curtains on the sides.

Removal of Sections from Evacuation Tube

The tube is taken from the deep freeze, and the vacuum is immediately checked, since the sections should warm up under vacuum. Because heat transfer is slow in the evacuated tube, warming will require an hour ordinarily, or the time can be reduced to 20 or 30 min by warming the outside with a warm wet towel. During the warm-up, the stopcock outlet must continue to be plugged or capped to prevent condensation of moisture inside. It cannot be too strongly emphasized that admission of even a few microliters of H_2O would ruin the sections. Opening thc vacuum tube while the sample holders are cold would be fatal to the sections as the result of condensation of moisture.

Dissection

As the previously cited tests of enzyme and metabolite stability show, to minimize changes that occur in air at room temperature, it is desirable to keep dry sections under vacuum as much as possible. A few sections are removed for dissection; the tissue holder is returned to the large storage tube, and the vacuum reestablished at once. This prevents exposure of the bulk of the samples for long periods to warm air. In the case of a few exceptionally sensitive enzymes and certain metabolites, it may be necessary to control strictly the period of exposure to room temperature even under vacuum.

The dissection is done freehand under a dissecting microscope. The sections on the stage are protected from drafts and the operator's breath by a piece of lint-free toweling taped to the front of the microscope. A hard plastic surface, such as Plexiglas (Lucite, Perspex), is fastened to the microscope stage. The plastic chosen is translucent (opalescent), so that the illumination provided from below is diffused. Illumination with a fiber light guide is especially satisfactory. This gives excellent visibility of structures within the frozen-dried material. In addition, cutting the tissue on a plastic surface rather than on glass prolongs the life of the knife edge. The plastic surface also facilitates dissection of very small samples, as will be described below.

To control static electricity, it is necessary to have a source of radiation on the microscope stage or suspended over it. A small amount of americium, such as used in the fish pole balance (Chapter 11), is satisfactory, or a polonium source can be used.

The frozen-dried material is held in place during dissection with a hair loop (Fig. 1) or in some cases with a hair point *(see Handling Dissected Samples)*. Dissection is accomplished by a series of short vertical cuts through the section. The knives are made of splinters of razor blades mounted on a flexible bristle (Fig. 1). Handles for razor blade knives are made from wooden dowel rod, 6 mm (1/4 in.) in diameter, cut in convenient lengths (15 cm) and tapered in a pencil sharpener. The ends and taper are sanded smooth, and the pointed tip is cut off to make a flat surface 2–2.5 mm in diameter. A hole about 1.5 mm in diameter is made in this surface with a wire drill. A piece of 18–20 gage copper wire (1.0–0.75 mm diameter) is sealed into this hole with epoxy resin or Duco cement. The wire is cut off 10 or 12 mm from the handle (Fig. 1).

The most satisfactory knives are made from splinters of thin, brittle stainless-steel blades. We know of no present source. The next best is brittle nonstainless-steel blades (e.g., Gillette "Super Blue"). These

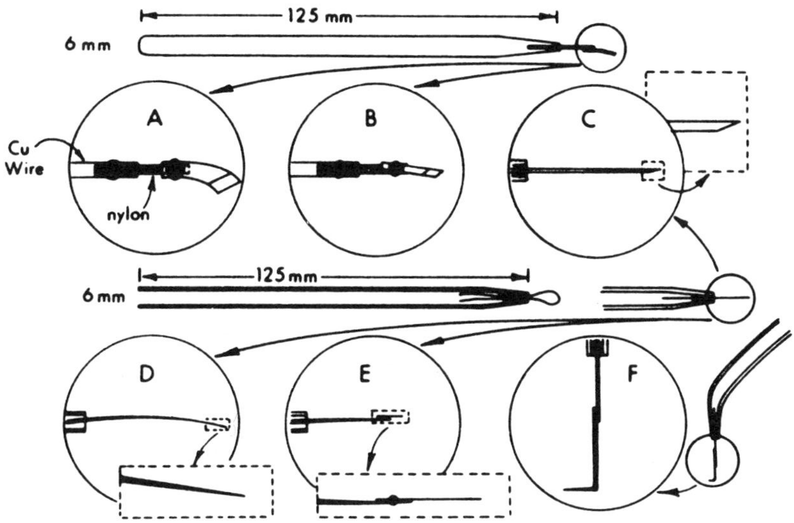

Fig. 1. Dissecting tools: razor blade knives, hair points, hair loop. A and B illustrate knives of two sizes. Hair point C is made from a human hair that is sharpened with a razor blade. Hair point D is a fine sable or camel's hair with a natural taper. E is a quartz-tipped hair point made from a fine camel's hair to which a 3–5 µm diameter quartz fiber is attached with epoxy cement. F is a special hair point for loading samples through oil in an oil well. A sufficiently stiff quartz fiber is attached with epoxy to a stiff hair (to prevent breakage). Note that the quartz fiber is drawn down to a tip beyond the bend.

must be protected from rusting. Most of the present thin blades on the market are not brittle, and points of knives made from these are easily bent during use. Nevertheless, these softer blades can be used for certain purposes: They can be cut with a sharp tin snips into very thin knives.

Splinters from brittle blades can be made by bending with the tip of needle nose pliers, or placing the blade on a hard plastic or smooth wood surface and applying pressure with a wood chisel. (Glasses or goggles must be worn in the latter case for safety from flying splinters.) The cutting edge is first broken off into a strip 5 or 6 cm wide, and broken into splinters at a 45° angle to the cutting edge. Fragments of blade 0.5–1.5 mm wide that resemble a scimitar in shape are most desirable to provide the proper angle when dissecting (Fig. 1). A short bristle is glued to the copper wire with epoxy resin. When the resin has

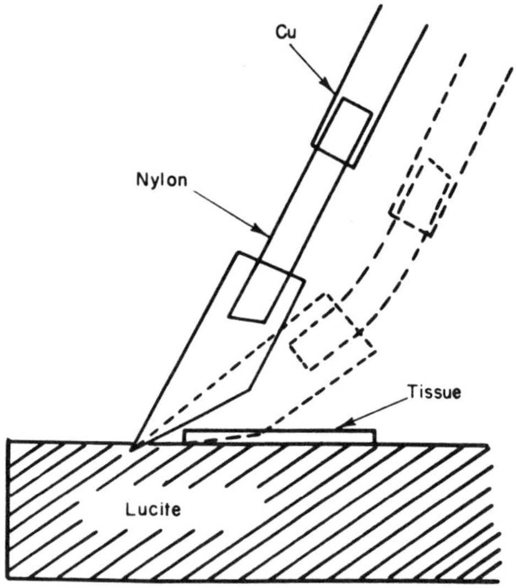

Fig. 2. Dissection technique to avoid tremor of the hand. *See text.*

set, the blade is fastened with epoxy to the free end of the bristle. Mounting the blade is easier if the resin is not too fluid. If necessary, it can be allowed to begin to set before mounting the blade. Bristles of different stiffness may be required for different purposes. The flexibility can also be adjusted by changing the free length of the bristle. Nylon bristles from a tooth brush are usually quite satisfactory.

The limiting factor in freehand dissection is tremor. For larger structures, this is sufficiently minimized by grasping the tools firmly with not more than 20 mm extending beyond the fingers. The side of the hand, the last three fingers, and if possible even the thumb are pressed firmly against the dissecting surface. The flexibility of the bristle mounting of the knife blade permits a dissection technique for further reduction of tremor and freehand dissection of very small objects. The tip of the razor blade is touched to the plastic surface enough to anchor it and stop the vibration (Fig. 2). The blade is exactly positioned by a slight lateral motion of the hand with the anchored tip as a pivot. When the position is correct, a slight downward pressure is exerted, causing the blade to rotate vertically, again with the tip as a pivot, making the desired cut in the dry tissue. In this manner, structures 15 or 20 μm in diameter may be readily dissected freehand.

When discrete layers from a structure, such as cerebellum or retina, are to be analyzed, the original microtome sections are usually cut at a small angle to the plane or the layers in order to make the layers appear as wide as possible. Consequently, unless the sections are exceedingly thin, the edge of a given layer may be in a slightly different position on one side of the tissue than the other. It may therefore be necessary to examine both sides of the section and discard the zone where adjacent layers overlap.

Handling Dissected Samples

After dissection, the samples are picked up and transferred with a short piece of hair mounted with Duco cement or similar adhesive in a pencil-shaped glass holder (Fig. 1). Such "hair points" are made from hair of different degrees of stiffness for different sample sizes or types of tissue. Human hair, extending 4–6 mm and sharpened by cutting at an acute angle, is suitable for a sample of 1–5 µg. The flexibility depends on the length of the hair and may be varied to suit the toughness of the tissue. A better hair point for smaller samples, down to 0.1 µg, is a fine hair with a natural taper, such as those in a sable or camel's hair paint brush. The tips of such hair taper to a diameter of 10 or 20 µm. A still better tip is provided by sealing to the end of a fine hair point a piece of quartz fiber 1–5 µm in diameter (Fig. 1). This is convenient for use with 0.1 µg samples, and is almost essential with samples weighing 0.01 µg or less. The tip of the hair is dipped in epoxy resin, leaving a very small amount on the hair. This is touched to a fine quartz fiber 5 or 10 mm long. Surface tension helps align the two fibers. When the resin has set, the quartz tip is cut off with scissors to a length of 0.5–1 mm.

Relatively large samples (over 2 or 3 µg) are lifted by pressing into the sample with a stiff sharp hair point to which the tissue then adheres. For smaller samples, it suffices just to touch the surface with the hair point or better to lift up the sample from underneath. Sometimes it is found easier to pick up the specimen than it is to disengage it subsequently, perhaps on a balance pan. This is a situation in which the fine quartz tip has great virtue, since a sample is easier to remove from a small tip than a large one. Static charge must be controlled with americium or polonium for satisfactory transfer of samples, particularly in the case of small sizes.

Identification and Histological Control

In most tissues, some study is required to recognize sufficient histological detail in frozen-dried material. With familiarity and a good light, however, a surprising amount of detail can be visualized. If the initial freezing is too slow, ice crystals will distort the material and make identification difficult. The thickness of the section may affect the ability to see and identify small structures. Slight differences in color are often helpful, and for this reason, white light is usually preferred for illumination.

As needed, sections are appropriately stained as a guide. The sections are placed in recorded order on a single slide lightly coated with histological glycerine-egg albumin. The sections can be flattened if necessary with hair loops and pressed down with the finger tip. The sections are then fixed and stained as desired. The accuracy of dissection may also be checked by mounting and staining the unused parts of the dissected section. Selected dry, unfixed sections may be stored for study and future reference. The sections are placed on a slide, a cover slip sealed at the edges with paraffin, and the slides stored at 4° to prevent deterioration.

Sample Carriers

Once the desired structures are dissected, the fragments are lined up on a tissue carrier for transfer to the balance and for temporary storage. It is convenient to design the carrier to hold 10–12 samples of the anticipated size (Fig. 3).

A piece of hardwood cut and shaped to be easily handled is used as the handle. This is glued to a segment of a glass microscope slide or, for smaller balances, a hemocytometer cover glass cut to the appropriate size. The glass is mounted at an angle that will facilitate insertion of the samples into the balance case for weighing. The hardwood handle is painted with black ink to reduce glare during weighing. The carrier with sections is placed in a Petri dish, so the sections will not be disturbed by air currents when moved. Handling at this stage is made easier if the Petri dish is turned upside down and if the lid (now the bottom) is shallower than the other part (Fig. 3). This permits the Petri dish and contents to be lifted without danger of pulling the two halves apart, which would suck in air and dislodge the samples. The Petri dishes should be set down gently, because if there is an electro-

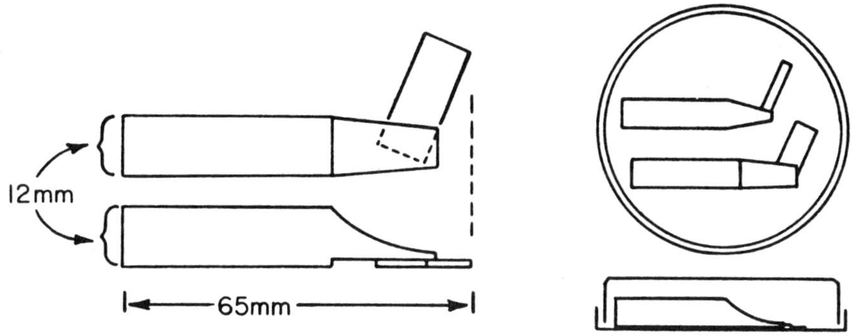

Fig. 3. Sample carriers. Two sizes are shown as well as the inverted Petri dish for protecting the samples from air currents. The body of the carrier is made of hard wood, the carrier surface is made of a glass strip cut from a microscope slide. For smallest carriers, the glass strip is cut from a hemocytometer cover.

static charge on the cover, a sharp jar may cause some of the samples to fly to the top. If this becomes a problem, it may be desirable to keep a small amount of americium (1–2 μCi) on metal in each dish.

Defatting Frozen-Dried Samples

In many tissues, there are regional differences in lipid content. It may be desirable to measure the lipid present, if for no other reason than to permit calculation of data on a fat-free dry weight basis. This is usually a truer basis physiologically than the total dry weight basis. An extreme example is presented by the nervous system, in which lipid can range in different regions from 20–75% of dry wt.

Samples weighing 1 μg or more are placed in a tube 4 or 5 mm id, and extracted with absolute ethanol by simply filling the tube a third full with the solvent and tapping just enough to dislodge the sample. After 10 min, the ethanol is carefully decanted in such a way as to leave the sample on the wall. The tube is then inverted for 30–60 s, and any fluid collected at the mouth is blotted off. The tube is immediately filled part way with hexane, the tube is tapped gently to be sure the sample is free, and the solvent is decanted as before. Hexane is added a second time to be sure all of the ethanol has been removed, and the last traces of solvent are aspirated carefully with a fine pipet. The section should come loose spontaneously or with a sharp tap. The sample can be removed from the tube by tapping onto a clean surface with the usual precaution to dissipate electrostatic charge.

Sticking will be caused by failure to use residue-free hexane, failure to wash away all of the ethanol, failure to use anhydrous ethanol, or leaving the ethanol on so long that it absorbs moisture from the air. Moisture can be easily absorbed if hexane is not added promptly after removal of the bulk of the ethanol. If the ethanol should dry, the section will be inevitably and irretrievably stuck. Sections ranging from 0.02 to 0.2 µg, because they are more difficult to see, are more easily extracted in an 8–10 mm glass tube of 2 or 3 mm bore sealed to a 50–60 mm length of tubing for a handle. The extraction procedure is the same as for larger samples except that the solvents are removed by aspiration with a fine-tipped pipet. Good visibility is maintained during aspiration to be sure the sample is not lost. Low-power magnification is helpful. These smaller samples may not fall out on tapping, but can be removed with a hair point.

Sample Volume

The sample volume can be measured provided the sections have been cut without serious compression. The volume is calculated from the thickness multiplied by the area. The fragment is placed on a microscope slide on a microscope stage, using a cover slip to keep the sample flat. The monocular eye piece is replaced with a right-angled prism. With a low-power objective and bright illumination, the image is projected on the wall and traced on a piece of paper. An ocular micrometer disk is projected at the same time and marked on the same paper to establish the magnification. The area of the fragment is measured with a planimeter.

Chapter 11

The Quartz Fiber Fishpole Balance

The fishpole balance is simply a quartz fiber mounted horizontally by one end. For large sizes, a light glass or quartz pan is attached to the free end. The sample is placed on the pan or the free end, and the displacement measured (Fig. 1).

This is an inelegant instrument compared to other balances. As ordinarily used, it is not capable of great precision. Any given balance is limited in usefulness to a narrow range. Nevertheless, the quartz fishpole balance is ideal for weighing small dry tissue samples, and it is capable of far greater useful sensitivity than any other mechanical balance presently available.

The great advantage of this simple balance is that it can be placed when necessary in a very small chamber and thereby be protected from the chief enemy of reproducibility: air currents. A balance of whatever design with sensitivity, for example, of ±10 pg, would be useless because of air currents if mounted in a balance case of 50–mL vol. However, a fishpole balance with this sensitivity can be mounted in a chamber of 0.2-mL vol where air currents are easily controlled.

Quartz is preferred to glass because of its greater strength and resistance to breakage, and because unlike most glasses it does not flow under tension. It will therefore retain its position indefinitely. A quartz fishpole balance that has been in continuous use for nearly 40 years shows no detectable change in position.

Measurement of Displacement and Useful Range

For most purposes, it is convenient to measure displacement of the balance tip on a micrometer ocular of a wide-angle stereomicroscope. The same microscope is used to supervise loading and unloading. A 5-mm ocular divided in 0.1-mm divisions has been found most satisfactory. (With a 10-mm ocular, loading is less convenient, since during loading the pan is near the zero point, which would be located at the extreme top of the field of view.)

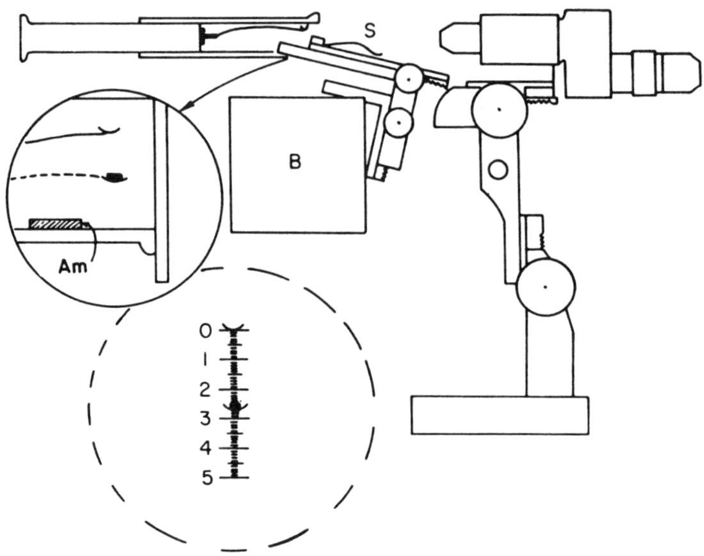

Fig. 1. Arrangement for loading and reading quartz fiber balance. The wooden sample holder is held by spring S to the double rack-and-pinion loading device shown (a modified stage micrometer). The loader is mounted on a 4 × 4 inch beam (B) that also provides a steady rest for the right hand with which samples are transferred to the balance tip. Transfers are supervised through the wide-angle microscope. The displacement of the balance tip is measured on the micrometer ocular in one eyepiece of the same microscope *(see* inset). The zero point is adjusted by means of the vertical rack and pinion of the microscope mount.

A 5-mm ocular scale represents a real distance that varies with the magnification of the objective, not with the magnification of the ocular. A 5-mm scale represents real distances of 5, 2.5, and 1.25 mm with 1, 2, and 4× objectives, and so on. For most purposes, a 1 or 2× objective is most satisfactory. Higher powers usually have inconveniently short working distances.

The scale of the 5-mm ocular can be read to a one-tenth of 1 small division or 0.01 mm. Since there is a 0.01-mm uncertainty at the zero point as well as at the point of deflection, the smallest deflection that can be measured to 2% is about 1 mm or one-fifth of the full scale. This limits the useful range to about fivefold for any given balance. Consequently, several balances may be needed if a wide range of sample sizes are to be analyzed. Nevertheless, to cover satisfactorily the range from 0.001 to 10 μg, only six balances would be required.

Choice of Suitable Fiber, Sensitivity as a Function of Length, and Diameter of the Fiber

For quartz fibers, the sensitivity can be adequately expressed by the formula:

$$S = \frac{L^3}{d^4} \text{ mm/}\mu\text{g} \qquad \text{[Eq. 11-1]}$$

where L is the length in millimeters and *d* is the diameter in micrometers. Thus the sensitivity increases as the cube of the length and inversely as the fourth power of the diameter. For example, a fiber 100 µm in diameter and 100 mm long would have a sensitivity of $10^6/10^8 = 0.01$ mm/µg or 100 µg/mm. The formula only applies strictly to small displacements, since as the fiber bends the horizontal distance from the point of attachment decreases.

Although the sensitivity increases as the cube of the length, there is for every fiber a length beyond which it would droop so far under its own weight as to be useless. For most purposes, this limit is reached when the droop (self-displacement of the tip) is >40 or 50% of its length. If the fiber balance requires a pan, it is usually desirable to keep the droop to 20–30%.

Figures 2 to 4 show the relationships between diameter, droop, and fiber length. These figures also show the sensitivity as a function of these three parameters and can serve as a guide to the choice of a suitable fiber for the sensitivity desired. For example, according to Fig. 2, a sensitivity of 2 µg/mm could be obtained with fibers ranging from 15–35 µm in diameter. A 15-µm fiber would be 25 mm long with droop equal to 5% of its length. A 35-µm fiber would be 115 mm long with droop equal to 50% of its length.

Although a suitable fiber could be chosen on the basis of its diameter, this becomes difficult to do accurately with the smaller sizes. A much more convenient method for fibers of all diameters is offered by the relationship seen in Figs. 2–4 between droop, length, and sensitivity. The fiber is suspended on a temporary mounting in a suitable glass container (to curtail air currents), and the displacement of the tip is estimated when the vessel is turned to shift the fiber axis from vertical to horizontal, or with the fiber horizontal, the vessel can be rotated 180° along the fiber axis, in which case the tip will move a distance equal to twice the droop. As an example, if a fiber 200 mm

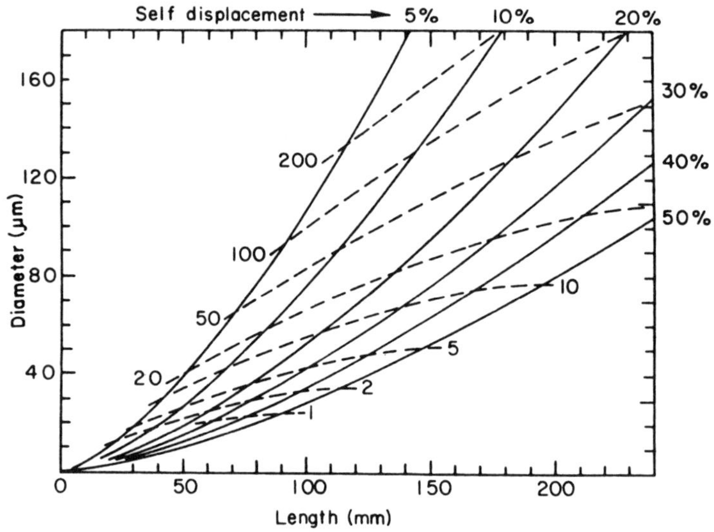

Fig. 2. Self-displacement (droop) and sensitivity of quartz fibers as a function of diameter and length [(corrected for deviations from formulas [Eq. 11-1] and [Eq. 11-2)]. Each solid line represents the relationship between diameter and length for a given percentage droop. For example, a fiber 80 μm in diameter at a length of about 135 mm would droop 20% or 27 mm. The droops are calculated for the condition in which the mounting is adjusted to bring the tip into line horizontally with the point of attachment. The dotted lines represent the sensitivity expressed as micrograms per millimeter of displacement. For example, a fiber 80 μm in diameter and 135 mm long would have a sensitivity of 20 μg/mm. This figure is also applicable to fibers under 0.18 μm in diameter. In this case, the ordinate scale would be 0–0.18 pm, the abscissa scale 0–2.4 mm, and the numerals on the sensitivity lines would represent picograms per millimeter of displacement. For example, a fiber 0.15 μm in diameter and 2.4 mm long would droop 30% of its length (0.7 mm) and have a sensitivity of 50 pg/mm of displacement.

long has a droop of 50%, it is about 80 μm in diameter and would be suitable for balances with sensitivity of 10 μg/mm or less (Fig. 2). If the droop is >50% of the length, a shorter segment of the fiber can be tested by mounting it so that only a portion of the total length projects past the support.

Instead of using the figures, the sensitivity can be calculated from the droop by the following formula:

$$S = \frac{(3 \times 10^6 D^2)}{L^5} = 3 \times 10^6 \left(\frac{D}{L}\right)^2 \left(\frac{1}{L^3}\right) \text{mm/}\mu\text{g} \qquad \text{[Eq. 11-2]}$$

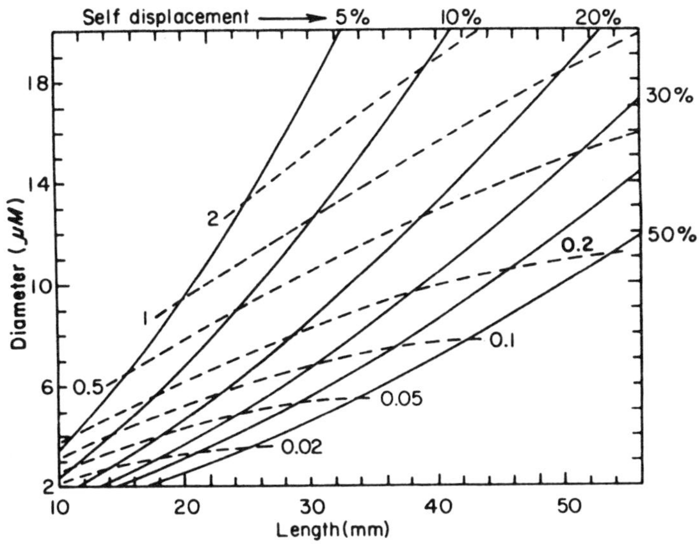

Fig. 3. Self-displacement sensitivity chart for fibers ranging from 2–20 μm in diameter (*see* Fig. 2).

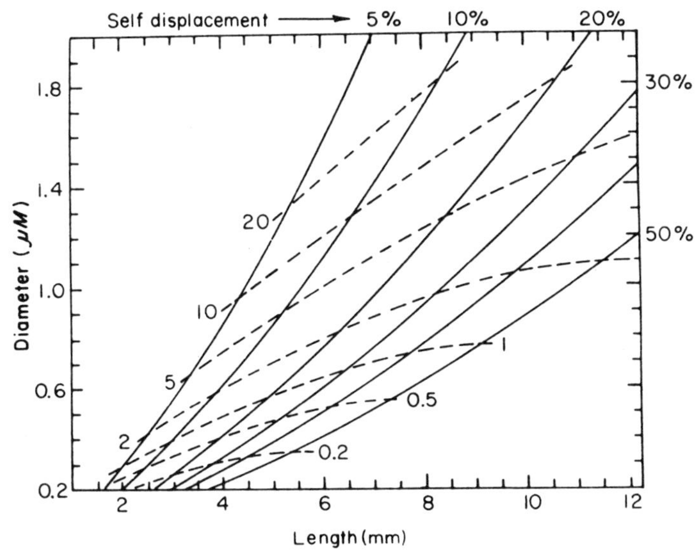

Fig. 4. Self-displacement sensitivity chart for fibers ranging from 0.2–2 μm in diameter (*see* Fig. 2). In this case, dotted lines represent sensitivity expressed as nanograms per millimeter of displacement.

where D is the droop in millimeters and L is also in millimeters. This applies strictly only to small droops, but the discrepancy is not serious with droops up to 50%. The error in using this formula with large droops tends to underestimate the sensitivity of the fiber.

With the use of Eq. (11-2) it is unnecessary to calculate the fiber diameter. For example, if a fiber 50 mm long droops 25 mm under its own weight, sensitivity is approximately:

$$3 \times 10^6 \times \left(\frac{1}{2}\right)^2 \times \frac{1}{50^3} = 6 \text{ mm/}\mu\text{g or } 0.17\ \mu\text{g/mm} \qquad \text{[Eq. 11-3]}$$

To make from this fiber a balance with a sensitivity of 0.4 µg/mm, since the sensitivity varies as the cube of the length (Eq. 11-1), the fiber is shortened to $50^3 \sqrt{(0.17/0.4)} = 38$ mm. Note the paradox that, because of the limitation on useful length set by droop, higher sensitivity can only be achieved ultimately by using *shorter* (and of course finer) fibers.

If 50% droop is set as a practical limit (D/L = 0.5), the maximum useful sensitivity for quartz fiber balances in general can be calculated as a function of the length. Substituting D/L = 0.5 in Eq. (11-2):

$$S = \frac{7.5 \times 10^5}{L^3} \text{ mm/}\mu\text{g} \quad \text{or } L = \frac{90}{\sqrt[3]{S}} \qquad \text{[Eq. 11-4]}$$

i.e., ultimate useful sensitivity for quartz fiber balances varies *inversely* as the cube of the length. This is illustrated in Figs. 2–4. A balance with a sensitivity of 10 µg/mm can be 200 mm long, but a balance with a sensitivity of 0.2 ng/mm can (for practical purposes) be no longer than 5.5 mm.

Balance Fibers

Commercially produced quartz fibers are available in sizes ranging down to 3 µm. They may also be easily made with a gas-oxygen torch, and at present, this is necessary for the smallest sizes. The method of preparing the fibers will depend on the size of the fiber desired. In the first method, one end of a piece of quartz tubing is fastened with adhesive tape to a 2–4 m length of rubber band or thin rubber tubing, which in turn is tied under tension to a stationary object so as to stretch the rubber when the quartz is held in the flame. The quartz is heated in the center until the lumen disappears. A small portion of the quartz is narrowed somewhat and then heated hotter until it is fluid, at which moment the tension on the rubber tubing is released. The heated quartz will be drawn into a long fiber. The fiber diameter depends on the amount of melted quartz, its temperature, and the tension and length of the rubber tubing. Fibers can easily be made in this way with diameters down to 10 of 15 μm (i.e., of a size suitable for balances capable of weighing 1 ± 0.01 μg or more).

A second method that will permit preparation of fibers of smaller diameter ulilizes the draft produced by the flame to pull the fiber. The quartz is first heated in a hot flame and pulled by hand to form a rod 0.5–1 mm diameter and 10–20 mm long. The narrow portion is then held nearly lengthwise in a large cool flame, where the rush of air will gradually thin the rod and then suddenly whip it out into a long fine filament. This will float in the air and can be trapped on a wooden club wrapped with black cloth. By varying the flame, smaller fibers of a wide range of sizes can be produced down to <0.2 μm in diameter.

A third method uses the weight of the quartz to pull the fiber. A rod of quartz is prepared from tubing as above. The rod is then suspended vertically 4 or 5 feet above the working surface. The narrow rod portion is heated just enough to soften the quartz and let it drop of its own weight, pulling a very fine fiber. This method is perhaps most useful for making balance fibers of 1–10 μm diameter.

The portion of the fiber selected for the balance ought to be of nearly uniform diameter. However, a slight uniform curvature is permissible, since the fiber can then be mounted with the concave edge upward. The result is to provide a more nearly straight fiber, when mounted, than if the fiber itself were perfectly straight.

Pans

Balances to weigh samples of 0.1 μg or more need pans to hold the samples. The pans are made from very thin fragments of Pyrex™ or quartz bubbles. Quartz is stronger than Pyrex™, but more difficult to blow. A compromise is Vycor™, which is almost pure quartz but melts at a somewhat lower temperature. To make the thin bubble, a piece of tubing is first closed at one end with heat, and then a fairly large portion of the closed end is nearly melted in the flame. The tubing is removed from the flame and quickly blown to produce a bubble thin enough to show interference colors. The bubble usually bursts, and the fragments are collected. The cuts necessary to reduce a fragment to an appropriate size and shape are made by placing it on a plastic surface and pressing straight down with the edge of a razor blade. The weight of the pan itself should not cause the fiber to bend by a distance of more than 20 or 30% of its length. This usually means that the pan should not weigh more than the load limit of the balance. No pan is needed with balances used to weigh samples of <0.1 µg, since such samples adhere well to the bare fiber tip. An adequate alternative to a pan for balances in the 0.05–2 µg capacity range (and easier to make) is a T at the fiber tip made by attaching with epoxy a short piece of quartz fiber of about the same diameter as the balance fiber.

Balance Case

The fiber is permanently mounted in a case to protect it from air currents and dust. The chamber must be of proper size in relationship to the balance to facilitate weighing and minimize air currents within the case. A balance case made from a glass syringe has been found most convenient (Fig. 1).

The size must be chosen according to the sensitivity of the balance, in order to keep internal air currents within tolerable limits. The proper sizes have been determined empirically. A 50-mL syringe is satisfactory for a balance of 5-µg capacity (1–5 µg range, ±0.01 µg). For balances of 0.5 µg, 0.05 µg, 5 ng, and 0.5 ng capacities, the chamber volume should not exceed 10, 2, 0.5, and 0.2 mL, respectively. (Since the two smallest balances will be very short, they can be accommodated in 2 and 1 mL syringes, respectively.)

To make the balance case, the closed end of the syringe barrel is cut off and the plunger reversed. The bottom end of the barrel may need to be ground out slightly with fine carborundum and water to admit the plunger. The flared end of the barrel is ground flat against a glass plate with carborundum and water to permit a nearly airtight closure by a glass cover.

A short piece of copper wire of suitable flexibility (20 gage, 0.75 mm diameter) is sealed to the end of the plunger with epoxy resin. The wire is mounted in such a way that 5–15 mm (depending on the size of the balance) extends straight out from the center (Fig. 5). For the more sensitive balances, to facilitate attachment of the fiber, a 5-mm

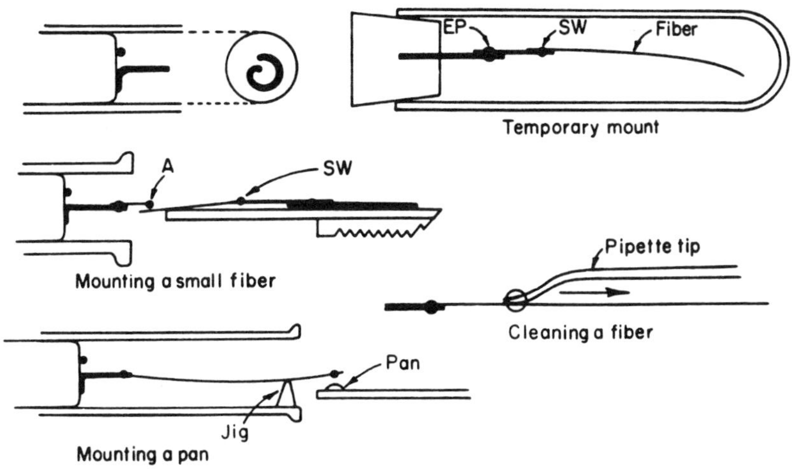

Fig. 5. Mounting quartz fibers and balance pans, and cleaning the fiber. At upper left is shown how a copper wire is bent to permit secure fastening (with epoxy cement) to the end of the syringe plunger. At upper right is a temporary mount for a quartz fiber. A fine steel wire is sealed to a stiff wire with epoxy (ep). This in turn is sealed to the fiber with a minute amount of sealing wax (sw) or similar material. (The sealing wax is first affixed to the steel wire; the wax is touched to the fiber tip, which is then trapped by momentarily melting the wax with an electrically heated wire held nearby.) A small fiber is mounted on the syringe barrel as shown. In the case illustrated, a short piece of fine steel wire has been fastened to the heavier copper wire with epoxy. A very small droplet of epoxy is placed on the tip (A). The quartz fiber tip is now raised until it touches A, at which time the epoxy is polymerized with heat *(see text)*. The other end of the fiber is released from its temporary mount by applying heat to the sealing wax. A pan is mounted in a similar manner as shown at lower left *(see text)*. Note the special bend near the tip of the pipet used to clean a fiber. This is to prevent the fiber from being drawn to the pipet shaft by capillarity. The diameter of the pipet is exaggerated in the figure, and the walls of the balance case are not shown.

length of fine steel wire (e.g., 38 gage) may be fastened to the copper, or in the case of the most sensitive balances, a short piece of stiff quartz fiber about 20 μm in diameter is used.

The balance is mounted rigidly on a heavy ringstand, and once the fiber is in place and calibrated, the orientation of the balance should not be changed without checking the calibration. Just under the open end of the balance, a firm support is provided on which to rest the hand when loading the sample. This support should not touch the balance support, to make sure that the balance position is not disturbed during loading. Loading samples is made easier by attaching to this same support a simple rack-and-pinion manipulator to hold the sample holder and move the samples in and out of the balance. This can be constructed from a microscope stage micrometer that is turned on edge and provided with a suitable spring clip (Fig. 1).

The balance closure is made with a piece of good-quality flat glass. A large or small microscope slide or portion thereof will usually suffice. Wire hooks or loops are sealed to the glass with epoxy cement to provide for attachment of long rubber bands to hold the glass plate in place. The rubber bands should be strong enough to provide firm closure, but at the same time permit easy removal of the plate without the slightest displacement of the balance. (Very weak long springs would be preferable to the rubber bands, which must be replaced at intervals.)

Mounting the Fiber

The quartz fiber is sealed to the copper or steel wire or stiff quartz with epoxy resin. Handling a fine fiber is made easier by the use of a temporary support. Convenient for this is a 5 mm length of fine wire (36–38 gage) that is permanently attached with epoxy cement to a larger wire (Fig. 5). A cement that softens with heat, such as De Khotinsky cement or sealing wax, is used to seal the fiber to the tip of the fine wire. The heat is provided by an electrically heated short piece of platinum or Nichrome wire brought nearby. The larger wire *(above)* can be inserted in a cork for easy manipulation of the fiber, or for storage in a tube and for testing its sensitivity.

The fiber is permanently mounted with the aid of a dissecting microscope. The plunger is extended through the syringe barrel until the wire or quartz support emerges. A small amount of epoxy resin is placed on the tip of the support. The free end of the quartz fiber is brought into contact with the resin with a hair point. For finer fibers, a rack-and-pinion device mounted on a ring stand may be used (Fig. 5). With the fiber properly aligned, the resin is polymerized by heat from the electrically heated wire held close by. The resin is approached with the hot wire until bubbles begin to form. The heating is continued until the bubbles do not shrink when the hot wire is temporarily withdrawn. The other end of the fiber is then released with heat from the temporary mounting. If necessary, any residual cement on the free end can be either burned off or removed with fuming HNO_3. The fiber is positioned by bending the wire support, and if necessary, the length is reduced with scissors to provide the desired sensitivity. The plunger should be positioned so that the natural curvature of the fiber, if any, is concave in the upward direction.

If a pan is needed, it is now attached upside down. A small amount of epoxy resin is placed on the tip of the fiber supported by a paper jig above the pan (Fig. 5). The pan is placed on a glass slide with its concave side down. The fiber is lowered to the pan by removing the paper jig or raising the glass slide, and the resin polymerized with heat from the hot wire.

The final adjustment of the balance is made so that the tip is 2–3 mm above the point of support to favor linearity of response. (Mounted in this way, as the fiber bends under a load, the horizontal lever arm will first lengthen slightly, and then shorten slightly.) To minimize disturbance from air currents, the tip should be drawn into the syringe a distance equal to the radius of the syringe and the syringe plunger taped in position. A piece of metal bearing 10–20 μCi of americium is placed in the bottom of the balance case to dissipate static electricity.

The distance of the tip from the top of the syringe should be recorded, since fibers may get dirty and droop further. This is particularly noticeable with very sensitive balances. When this occurs, the fiber should be cleaned *(see Cleaning the Balance Fiber)* so that the original position, and therefore the original calibration is regained.

Lighting and Viewing

Good lighting is required with a minimum of heat. With larger balances, is can easily be accomplished with fluorescent lamps. With smaller balances, the fiber, especially when new and clean, may be difficult to see clearly. A fiber light guide provides excellent illumination with usually negligible heat.

Weighing

The zero point is adjusted. The balance case is opened. With the rack-and-pinion, the samples on the holder are moved into the balance case just under the fiber tip. A sample is loaded onto the balance pan or tip with a hair point, and the holder withdrawn. The glass cover is replaced, and a reading made.

Loading a balance fiber that does not have a pan is not as difficult as it might seem. The sample is picked up from the carrier with a fine hair point, preferably quartz-tipped *(see* Chapter 10). The sample is touched to the fiber tip as it floats freely in the air. The touching is repeated several times, if necessary, perhaps with slightly different orientation, until the transfer is accomplished. The process is repeated to remove the sample after the weighing has been made. Samples of 0.02 µg or less will adhere either to the tip of the hair or the tip of the balance, and will almost never fall off. A sharp tap of the balance case might, however, cause a sample to fall. Good control of electrostatic charge makes for easier transfer.

In general, it is easier to load and unload samples of all sizes if air currents in the neighborhood are also well controlled so that the balance tip remains reasonably stationary when the case is open. *(see* Chapter 10 for suggestions on screening the room.) On this account, it is useful to shield the balance from the operator's breath by draping the front of the microscope head with a piece of cloth. Ordinarily readings can be made within 10–20 s after closing the case, and should be reproducible to 0.01 mm or at most 0.02 mm on the micrometer ocular (0.005–0.01 mm real distance with a 2× objective). If there is a slow drift to a stable position, there may be insufficient americium in the balance, or the operator may be wearing clothing particularly likely to generate static electricity. Other sources of instability could be too much heat from the lighting, poor closure of the chamber and a drafty room, unstable mounting of the balance, or wobble in the microscope. In any event, the zero point should be checked frequently. Often there is a slight shift in the zero point during the first few weighings—possibly because of heat from the hand.

A good balance will give a nearly linear relationship between weight and deflection, although there must, because of the nature of the fiber balance, always be some departure from linearity, since the horizontal distance between the two ends must change as the fiber bends. It is easily possible to keep the nonlinearity to within 2–5%. Often the difference can be ignored, but for some purposes, it may be worthwhile to make a correction. In any event, a new balance should be checked for linearity *(see Calibration of Balances and Testing Linearity).*

If the pan is large in relation to the length of the fiber, reproducibility will be increased slightly if the sample is always placed near the center of the pan. Since the sample actually sticks to the surface of the pan, the length of the lever arm is determined by the center of gravity of the sample, not by the point at which the sample happens to touch the pan. (The displacement of the tip varies as the square of the horizontal distance of the sample from the mounting. Thus, a 1% difference in this distance will make a 2% difference in tip displacement.)

Correction of Weight for Adsorption of Gases and Moisture

Porous frozen-dried tissue has a large surface area that adsorbs O_2, N_2, and moisture. When a 20 µm thick sample is taken out of a vacuum, this adsorption reaches equilibrium in a few seconds. In the case of animal tissue, adsorbed O_2 and N_2 add 2% to the weight. The amount of moisture adsorbed is proportional to the humidity and is equal to about 1% for each 10% humidity. With 50% humidity, therefore, a tissue sample will weigh approx 7% above its true dry wt. This correction can be incorporated into the balance factor. (Dry tissue rich in lipids will increase somewhat in weight if left in room air for several days, probably as the result of lipid oxidation.)

Calibration of Balances and Testing Linearity

Balances of 0.5-µg capacity or larger can be calibrated colorimetrically using *p*-nitrophenol crystals. The deflections owing to single crystals of suitable size are measured. Each is then dissolved in a measured volume of carbonate buffer (50 m*M* Na_2CO_3, 50 m*M* $NaHCO_3$) in the proportion of 1 mL for a 1–6 µg crystal. Readings are made at 400 nm in the spectrophotometer, and the weights are calculated from readings made with appropriate standards. For example, crystals ranging from 0.5 to 3 µg in size are dissolved in 500 µL to give OD of 0.123–0.8 at 400 nm. A stock solution containing 1 mg/mL is prepared from the same bottle of *p*-nitrophenol. This is diluted 200-fold with the carbonate buffer (giving a 5 µg/mL solution), and the reading obtained is used to calculate the weight of the single crystal samples.

Balances of smaller capacity can be calibrated fluorometrically with crystals of quinine hydrobromide. (Quinine sulfate crystals are unsuitable, because they are flat and tend to stick to flat surfaces, including that of the balance pan.) The fluorescence of the quinine crystal dissolved in 1 mL of 0.1*N* H_2SO_4 is read against an appropriate standard.

Quinine at a concentration of 0.001 µg/mL will yield measureable fluorescence. However, it is difficult to manipulate crystals smaller than about 0.01 µg.

Frozen-dried tissue samples, being light and soft, are much easier than crystals to pick up and to manipulate on and off the balance pan or fiber. Therefore, once a relatively large balance has been calibrated, the easiest procedure is to calibrate a new balance by comparison with the old one. For example, suppose a balance with a capacity of 5 µg has been calibrated, and it is desired to calibrate a balance with a capacity of about 1 µg. A piece of tissue weighting 3 or 4 µg is cut into four pieces of nearly equal size (*see* Chapter 10). The four pieces are weighed together on the larger calibrated balance, and each piece is then weighed separately on the smaller balance. From the sum of the weights on the smaller balance its sensitivity is readily calculated. This second balance, can in turn be used in the same manner to calibrate a still smaller balance and so on.

Frozen-dried tissue samples can also be used to assess the linearity of a balance. A sample that gives a deflection slightly less than full scale is divided into four or five nearly equal parts. These are weighed all together, then singly, and in combinations of two and three. For example, suppose there are four pieces that weighed together give deflections of 4.00 divisions and this has been determined to equal 4.00 µg (i.e., the sensitivity in this region is 1.00 µg/division). The individual samples a, b, c, and d give deflections of 0.9, 1.10, 1.15, and 1.05 divisions, which total 4.20 divisions. Therefore, in the region of the first division on the balance, the sensitivity is 4.00 µg/4.20 divisions of 0.95 µg/division. When samples a and b are weighed together, they give a deflection of 1.98 divisions. Therefore, in the region of two divisions, the sensitivity is $0.95(0.90 + 1.10)/1.98 = 0.96$ µg/division, and so on.

Once standardized, the sensitivity of a quartz fiber balance has not been found to change even after years, unless the fiber becomes coated with dirt. Nevertheless, it may be wise to check the sensitivity occasionally. For this purpose, permanent balance standards can be prepared with gold leaf. These standards can also be used to calibrate new balances. The piece of gold leaf of suitable size is crumpled slightly to make it easier to pick up and less likely to stick to a smooth surface. The weight is determined on a calibrated balance, and it is then stored where it is unlikely to collect a film of dust. A convenient storage vessel is one shaped like the sample holders for dissected tissue samples (Chapter 9, Fig. 1), except that a piece of aluminum replaces the glass. A shallow depression somewhat larger than the piece of gold leaf is drilled in the metal toward the front. The gold leaf is stored in this hole covered with a small piece of glass that is held in place with a clip. Only weights of 1–3 μg have been used, but it is probable that smaller weights could be handled and stored satisfactorily.

Cleaning the Balance Fiber

In spite of nearly airtight closure of a balance, the fiber will ultimately become covered with a thin film and droop further. To clean the fiber, a small droplet of cleaning fluid is formed at the end of a very fine-tipped pipet. This droplet is carefully touched to the fiber near the mounting (Fig. 5) and is drawn forward the length of the fiber until the fiber escapes from the droplet. This may be repeated, if neccssary. The dirt film can usually be removed by successive rinsings with 20% ethanol in H_2O, 95% ethanol, and finally hexane. With care, this can be done without dislodging or breaking the pan. The ethanol is added to the H_2O for the first rinse to lower the surface tension, which might cause a delicate fiber to be pulled off.

If these solutions do not clean the fiber, fuming HNO_3 may be used as a last resort, but this will usually take off the pan. Cleaning is therefore easier (and required more often) with balances small enough not to need a pan. This is true except with exceedingly sensitive balances (capacities under micrograms), for which great care must be exercised not to remove the fiber along with the film of dirt.

CHAPTER 12

Histochemical Analyses

Introduction

Analyses of histological samples are made by the same basic principles described for larger specimens. Samples of course will generally be small, and may become very minute as the analyst's ambition grows to study smaller and smaller structures. This implies a demand for an enormous range of sensitivity for studies, all of which could be denoted as quantitative histochemistry or cytochemistry. Consequently, in adapting methods to histochemical purposes, certain modifications must be introduced depending on the sample size and the substance concerned. To measure lactate dehydrogenase in a 2-μg sample from the molecular layer of cerebellum requires a million times less sensitivity than to measure lactate in a 2-ng portion of an anterior horn cell. To measure the lactate dehydrogenase, it is easy to provide sufficient sensitivity, but for reasons of convenience, the reaction is carried out at high tissue dilution. This makes it necessary to be concerned about the effects of dilution on enzyme stability. To measure the lactate requires not only great sensitivity, but also the use of very small volumes for the reaction in order to circumvent blank problems.

Enzyme Stability at High Dilution

Primarily because of surface denaturation, most enzymes become unstable at high dilution. This is particularly significant when the ratio of surface to volume is large, as it is when the volume is small. The addition of an excess of inert protein (e.g., serum albumin) crowds out most of the enzyme from the surface and protects it. A rule of thumb is that the total protein content should be 0.02% in volumes down to 2 or 3 μL and 0.05–0.1% in the much smaller volumes sometimes used in oil wells. A 1-μg dry tissue sample in a 10-μL vol constitutes a 0.005–0.01% protein solution.

Addition of Sample and Reagent

The introduction of the dry sample into the reagent in a small test tube is a critical step, particularly in the case of a small sample in an enzyme assay. When the sample becomes wet, it becomes more difficult to visualize, and it could be left stranded unnoticed on the wall above the meniscus. Every effort should be made to see the sample after it is in the reagent. It is important to have good lighting and clear vision. The operator may benefit from special glasses that permit close examination, or a wide-field microscope may be used. Another more serious problem with small samples is discussed below.

Dry Loading

Relatively large samples (1 µg or more) are transferred to the bottom of a dry tube. Usually, the sample is placed inside the mouth of the tube and then dislodged with sharp taps, while the tube is in a vertical position. Sometimes electrostatic charges make it difficult to make the sample fall to the bottom. In this event it is helpful not to handle the tubes more than necessary and to keep an α-particle radiation source near the samples being loaded. When reagent is added to the sample, in order to prevent the possibility of the sample attaching itself to the pipet tip, it is helpful to keep the tip away from the sample when delivery is started and to keep the tip in the extreme upper part of the meniscus as delivery continues.

In an enzyme assay, it is unnecessary and undesirable to mix the sample after reagent has been added. In addition to the danger of washing the sample up onto the wall, violent mixing could result in surface denaturation. Repeated tests have shown that, with volumes of reagent even as large as 40 µL, diffusion is adequate to bring substrate to the enzyme. This is only true for reagent added to the porous frozen-dried material. If a second reagent solution were to be added to the first reagent, mixing would, of course, be required. If in spite of precautions a section should be stranded above the reagent, gentle tapping with the finger could be used to swirl the liquid gently and wash the sample down.

In a typical enzyme assay, with 1-µg samples or larger, the tubes with sections are lined up in a rack at room temperature. The tubes are identified by their position in the rack. (Until reagent is added, the tubes must not be chilled because of the danger of condensation of moisture from the air.) As each tube receives the reagent it is transferred to the same position in a second rack in an ice bath. When all tubes, standards, and blanks are ready, this second rack is transferred to the incubation bath (usually 38°) for the required period and then returned to the ice bath. If a second solution is used to stop the reaction, this is added at about the same pace as that used for the first reagent addition. Consequently, all samples are exposed to the first reagent for the same total time at 0° (as well as at 38°), and therefore, the enzyme will be given the same opportunity for action in all samples. If the reaction is instead to be stopped by heat, allowance may have to be made for the fact that the first samples were exposed to reagent for a longer time at 0° than the last samples. (It is highly desirable to know the relative rate at 0° for the enzyme concerned. This can vary from 2 to 40% of that at 38°.) For samples smaller than 0.2-μg dry wt and assays that require reagent volumes smaller than 20 μL, we strongly recommend substitution of oil wells for tubes (*see* next section).

Oil Well Technique

To utilize the full sensitivity of enzymatic cycling, it is usually necessary to carry out the first analytical steps in very small volumes. It is difficult to work with volumes of <1 μL in open test tubes without serious evaporation. By working under oil in holes drilled in a block of hydrophobic plastic, aqueous volumes as small as 1 nL can be employed with little or no evaporation *(1–3)*. According to the demands of the particular analysis, successive additions can be made to the sample, as in any test tube, except that there is greater latitude in regard to volume. For example, an initial vol of 5 nL may be allowed to grow to 10 μL within the oil well. Usually, all but the last analytical steps are carried out in the oil well. Heating steps can be employed, with temperatures approaching 100° without evaporation.

Composition of Oil Well Rack

It is essential that the oil well be hydrophobic; otherwise, droplets will wet the surface and spread, making mixing with subsequently added reagents difficult or impossible. Teflon™ is nearly ideal in that it is both hydrophobic and chemically inert. Its one disadvantage is its opalescence, which decreases visibility. Polyethylene is better in this respect and may be used, but destructive reagents must not be employed for cleaning. Siliconed glass has been tried, but alkali will remove the silicone coating. Therefore, no analytical step requiring alkalinization can be employed. This is a great disadvantage. Polycarbonate (Lexan™) has been used successfully and also works best if a silicone coating is used. However, alkaline solutions used during the analyses do not spread. Although a degree of trouble is involved in washing and preparing the polycarbonate wells, the advantage of increased visibility over Teflon™ may be worth the effort.

Teflon™ or Polycarbonate Oil Well Racks

The size of the Teflon™ or Lexan™ racks and of the wells may be varied as needed. A convenient rack consists of 20 × 120 mm block, 5 mm thick, drilled with 60 holes of 3 mm diameter. In the case of Teflon™, a flat-ended drill is used to create a flat-bottomed well deep enough so that only a very thin translucent layer remains. The thickness of the bottom wall is not so critical with Lexan™. The Teflon™ rack may be polished with a piece of fine sandpaper wrapped around a metal rod slightly smaller than the diameter of the well. The polycarbonate wells can be freed of burrs with acetone treatment. A drop of acetone is added to each well individually with a Pasteur pipet and aspirated within seconds. For very small tissue samples, there is some advantage in using thinner racks, 3/32 inch or 2.5 mm thick, with correspondingly shallower wells.

If the oil wells are to be siliconized, they are placed in a vacuum dessicator with 1–2 mL of silicone (Pierce™) in a watch glass, and a slight vacuum is created using a pump or vacuum line. The wells are left under vacuum for about 5 min, and the dessicator opened under a hood. The wells are transferred to another dessicator and put under vacuum for a few minutes to remove the silicone fumes. The wells are removed and stored covered with foil until use.

Oil

A mixture of 40% hexadecane and 60% USP light mineral oil (specific gravity, 0.83–0.86) has been found to be satisfactory. The viscosity of oil used must be great enough to protect the fluid droplet from access of CO_2 from the air and still permit the droplet to fall rapidly to the bottom of the well when delivered from a pipet. Too viscous an oil adheres to the tip of the pipet to a disturbing degree.

To ensure that the oil is free of disturbing impurities, the following washing procedure is recommended. The oil mixture is shaken vigorously in a separatory funnel with 2 vol of 1*M* NaOH, and the oil and water phase separated. The oil is then washed with H_2O, 0.02*M* HCl, and repeated changes of water until the washings are neutral. The oil is finally centrifuged at 10,000 rpm, and the oil phase removed from the water phase. The last traces of water are removed by drying the oil in a vacuum dessicator. Unless the oil is cleaned, alkaline droplets may react with the oil during heating steps, and the droplets may spread out and be difficult to see. If it is necessary to work at 0°, light mineral oil alone is substituted for the hexadecane-mineral oil mixture, which solidifies below 10°.

Heating and Cooling Teflon™/Polycarbonate Racks

Because the plastics are good thermal insulators, the time of heating and cooling the racks must be made longer than that for tubes immersed in water. It takes about 10 min for the rack to reach the ambient temperature. Heating is most easily done in an oven of the desired temperature, or the rack may be placed in a metal pan covered with foil and floated in a water bath. The foil is to protect the rack from any condensation of water. If it is necessary to make additions or remove samples at 0°, the racks may be cooled by placing them on a plastic bag filled with crushed ice in a Petri dish. As said above, when low temperatures are needed, mineral oil alone is used because the hexadecane–oil mixture solidifies.

Cleaning Oil Well Racks

Whether the racks are Teflon™ or polycarbonate, the bulk of the oil is removed by inverting the rack and tapping it out. In the case of Teflon™, any remaining oil is rinsed out with acetone, and the racks heated for 15 min in 0.5*M* ethanolic KOH or NaOH *in a water bath in a hood at no higher than 70°. (Fire hazard!)* The racks are carefully rinsed with tap water, and then deionized water and dried in an oven at 100°.

After tapping out the polycarbonate racks, they are rinsed with fresh petroleum ether. The racks are then placed in a beaker of Alconox™ solution, and the beaker placed in a boiling water bath for 15–20 min. After cooling, the tubes are rinsed and dried as described for the Teflon™ racks.

Reagent Containers

During the analytical procedure, it is convenient to have the reagcnt in a container that can be viewed in the microscope along with the rack. A convenient holder is another plastic rack with larger wells in which the reagent can be stored under oil. Another simple solution is to place 1 mL or less of the reagent in a 10 × 75 mm tube. The tube is placed horizontally on a suitable holder, such as one of the pipet racks described in Fig. 3, Chapter 3. Surface tension will prevent the fluid from running out, and the reagent is easily accessible.

Pipets and Oil Well Pipeting

Constriction pipets of a special design are made for use with oil wells (Fig. 1, Chapter 3). A finer tip than usual is needed, since the volume must be removed by scraping it off at the oil surface, rather than by transferring it with the aid of surface tension to a glass surface. If the tip is too blunt, none or only part of the volume will be removed. Quartz rather than glass is used, because its better heat conductivity makes it easier to make a uniform constriction, and its greater strength permits a sturdier sharp tip. The tip is drawn in a slender taper to a fine point and bent at an angle of 45°, 5–6 mm above the orifice. The bend makes it possible to pipet into the oil with the hand steadied by resting on the working surface. The constrictions must be well balanced (Chapter 3), because if air should be delivered after the liquid, the bubble will splatter and ruin the sample. If delivery is too rapid, the liquid may break up into several droplets.

As with the usual constriction pipet, the tip should not be dipped further than necessary into the fluid to be pipeted. Since there is no hydrophilic surface against which to deliver the fluid, pipeting is made directly into the oil. By withdrawing the pipet from the oil, the droplet is stripped from the tip and sinks to the bottom of the well. If a second volume of reagent is added, the pipet is touched to the surface of the previous droplet, and the fluid delivered into it (Fig. 1).

There is frequently a tendency for droplets to remain suspended from the oil surface. If so, it is essential that they be dislodged; otherwise, disastrous evaporation will occur through the point of contact.

Microscope and Working Stage

All the operations are carried out under low magnification (0.5× or 1× objective, 10× oculars) of a stereo dissecting microscope. A vari-

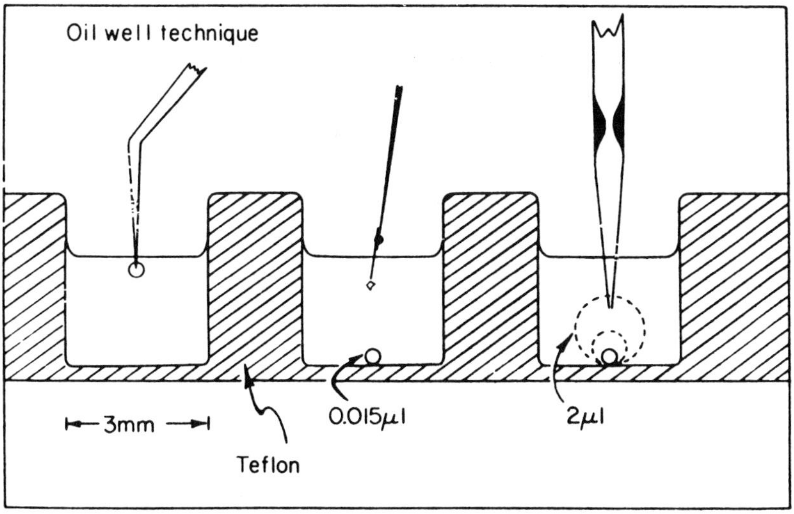

Fig. 1. Steps in an analysis in oil wells. The hair point shown would be improved if the tip had a short foot (*see* Fig. 1, Chapter 10). The volumes indicated are completely arbitrary.

able magnification (Zoom) microscope with long focal distance is especially useful. The microscope should be mounted on an arm projecting over the working surface.

The rack is illuminated by transmitted light from below. A suitable working surface and light are provided by a wooden stand in which a 90° glass prism is fitted to project the light through the rack (Fig. 2). Visibility is further improved if the illumination is restricted to a single well, thereby reducing glare from the rest of the rack. The stage is covered with a sheet of dark plastic metal drilled with a hole the size of an oil well (Fig. 2). The light entering obliquely is focused on this opening.

Static electricity is eliminated by suspending a piece of metal with americium or polonium on a stand above the rack during analyses. The static must be dissipated, or the aqueous droplets may migrate capriciously or even disintegrate into smaller droplets.

Dust must be prevented from falling into the oil wells. Except when additions are being made, the oil rack should be covered. Except for the least demanding analyses, it is advisable to suspend a shelf of clear Plexiglas™ over the wooden stand just below the microscope. This

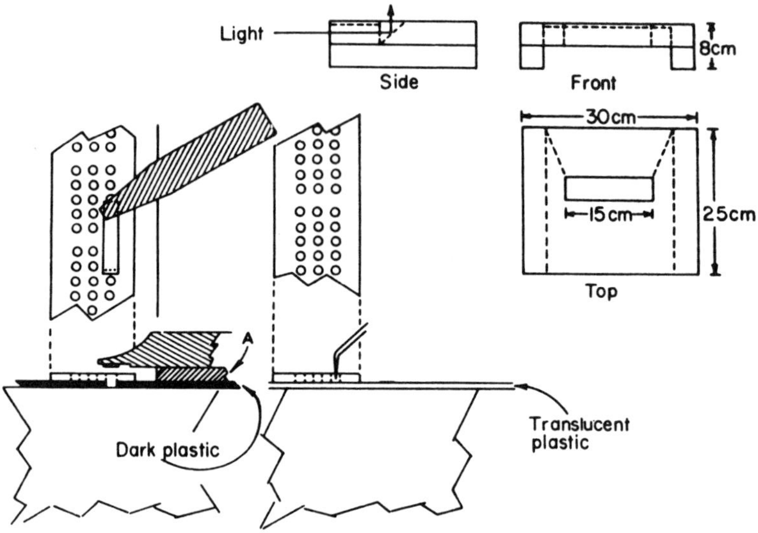

Fig. 2. Platform for oil well work. Light is transmitted from below by means of a right-angle prism or mirror. When loading samples, as shown on the left, the surface is completely covered by a sheet of dark plastic to absorb most of the light, except where a hole is drilled with a diameter about equal to that of the oil wells. The right half of the platform is covered by a smooth plate (A) somewhat thicker than the oil well rack. This is fixed firmly in place with tape and permits the oil well rack to slide under the sample holder. When adding reagent, as shown in the middle, plate A is not needed, and better visibility is provided by covering the platform with a sheet of translucent plastic to illuminate the reagent vessel as well as the oil wells.

protects the oil wells from dust (and dandruff!) during analytical manipulations.

Addition of Sample Reagent

Two different procedures have been used to bring the sample and reagent together. If the volume of the initial step is >0.1 µL, the tissue sample can be placed in an empty well, the reagent added, and the droplet immediately covered with oil from an eye dropper.

An alternative and better procedure of general applicability and almost essential with smaller volumes was kindly suggested to us by J.-E. Edstrom. Oil is added first, followed by the reagent droplet, and then the sample is introduced into the droplet through the oil (Fig. 1). It is guided or pushed through the oil with a special quartz-tipped hair point until it makes contact with the reagent. During the passage through the oil, the porous tissue becomes soaked with the oil, but this is displaced upon contact with the aqueous reagent, on which the sample rapidly spreads.

In the case of very small samples, the introduction of the sample through the oil can be difficult at first. Therefore, this step deserves elaboration. The hair point used to guide the sample through the oil is constructed with a fine foot (Fig. 1; Chapter 10; notice the angle of the shaft). The sample is picked up from the carrier with the tip of this foot and pushed into the oil where actual contact is lost. Thereafter, the foot acts as a kind of fan to wave the sample down to the droplet. Once real contact is made with the droplet, the samples of most tissues will spread over the surface (muscle is an exception). If this sign is not observed, the sample can be gently pushed into closer contact. If possible, however, the tip is kept free of the droplet, otherwise there is danger that part of the sample will adhere to the tip. As the sample is pushed or fanned through the oil, it is followed by focusing down with the microscope. If the sample falls to the bottom of the oil well without making contact with the droplet, it may become invisible.

The sample, while in the oil, becomes translucent because of impregnation with oil and may be somewhat difficult to visualize. This is made worse by the fact that the oil meniscus acts as a negative lens, causing the sample to appear to shrink, more and more, the further it gets from the surface. If the volume of oil is kept to a minimum, however, this lens effect is decreased, and the distance through which the sample must be pushed is diminished. (The substitution of shallower oil wells was mentioned above.) If too much difficulty is experienced, the oil volume can be reduced until it only barely covers the aqueous droplet. After sample addition, oil is added to all wells. (Reducing the percentage of hexadecane or eliminating it altogether may improve visibility by increasing the difference in refractive index between sample and oil.)

Loading either directly or through the oil appears to give the same results. Glucose, glucose-6-P, glycogen, and a number of enzymes have been specifically tested in this regard. Nevertheless, the possibility should be borne in mind that some compounds or certain enzymes might be affected by direct exposure to the oil. Cleaning the oil as recommended above should help eliminate possible harmful impurities.

Special Considerations

The presence of the oil–water interface creates certain problems. Inactivation of enzymes may be more likely to occur during incubation; high concentrations of bovine plasma albumin (1 mg/mL) have been found to prevent this in situations tested. If a cycling step is carried out in the oil well, the high concentrations of enzymes used form a precipitate during heating that makes the samples difficult to transfer. In this case, the reaction is stopped by adding NaOH to give a 0.1 or 0.2*M* concentration followed by heating 10 min at 90°. The alkali keeps the protein in solution, and the sample can be easily pipeted.

Sample volumes much greater than 10 µL so fill the standard-sized wells that there is danger of breaking through the oil surface with consequent evaporation. When the analysis demands that the volume be expanded beyond 10 µL, the samples can be transferred to ordinary glass test tubes for subsequent steps. In the majority of cases, however, the sample aliquot can be transferred directly from the oil well to the fluorometer tube for the final step and for reading in a vol of 1 mL.

PART 4

Appendix and References

Appendix

Acid-Base Made Easy

Some Definitions

LOGARITHM (TO THE BASE 10)

A logarithm is an exponent. It is a number that, if used as an exponent to 10, will give the number whose log it is.

EXAMPLES

$10^3 = 1000$	$\log 1000 = 3$
$10^{-3} = 0.001$	$\log 0.001 = -3$
$10^{0.3} = 2$	$\log 2 = 0.3$
$10^{-0.3} = 1/2 = 0.5$	$\log 0.5 = -0.3$

Remember, $\log N + \log M = \log N \times M$, $\log N - \log M = \log N/M$, $\log 10^{-N} = \log 1/10^N$, $\log \sqrt{N} = 1/2 \log N$. *(See also* A Homemade Log Table at the end of this section.)

PH

pH is the negative logarithm of the hydrogen ion *concentration* (i.e., $pH = -\log [H^+]$ or 10^{-pH} $[H^+]$). (Note *concentration is always molar.)*

EXAMPLES

$[H^+] = 0.1M$	$0.1 = 10^{-1}$	$pH = 1$
$[H^+] = 0.0001M$	$0.0001 = 10^{-4}$	$pH = 4$
$[[H^+] = 0.5M$	$0.5 = 5 \times 10^{-1}$	$pH = -(0.7 - 1) = 0.3$
$[[H^+] = 0.0002M$	$0.0002 = 2 \times 10^{-4}$	$pH = -(0.3 - 4) = 3.7$
$[H^+] = 5 \times 10^{-9}M$		$pH = -(0.7 - 9) = 8.3$

ACID

Acid is any compound that can lose a hydrogen ion.

EXAMPLES

CH_3COOH, CH_3CH_2OH, $CH_3NH_3^+$, NH_4^+

Note "strong acids," such as HCl, really do not fit this definition, *see* under Conjugate Base.

BASE

Base is any compound than can gain a hydrogen ion.

EXAMPLES

$$CH_3COO^-, CH_3CH_2O^-, CH_3NH_2, NH_3$$

CONJUGATE BASE (OR ACID)

This is the base (or acid) in the following general equation:

$$\text{acid} \leftrightharpoons \text{base} + H^+ \qquad \text{[Eq. A-1]}$$

i.e., every acid has its conjugate base (and every base has its conjugate acid). Strong acids and bases, such as HCl and NaOH, really do not fit this formulation. HCl in solution has already lost its H^+ (i.e., it is completely dissociated and can be written H^+Cl^-). NaOH is really Na^+OH^-, and the real base is OH^-, which qualifies because it *can* gain a H^+:

$$OH^- + H^+ \leftrightharpoons H_2O \qquad \text{[Eq. A-2]}$$

BUFFER

Buffer is a mixture of a conjugate acid and base. It is a buffer (i.e., a pH buffer), because if H^+ is added most of H^+ will combine with the base. Eq. (A-1)], lessening the increase in H^+ concentration that would otherwise result. Similarly, if H^+ is removed from such a mixture, more will be formed from the acid [Eq. (A-1)], lessening the decrease in H^+ concentration that would otherwise result.

p*K*S AND THE HENDERSON-HASSELBALCH EQUATION

The more readily an acid loses H^+, the stronger an acid it is. This strength is measured by its *dissociation constant.*

$$K = \frac{[\text{base}][H^+]}{[\text{acid}]} = \frac{[B][H^+]}{[A]} \qquad \text{[Eq. A-3]}$$

When the acid is 50% dissociated, base and acid are equal and $K = [H^+]$. For example, the dissociation constant of acetic acid is $2.5 \times 10^{-5}M$. Therefore, in a mixture of equal molar concentrations of acetic acid and sodium acetate $[H^+] = 2.5 \times 10^{-5}M$ or pH $= -(0.4 - 5) = 4.6$.

Similarly, the dissociation constant of NH_4^+ is 5×10^{-10}. Therefore, in a mixture of equal molar concentrations of ammonium hydroxide ($NH_3 \cdot H_2O$) and $NH_4^+Cl^-$, $[H^+] = 5 \times 10^{-10}$ or pH $= -(0.7–10) = 9.3$.

Because of this relationship between *K* and $[H^+]$, and by analogy with pH, *the negative log of* K *is called* p*K*. It is convenient to replace *K* and $[H^+]$ in Eq. [A-3] with pH and p*K* by taking the log of each side.

$$\log K = \log \frac{[B][H^+]}{[A]} = \log \frac{[B]}{[A]} + \log [H^+] \quad \text{[Eq. A-4]}$$

rearranging:

$$-\log [H^+] = -\log K + \log \frac{[B]}{[A]} \text{ or } pH = pK + \log \frac{[B]}{[A]} \quad \text{[Eq. A-5]}$$

This is the famous Henderson-Hasselbalch equation. This equation was originally written:

$$pH = pK + \log \frac{[Salt]}{[Acid]} \quad \text{[Eq. A-6]}$$

This was before Brönsted proposed the definition of acids and bases given above. Equation [A-6] is harder to use than Eq. [A-5], and is not quite exact.

From the p*K* of an acid and the Henderson-Hasselbalch equation, it is easy to calculate the pH for any ratio of acid to base.

Base/acid	pH	Base/acid	pH
100	p*K* + 2	2	p*K* + 0.3
1/100	p*K* – 2	1/2	p*K* – 0.3
20	p*K* + 1.3	1	p*K*
1/20	p*K* – 1.3	200	p*K* + 2.3
10	p*K* + 1	5000	p*K* + 3.7
1/10	p*K* – 1	10,000	p*K* + 4

It is easy to make this calculation in your head, as accurately as there is usually any need, especially if you memorize the log of 2 and 1.25 (0.3 and 0.1). *(See* A Homemade Log Table *below.)* You can even forget whether it is base/acid or acid/base in the H-H equation if you simply remember that, if *base* predominates, the pH is on the basic side of the p*K*, and if *acid* predominates, the pH is on the acid side of the p*K*.

However, if you prefer, Table 1 makes the calculation for you and gives the percentage of the buffer that is in the basic form. Notice it takes three times more acid or base to make a pH change of 0.1 unit near the p*K* than it does 1 pH unit away. At 1.3 units from p*K*, a 1% shift from acid to base makes a 0.1 unit pH change. Thus, the useful pH range is limited to about 1 pH unit on either side of the p*K*.

Table 1
Base:Acid Ratios to Give Desired pH

pH	Base/acid	Base as % of A + B	pH	Base/acid	Base as % of A + B
pK + 1.0	10	91	pK – 1.0	10	9
pK + 0.9	8	89	pK – 0.9	8	11
pK + 0.8	6.3	86	pK – 0.8	6.3	14
pK + 0.7	5	83	pK – 0.7	5	17
pK + 0.6	4	80	pK – 0.6	4	20
pK + 0.5	3.2	76	pK – 0.5	3.2	24
pK + 0.4	2.5	71	pK – 0.4	2.5	29
pK + 0.3	2	67	pK – 0.3	2	33
pK + 0.2	1.6	61	pK – 0.2	1.6	39
pK + 0.1	1.25	56	pK – 0.1	1.25	44
pK + 0.0	1.0	50	pK – 0.0	1.0	50

Making a Buffer

A buffer mixture can be made by either (a) mixing the free acid or base with its salt, or (b) adding a calculated amount of strong acid or strong base to the free base or free acid. For example, a pH 4.6 buffer can be made by making the solution 50 mM in acetic acid and 50 mM in sodium acetate, or it can be made by making the solution 100 mM in acetic acid and 50 mM in NaOH. This ignores the fact that some H^+ had to be produced ($10^{-4.6}M$) from the acetic acid. Therefore, the base (acetate) is in fact slightly greater than the added sodium or NaOH. In this case, the difference is only 1 part in 2000. Only below pH 3 and above pH 11, and with low buffer concentrations, does this ever become significant.

Mixture of Two or More Acid-Base Pairs

The H-H equation must apply simultaneously to all acid-base pairs in a given solution, i.e.:

$$\mathrm{pH} = \mathrm{pK_1} + \log \frac{[\mathrm{B_1}]}{[\mathrm{A_1}]} = \mathrm{pK_2} + \log \frac{[\mathrm{B_2}]}{[\mathrm{A_2}]} \text{ etc.} \qquad \text{[Eq. A-7]}$$

Take as an example a mixture of acetic acid, acetate (pK = 4.6), $Tris^+$, and Tris base (pK = 8.1). At pH 7:

$$\log [\mathrm{B_1}]/[\mathrm{A_1}] = \log [\mathrm{Ac^-}]/[\mathrm{HAc}] = 2.4 = \log 500/1 \text{ and}$$
$$\log [\mathrm{B_2}]/[\mathrm{A_2}] = \log [\mathrm{Tris}]/[\mathrm{Tris^+}] = -1.1 = \log 1/12.5 \qquad \text{[Eq. A-8]}$$

DISSOCIATION CONSTANT OF H_2O

There is one acid that must not be forgotten; this is H_2O. It is a very weak acid; its p*K* is 15.7. That means that:

$$\frac{[H^+][OH^-]}{[H_2O]} = 10^{-15.7}\ M \text{ and } pH = 15.7 + \log \frac{[OH^-]}{[H_2O]} \qquad \text{[Eq. A-9]}$$

Because the concentration of H_2O is practically constant in most aqueous solutions (55.5*M*), it has become customary to merge its concentration with the dissociation constant:

$$[H^+][OH^-] = 55.5 \times 10^{-15.7} = 10^{-14} \qquad \text{[Eq. A-10]}$$

In the logarithmic form, this becomes:

$$pH + pOH = 14 \quad \text{or} \quad pOH = 14 - pH \qquad \text{[Eq. A-11]}$$

where pOH is the negative log of $[OH^-]$.

PH OF FREE ACIDS AND BASES

The pH of a weak acid (or base) in the absence of added conjugate base (or acid) is easily calculated in all practical situations. When an acid is dissolved in H_2O, $A \leftrightarrow B + H^+$ (i.e., $B = H^+$). Equation [A-3] rearranged becomes $[B][H^+] = K[A]$, or in this case $[H^+]^2 = K(A)$, and $[H^+] = \sqrt{K[A]}$, or:

$$pH = 1/2(pK - \log[A]) \qquad \text{[Eq. A-12]}$$

For example, 0.1*M* acetic acid has pH = 1/2(4.6 + 1) = 2.8. A 0.01*M* solution (–log [A] = 2) would have a pH of 3.3. The p*K* of NH_4^+ is 9.3. A 50 m*M* solution of NH_4Cl has pH = 1/2(9.3 + 1.3) = 5.3. The pH of a 1*M* solution would be 4.65.

When a free base is dissolved in H_2O, it removes H^+, forming hydroxyl ion: $B + H_2O \leftrightarrow A + OH^-$. Here $[A] = [OH^-] = 10^{-14}/[H^+]$ *(see above)*. Equation [A-3] rearranged becomes $[H^+]/[A] = K/[B]$. Substituting for [A]:

$$[H^+]^2/10^{-14} = K/[B],\ [H^+]^2 = 10^{-14}\ K/[B], \text{ and } [H^+] = \sqrt{10^{-14}K/[B]}, \text{ or}$$

$$pH = 1/2(14 + pK + \log[B]) \qquad \text{[Eq. A-13]}$$

For example, 0.2*M* Tris base has pH = 1/2(14 + 8.1 – 0.7) = 10.7. A 0.1*M* sodium acetate solution has pH = 1/2(14 + 4.6 – 1) = 8.8. (An easy way to remember Eqs. [A-12] and [A-13] is that an acid has a pH halfway between its p*K* and the pH it would have if it were a strong acid, and a base has a pH halfway between its p*K* and the pH it would have if it were a strong base.)

Equations [A-12] and [A-13] are not strictly true, because they ignore the decrease in A or in B required to form H^+ in one case and OH^- in the other. The difference is exceedingly small, except with low concentrations of acids having p*K*s below 3 or bases with p*K*s over 11.

Some Complications with p*K*s

We have been talking about p*K*s and dissociation constants as though they were invariant. Actually, they are affected by temperature and salt concentration (ionic strength). Increasing temperature and increasing ionic strength increase the dissociation constants (decrease the p*K*s). Tables of p*K*s are often calculated for 25° for "infinite dilution" (zero ionic strength). An observed p*K* is therefore, always lower than these "true" p*Ks,* and should properly be called an apparent p*K* and written p*K*'.

Ionic strength has a much greater effect if the dissociation involves a separation of charges than if it does not. Consider the three dissociations:

$$RNH_3^+ \leftrightharpoons RNH_2 + H^+$$

$$RCOOH \leftrightharpoons RCOO^- + H^+ \quad \text{[Re. A-1]}$$

$$RPO_2H^- \leftrightharpoons RPO_2^{2-} + H^+$$

Ionic strength will have little effect in the first case, a moderate effect in the second, and a large effect in the third. (Increasing ionic strength by increasing the dielectric constant of the solution, shields the charges from each other, making it easier for them to come apart.)

As examples, the true p*K* of acetic acid is given as 4.76, but a 0.1*M* acetate buffer has a p*K*' near 4.6. The pK_2 of a phosphate buffer is 7.2, a 0.1*M* phosphate buffer has a pK_2' of 6.8, and in a 1*M* phosphate buffer the pK_2' is around 6. (The pK_3 is even more sensitive to ionic strength.) True p*K*s are given in Table 2, and in some cases, ionic strength effects may need to be taken into account.

There is one final complication in regard to p*K*s. We have been considering NH_4^+, for example, as an acid with p*K* = 9.3 (i.e., a dissociation constant of 5×10^{-10}. Unfortunately, in preBrönsted days, NH_3 was regarded in solution as NH_4OH, which dissociated as $NH_4OH \leftrightarrow NH_4^+ +$

OH^-. Measurements of OH^- concentration indicated a dissociation constant of 2×10^{-5}; or pK = 4.7. To distinguish these, one is written pK_a (for the acid dissociation constant); the other is written pK_b (for the "base dissociation constant"). It is clear that, since pOH + pH = 14, pK_a + pK_b = 14. pK_bs are still to be found in handbooks, often written simply as pK, and it is not always clear whether pK_a or pK_b is meant.

A Homemade Log Table

All the logarithms needed for practical acid-base aspects of most analytical work can be carried in your head, or figured out quicker than you can look up a log table. All you need to remember is that the log of 2 = 0.3 (actually 0.30103) and the log of 1.25 = 0.1 (actually 0.0969). Knowing the log of 2, it is clear that log 4 = 0.6 and log 8 = 0.9. Similarly, log 5 = 0.7 (i.e., log 10 – log 2), and log 2.5 = 0.4 (i.e., log 5 – log 2). We are only missing 0.2, 0.5, and 0.8, which we obtain as the following tabulation:

log 1 = 0.0
log 1.25 = 0.1 (memory)
log 1.6 = 0.2 (log 2 – log 1.25)
log 2 = 0.3 (memory)
log 2.5 = 0.4 (log 10 – log 4)
log 3.2 = 0.5 (log 4 – log 1.25)
log 4 = 0.6 (log 2 + log 2)
log 5 = 0.7 (log 10 – log 2)
log 6.3 = 0.8 (log 5 + log 1.25)
log 8 = 0.9 (log 4 + log 2)
log 10 = 1.0

If you need to be closer than 0.1 pH unit, you can interpolate from this, e.g., log 9 is close to 0.95 (0.954). This list and the Henderson-Hasselbalch equation, both in your head, will save time and, more importantly, avoid blunders.

Statistical Shortcuts

This section is aimed at the "casual statistician" (Snedecor). Rarely, for his purposes, does the situation require or justify elaborate statistical treatment.

We are concerned here only with means, standard deviations, standard errors (of the mean), and standard errors of differences.

First, the orthodox calculations:

$$\text{Standard deviation (St. dev.)} = \sqrt{\frac{\text{sum of dev}^2}{n - 1}} \qquad \text{[Eq. A-14]}$$

where dev is the individual deviation from the means and *n* is the number of samples.

$$\text{Standard error (of the mean) (SE or SEM)} = \sqrt{\frac{\text{sum of dev}^2}{n(n-1)}} = \frac{\text{st dev}}{\sqrt{n}} \quad \text{[Eq. A-15]}$$

$$\text{Standard error of the difference (SED)} = \sqrt{SE_1^{\,2} + SE_2^{\,2}} \quad \text{[Eq. A-16]}$$

There is a longer way of calculating SED, but it has no clear advantage. Now the shortcuts.

1. For most purposes, calculation of SE to within 10% is all that is worthwhile. For example: 110 ± 10 (SE) says that there is a 1 out of 3 chance that the true mean lies outside of the range 90–110. What virtue could there possibly be in calculating that the SE is actually 10.3?
2. Conversely, calculation of the *mean* to closer than 10% of the SE is seldom worthwhile. In the previous example, what possible use would there be in calculating that the average is actually 110.4? If, however, the standard error was 1 instead of 10, it might be useful to calculate the mean more closely, say ± 0.1.
3. An approximate value for the SE is:

$$SE = \frac{\text{range of values}}{n} \quad \text{[Eq. A-17]}$$

For example, if there are 10 values ranging from 80 to 120, SE = 40/10 = 4. This gives a surprisingly close estimate of the SE (usually within 15%). It is a valuable check for errors in more exact calculations. (Note, however, that with $n > 12$ or 13, this gives erroneous small estimates of SE).

4. Similarly, because $SE = SD/\sqrt{n}$:

$$\text{st. dev.} = SE\sqrt{n} = \frac{\text{range}}{\sqrt{n}} \quad \text{[Eq. A-18]}$$

5. For more orthodox calculations of means, SEs, and SDs, individual values and individual *deviations from the mean* can usually be rounded off with great saving of time. What guidelines are there to safe rounding off?

It is usually completely safe to round off individual values to within 10% of the range. For example, there are 10 numbers ranging from 32.2 to 43.6. The range is 11.4. Therefore, the numbers can be rounded off to the nearest whole number. No number will be changed by more than 0.5. This cannot increase the range by more than twice 0.5 or 1 (i.e., 9%).

Therefore, this will not increase the SE or the SD by more than 9%, as the above shortcut method for calculating these statistics indicates.

Rounding off to within 10% of the range will have a completely negligible effect on the mean. Thus, in the above case, the average number is changed by only 0.25, and changes are randomized plus and minus.

Table 2
Dissociation Constants for Acids and Bases Used in Making Buffers

Acid or base	pK_a	Acid or base	pK_a
Oxalic acid, K_1	1.21	4-Phenylimidazole	6.13
β-glycerophosphoric, K_1	1.47	MES[a,h]	6.15
Histidine, K_1	1.73	β-glycerophosphoric, K_2	6.19
Maleic, K_1	1.83	Cacodylic	6.19
Phosphoric, K_1	1.96		
Aspartic, K_1	2.14	2-Phenylimidazole	6.53
Glutamic, K_1	2.14	4-Hydroxyimidazole	6.53
Arsenic, K_1	2.25	Arsenic, K_2	6.77
Lysine, K_1	2.25	Imidazole	7.07
Glycine, K_1	2.35	Phosphoric, K_2	7.12
Phthalic, K_1	2.92	BES[b,h]	7.15
Tartaric, K_1	3.00	MOPS[c]	7.2
Citric, K_1	3.05	2,4,6-Collidine	7.40
Fumaric, K_1	3.10	TES[d,h]	7.5
Glycylglycine, K_1	3.10	HEPES[e,h]	7.55
Malic, K_1	3.40	4-Methyimidazole	7.65
Formic	3.76	Triethanolamine	7.9
Aspartic, K_2	3.87	2-Methyimidazole	8.00
Lactic	3.87	Glycylglycine, K_2	8.07
Barbituric	4.00	Tris[f]	8.1
Glutamic, K_2	4.09	Tricine[g,i]	8.15
Tartaric, K_2	4.13	Pyrophosphoric, K_2	8.22
Succinic, K_1	4.13	Hydrazine, K_2	8.23
Oxalic, K_2	4.22	2,4-Dimethylimidazole	8.48
Fumaric, K_2	4.53	2-Amino-2-methyl-1.3 propanediol	8.78
Citric, K_2	4.75	Lysine	8.92
Acetic	4.76	Histidine, K_3	9.14
Malic, K_2	5.00	Ammonium hydroxide	9.23
Benzoic	5.20	Glutamic acid, K_3	9.43
Phthalic, K_2	5.57	Glycine, K_2	9.75
Succinic, K_2	5.57	Aspartic acid, K_3	9.75
Maleic, K_2	5.75	2-Amino-2-methyl-1-propanol	9.90
Pyrophosphoric, K_3	5.77	Carbonic acid, K_2	10.36
Uric	5.83	Lysine, K_3	10.54
Citric, K_3	6.00	Arsenic, K_3	11.6
Histidine, K_2	6.10	Phosphoric, K_3	12.24

[a]MES, 2-(*N*-morpholino)ethanesulfonic acid.
[b]BES, *N,N-bis*(2-hydroxyethyl)-2-aminoethanesulfonic acid.
[c]MOPS, 3-*N*-(Morpholino)propanesulfonic acid.
[d]TES, *N-tris*(hydroxymethyl)methyl-2-aminosulfonic acid.
[e]HEPES, *N*-2-hydroxyethylpiperazine-*N*'-2-ethanesulfonic acid.
[f]*Tris*, *Tris*(hydroxymethyl)aminoethane.
[g]Tricine, *N-tris*(hydroxymethyl)methylglycine.
[h]At 20°, Good et al., 1966.
[i]Good, 1962.

Table 3
Preparation and Storage of Some Common Stock Solutions[a]

Compound	Storage conditions		Comments
	Powder	Solution	
ATP · Na_2, 551	–20°, desiccant	In H_2O, bring to pH 7 with 2 equiv. NaOH, –20°	No loss after months in storage
ADP · Na_2,471	–20°, desiccant	In H_2O, bring to pH 7 with an equiv. NaOH, –20°	Usually about 0.5–1% ATP on a molar basis, sometimes contains $NADP^+$; $NADP^+$ can be destroyed by bringing solution to pH 12 with NaOH; heat 10 min at 60° and neutralize with HCl
5'-AMP · Na, 369	–20°, desiccant	In H_2O, –20°	–
Albumin, bovine plasma crystalline	4°, desiccant	In H_2O, –20°	Some lots contain citrate and isocitrate
Citric acid · H_2O, 210	4°, desiccant	In H_2O, –20°	–
Creatine-H_2O, 149	RT	In H_2O, –20°	–
Creatine-P · Na_2, 255	–20°, desiccant	In H_2O, –20°	No loss after 60 min at 0° in 0.6*M* $HClO_4$; or at –10° in 3*M* $HClO_4$; no loss after 30 d and probably much longer at pH 8 and –50°; stable to alkali
Dihydroxyacetone-P, Li salt, 182	–20°, desiccant	Store in buffer at pH 3, or up to 0.01*M* acid at –20°	No loss after 2 h at 60° in 0.3*M* $HClO_4$ or at pH 3.0; unstable above pH 8.0

(continued)

Table 3 *(continued)*

Compound	Storage conditions		Comments
	Powder	Solution	
Fructose-1,6-$P_2 \cdot Na_4$ 428	RT	In H_2O, –20°	–
Fructose-6-P · Na 282	RT	In H_2O, –20°	–
Fumaric acid · Na_2, 160	RT	In H_2O, –20°	–
α-D-Glucose-1-P · Na_2, 301	RT, desiccant	In H_2O, –20°	Contaminated to varying degrees with glucose-1,6-P_2; hydrolyzed to glucose in 0.1*M* HCl after 10 min, 100°
α-D-Glucose-1,6-P_2 · K_4 492	RT, desiccant	In H_2O, –20°	Can be standardized by hydrolyzing in 0.1*M* acid at 100° for 10 min and for analyzing glucose-6-P
α-D-Glucose-6-P · Na_2 304	RT	In H_2O, –20°	–
Glutamic acid · Na, 169	RT, desiccant	In H_2O, –20°	–
Glyceraldehyde-3-P, acetal · Ba	RT, desiccant	In acid, pH 1 to 3 at –20°	Remove acetal and Ba^{+2} with cation-diethyl exchange resin, or by adding a slight excess of H_2SO_4 and heating at 60° for 60 min; $BaSO_4$, or resin is removed by centrifugation; no loss after 2 h at 60° in 0.3*M* $HClO_4$ or pH 3; unstable above pH 8
α-Glycero-P · Na_2, 194	RT	In H_2O, –20°	–

Glycogen	RT, desiccant	In H_2O, –20°	Glycogen forms an aggregate on freezing that can be dispersed by heating 10 min at 60°; the storage of weak solutions (1 m*M* or less) is not recommended since there appears to be some loss after freezing and thawing
DL-Isocitrate · 3Na 258	RT, desiccant	In H_2O, –20°	May contain some of the allo isomer
2,3 Diphospho-D-glyceric acid pentacydohexylammonium, 762	RT, desiccant	In H_2O, –20°	–
α-Ketoglutaric acid, 146	RT	In H_2O, –20°, at pH 8, –50°	At pH 8, the α-ketoglutarate is stable for 2 wk at –20° and at least 2 mo at –50°; longer storage results in some loss
Lactate · Li, 96	RT	In H_2O, –20°	Many lactic acid or lactate preparations may contain pyruvate, which can be removed by heating with a slight molar excess of H_2O_2
Malic acid, 134.1	RT	In H_2O, –20°	–
β-NAD^+, free acid 663	–20°, desiccant	In H_2O, –20°	Stable indefinitely
β-NADH · Na_2, 708	–20°, dark, desiccant	5–10 m*M* in carbonate buffer (80 m*M* Na_2CO_3, 20 m*M* $NaHCO_3$); *see* Chapter 1	*See* Chapter 1; the solution is slowly oxidized to NAD^+ if stored at –20° or above; preformed NAD^+ can be destroyed by heating in the carbonate buffer at 100° for 5 min; may contain 5–15% 5'-AMP on a molar basis (*see* AMP method, Chapter 6)

(continued)

Table 3 *(continued)*

Compound	Storage conditions: Powder	Storage conditions: Solution	Comments
$NADP^+ \cdot Na$, 765	–20°, desiccant	In H_2O, –20°	Stable indefinitely
$NADPH \cdot 4Na$, 833	*See* NADH	*See* NADH	May contain $NADP^+$, which can be removed by heating (*see* NAD^+) slowly oxidized to $NADP^+$ at 4°
Oxaloacetic acid, 132	–20°, desiccant	In 0.5*M* HCl, –50°	Unstable except in strong acid; there is 5% loss at pH 7 after 30 min at 25°
2-P-enolpyruvic acid, 3Na, 234	RT, desiccant	In H_2O, –50°	There is some decomposition to pyruvate when stored in solution at –20°
6-P-gluconate · 3Na, 342	RT, desiccant	In H_2O, –20°	Stable for 2 h at 60° in 0.6*M* $HClO_4$ or neutral solution
3-P-glycerate · Na, 230	RT, desiccant	In H_2O, –20°	–
2-P-glycerate · Na, 274	RT	In H_2O, –20°	–
Pyruvic acid, 88.1, or pyruvate Na, 110.1	–20°, desiccant	In H_2O, –50°	May polymerize on storage, which may be reversed in 0.1*M* NaOH at 100° for 1–2 min; if stored in Tris buffer, pH 8, there is a 40% loss in month at 7°, 30–100% loss in 3 d at –18°
Uridine-5-diphospho-glucose · Na, 610	–20°, desiccant	In H_2O, –20°	–

[a]The formula weight is given after the name. Many of these compounds contain variable amounts of H_2O of crystallization, usually recorded on the label. RT is room temperature.

Table 4
Normality of Common Concentrated Acids and Bases

	% by Wt	SG	Normality	Mol wt
Acetic acid	100	1.05	17	60.1
Formic acid	89	1.20	23	46.1
HCl	36	1.18	12	36.5
HNO_3 (fuming)	90	1.5	21	63.0
HNO_3 (constant boiling)	68	1.41	15	63.0
H_2SO_4	96	1.84	36	98.1
H_3PO_4	85	1.70	44	98
$HClO_4$	70	1.67	11.65	100.5
HBr	48	1.50	9	80.9
HI	57	1.70	7.6	128
NH_3 solution	28	0.898	15	17
Hydrazine hydrate	64	1.03	–	32

Table 5
An Abbreviated List of Atomic Weights

Aluminum	Al	27.0
Antimony	Sb	121.8
Arsenic	As	74.9
Barium	Ba	137.4
Bismuth	Bi	209.0
Boron	B	10.8
Bromine	Br	79.9
Cadmium	Cd	112.4
Calcium	Ca	40.1
Carbon	C	12.0
Cesium	Cs	132.9
Chlorine	Cl	35.5
Chromium	Cr	52.0
Cobalt	Co	58.9
Copper	Cu	63.5
Fluorine	F	19.0
Gold	Au	197.0
Hydrogen	H	1.008
Iodine	I	126.9
Iron	Fe	55.8
Lanthanum	La	138.9
Lead	Pb	207.2
Lithium	Li	6.9
Magnesium	Mg	24.3
Manganese	Mn	54.9
Mercury	Hg	200.6
Molybdenum	Mo	96.0
Nickel	Ni	58.7
Nitrogen	N	14.0
Oxygen	O	16.0
Phosphorus	P	31.0
Platinum	Pt	195.1
Potassium	K	39.1
Selenium	Se	79.0
Silicon	Si	28.1
Silver	Ag	107.9
Sodium	Na	23.0
Strontium	Sr	87.6
Sulfur	S	32.1
Tellurium	Te	127.6
Tin	Sn	118.7
Tungsten	W	183.9
Uranium	U	238.1
Zinc	Zn	65.4

References

Chapter 1

1. Negelein, E. and Haas, E. (1935) Über die Wirkungweise des Zwischenferments. *Biochem. Z.* **282,** 206–220.
2. Greengard, P. (1956) Determination of intermediary metabolites by enzymic fluorimetry. *Nature (London)* **178,** 632–643.
3. Kaplan, N. O., Colowick, S. P., and Barnes, C. C. (1951) Effect of alkali on diphosphopyridine nucleotide. *J. Biol. Chem.* **191,** 461–472.
4. Lowry, O. H., Roberts, N. R., and Kapphahn, J. I. (1957) The fluorometric measurement of pyridine nucleotides. *J. Biol. Chem.* **224,** 1047–1064.
5. Lowry, O. H. and Carter, J. G. (1974) Stabilizing the alkali-generated fluorescent derivatives of NAD and NADP. *Anal. Biochem.* **59,** 639–642.
6. Lowry, O. H., Passonneau, J. V., and Rock, M. K. (1961) The stability of pyridine nucleotides. *J. Biol. Chem.* **236,** 2756–2759.
7. Burch, H. B., Bradley, M. E., and Lowry, O. H. (1967) The measurement of triphosphopyridine nucleotide and the role of hemoglobin in producing erroneous triphosphopyridine nucleotide values. *J. Biol. Chem.* **242,** 4546–4554.

Chapter 3

1. Levy, M. (1936) Studies on enzymatic histochemistry XVII. A micro kjeldahl estimation. *C. R. Trav. Lab. Carlsberg, Ser. Chim.* **248,** 101–110.
2. Peters, J. P. and Van Slyke, D. D. (1932) *Quantitative Clinical Chemistry,* vol. 2 (Williams & Wilkins, Baltimore, MD), p. 4.

Chapter 4

1. Richter, D. and Dawson, R. M. C. (1948) Brain metabolism in emotional excitement and in sleep. *Amer. J. Physiol.* **154,** 73–79.
2. Wollenberger, A., Ristau, O., and Schoffa, G. (1960) Eine einfache technik der extrem schnellen abkühling grösserer gewebestücke. *Pflueger's Arch. Gesamte Physiol. Menschen Tiere.* **270,** 399–412.
3. Kerr, S. E. and Ghantus, M. (1937) The carbohydrate metabolism of the brain. III. On the origin of lactic acid. *J. Biol. Chem.* **117,** 217–225.
4. Ferrendelli, J. A., Gay, M. H., Sedgwick, W. G., and Chang, M. M. (1972) Quick-freezing of the murine CNS: Comparison of regional cooling rates and metabolite levels when using liquid nitrogen or Freon-12. *J. Neurochem.* **19,** 979–987.
5. Nelson, S. R., Lowry, O. H., and Passonneau, J. V. (1966) in *Head Injury Conference Proceedings* (Caveness, W. F. and Walker, A. E., eds.), Lippincott, Philadelphia, PA, p. 444.
6. Bessey, O. H., Lowry, O. H., and Love, R. H. (1949) The fluorometric measurement of the nucleotides of riboflavin and their concentration in tissues. *J. Biol. Chem.* **180,** 755–769.

7. Burch, H. B., Bradley, M. E., and Lowry, O. H. (1967) The measurement of triphosphopyridine nucleotide and the role of hemoglobin in producing erroneous triphosphopyridine nucleotide values. *J. Biol. Chem.* **242,** 4546–4554.

Chapter 5

1. Cha, S. and Cha, C.-J. M. (1965) Kinetics of cyclic enzyme systems. *Mol. Pharmacol.* **1,** 178,179.
2. Hintz, C. S., Chi, M. M.-Y., and Lowry, O.H. (1980) Correcting potential defect in an enymatic cycle for NADP. *Anal. Biochem.* **128,** 186–190.
3. Lowry, O. H., Passonneau, J. V., Schulz, D. W., and Rock, M. K. (1961) The measurement of pyridine nucleotides by enzymatic cycling. *J. Biol. Chem.* **236,** 2659–2756.
4. Chi, M. M.-Y., Lowry, C. V., and Lowry, O. H. (1978) An improved enzymatic cycle for nicotinamide adenine dinucleotide phosphate. *Anal. Biochem.* **89,** 119–129.
5. Kato, T., Berger, S. J., Carter, J. A., and Lowry, O. H. (1973) An enzymatic cycling method for nicotinamide-adenine dinucleotide with malic and alcohol dehydrogenases. *Anal. Biochem.* **53,** 86–97
6. Breckenridge, B. McL. (1964) The measurement of cyclic adenlyate in tissues. *Proc Natl. Acad. Sci. USA* **52,** 1580–1586. .
7. Goldberg, N. D., Dietz, S. B., and O'Toole, A. G. (1969) Cyclic guanosine 3',5' monophosphate in mammalian tissues and urine. *J. Biol. Chem.* **224,** 4458–4466.

Chapter 6

1. Kornberg, A. and Pricer, W. E., Jr. (1951b) Enzymatic phosphorylation of adenosine and 2,6-diaminopurine riboside. *J. Biol. Chem.* **193,** 481–495.
2. Pfleiderer, G., Grein, L., and Wieland, T. (1955a) Specific determination of L-alanine and L-glutamine by means of glutamic-pyruvic transaminase. *Ann. Acad. Sci. Fenn. Ser. A2,* **60,** 381–388.
3. Pfleiderer, G., Gruber, W., and Wieland, T. (1955b) Eine enzymatische bestimmung der L-asparaginsäure. *Biochem. Z.* **326,** 446–450.
4. Kornberg, A. (1950) Reversible enzymatic synthesis of diphosphopyridine nucleotide and inorganic pyrophosphate. *J. Biol. Chem.* **182,** 779–793.
5. Moellering, H. and Gruber, G. (1966) Determination of citrate with citrate lyase. *Anal. Biochem.* **17,** 369–376.
6. Thorn, W., Pfleiderer, G., Frowein, R. A., and Ross, I. (1955) Stoffwechselvorgänge im Gehirn bei akuter Anoxie, akuter Ischämie und in der Erholung. *Pflueger's Arch. Gesamte Physiol. Menschen Tiere* **261,** 334–360.
7. Slein, M. W. (1950) Phosphomannose isomerase. *J. Biol. Chem.* **186,** 753–761.
8. Slater, E. C. (1953) Spectrophotometric determination of fructose-1:6-diphosphate, hexose monophosphates, adenosinetriphosphate and adenosine-diphosphate. *Biochem. J.* **53,** 157–167.
9. Vishniac, W. and Ochoa, S. (1952) Fixation of carbon dioxide coupled to photochemical reduction of pyridine nucleotides by chloroplast preparations. *J. Biol. Chem.* **195,** 75–96.

10. Slein, M. W., Cori, G. T., and Cori, C. F. (1950) A comparative study of hexokinase from yeast and animal tissues. *J. Biol. Chem.* **186,** 763–780.
11. Paladini, A. C., Caputto, R., Leloir, L. F., Trucco, R. E., and Cardini, C. E. (1949) The enzymatic synthesis of glucose-1,6-diphosphate. *Arch. Biochem.* **23,** 55–66.
12. Albers, R. W., Koval, G. McKahnn, G., and Ricks, D. (1961) in *Regional Neurochemistry* (Kety, S. S. and Elkes, J., eds.), Pergamon, Oxford, p. 340.
13. Bublitz, C. and Kennedy, C. P. (1954) Synthesis of phosphatides in isolated mitochondria III. The enzymatic phosphorylation of glycerol. *J. Biol. Chem.* **211,** 951–961.
14. Ochoa, S. (1948) Biosynthesis of tricarboxylic acids by carbon dioxide fixation. III. Enzymatic mechanisms. *J. Biol. Chem.* **174,** 133–157.
15. Noll, F. (1966) Methode zur Quantitativen Bestimmung von L(+)-Lactat mittels Lactat-Dehydrogenase und Glutamat-Pyruvate Transaminase. *Biochem. Z.* **346,** 41–49.
16. Kubowitz, F. and Ott, P. (1943) Isoliering und Kristallisation eines Gärungsfermente aus Tumoren. *Biochem. Z.* **314,** 94–117.
17. Strominger, J. L., Maxwell, E. S., and Kalckar, H. M. (1957) *Meth. Enzymol.* **3,** 974.
18. Strominger, J. L. (1955) Enzymic synthesis of guanosine and cytidine triphosphates: A note of the nucleotide specificity of the pyruvate phosphokinase reaction. *Biochim. Biophys. Acta* **16,** 616,617.
19. Noda, L. and Kuby, S. A. (1963) *Meth. Enzymol.* **6,** 223.
20. Lowry, O. H., Schulz, D. W., and Passonneau, J. V. (1964) Effects of adenylic acid on the kinetics of muscle phosphorylase *a. J. Biol. Chem.* **239,** 1947–1953.
21. Barbehenn, E. K., Law, M. M-Y., Brown, J. G., and Lowry, O. H. (1976) Measurement of 5'adenylic acid by stimulation of phosphorylase *a. Anal. Biochem.* **70,** 554–562.
22. Lienhard, G. E. and Secemski, I. I. (1973) P^1,P^5-Di(adenosine-5')penta-phosphate, a potent inhibitor of adenylate kinase. *J. Biol. Chem.* **248,** 1121–1123.
23. Lust, W. D., Feussner, G. K., Barbehenn, E. K., and Passonneau, J. V. (1981) The enzymatic measurement of adenine nucleotides and P-creatine in picomole amounts. *Anal. Biochem.* **110,** 258–266.
24. Manchester, J. K., Chi, M. M.-Y., Carter, J., Pusateri, M. E., McDougal, D. B., and Lowry, O. H. (1990) Measurement of 2-deoxyglucose and 2-deoxyglucose-6 phosphate in tissues. *Anal. Biochem.* **185,** 118–124.
25. Baranowski, T. (1963) in *The Enzymes* (Boyer, P. D., Lardy, H., and Myrback, K., eds.), rev. ed., vol. 7, Academic, New York, p. 63.
26. Velick, S. F. and Furfine, C. (1963) in *The Enzymes* (Boyer, P. D., Lardy, H., and Myrbäck, K., eds.), vol. 7, Academic, New York, p. 243.
27. Massey, V. and Alberty, R. A. (1964) Ionisation constants of fumarase. *Biochim. Biophys. Acta* **13,** 354–359.
28. Ray, W. J. Jr. and Roscelli, G. A. (1964) A kinetic study of the phosphoglucomutase pathway. *J. Biol. Chem.* **239,** 1228–1236.

29. Frieden, C. (1959a) Glutamic dehydrogenase II. The effect of various nucleotides on the association-dissociation and kinetic properties. *J. Biol. Chem.* **234,** 815–820.
30. Frieden, C. (1959b) Glutamic dehydrogenase III. The order of substrate addition in the enzymatic reaction. *J. Biol. Chem.* **234,** 2891–2896.
31. Matschinsky, F. M. (1964) Personal communication.
32. Passonneau, J. V. and Lauderdale, V. R. (1974) A comparison of three methods of glycogen measurement in tissues. *Anal. Biochem.* **60,** 405–412.
33. Lust, W. D., Passonneau, J. V., and Crites, S. K. (1975) The measurement of glycogen in tissues by amylo-α-1,4-α-1,6-glucosidase after the destruction of preexisting glucose. *Anal. Biochem.* **68,** 128–131.
34. Passonneau, J. V. and Rottenberg, D. A. (1973) An assessment of methods for measurement of glycogen synthetase activity including a new direct one-step assay. *Anal. Biochem.* **51,** 405–412.
35. de Azeredo, F. A. M., Feussner, G. K., Lust, W. D., and Passonneau, J. V. (1979) An enzymatic method for the measurement of GTP and GDP in biological samples. *Anal. Biochem.* **95,** 512–519.
36. Kaufman, F. and Alivisatos, S. G. A. (1955) Purification and properties of the phosphorylating enzyme from spinach. *J. Biol Chem.* **216,** 141–152.
37. Cha, S. and Parks, J. E., Jr. (1964) Succinic thiokinase. II Kinetic studies; initial velocity, product inhibition, and effect of arsenate. *J. Biol. Chem.* **239,** 1968–1977.
38. Olson, J. A. and Anfinsen, C. B. (1953) Kinetic and equilibrium studies on crystalline L-glutamic acid dehydrogenase. *J. Biol. Chem.* **202,** 841–856.
39. Hintz, C. S., Chi, M. M.-Y., Fell, R. D., Ivy, J. L., Kaiser, K. K., Lowry, C. V., and Lowry, O. H. (1982) Metabolite changes in individual rat muscle fibers during stimulation. *Am. J. Physiol.* **242,** C218–C228.
40. Krebs, H. A. (1953) Equilibria in transamination systems. *Biochem. J.* **54,** 82–86.
41. Fawaz, E. N., Roth, L., and Fawaz, G. (1966) The enzymatic estimation of inorganic phosphate. *Biochem. Z.* **344,** 212–214.
42. Pontremoli, S., de Flora, A., Grazi, E., Mangiarotta, G., Bonsignore, A., and Horecker, B. L. (1961) Crystalline D-gluconate-6-phosphate dehydrogenase. *J. Biol. Chem.* **236,** 2975–2980.
43. Outlaw, W. H., Jr. and Lowry, O. H. (1979) Measurement of 10^{-7} to 10^{-12} mol of potassium by stimulation of pyruvate kinase. *Anal. Biochem.* **92,** 370–374.
44. Cook, G. A., O'Brien, W. E., Wood, H. G., King, M. T., and Veech, R. L. (1978) A rapid, enzymatic assay for the measurement of inorganic pyrophosphate in animal tissues. *Anal. Biochem.* **91,** 557–565.
45. Jones, M. G. K., Outlaw, W. H., Jr., and Lowry, O. H. (1977) Enzymic assay of 10^{-7} to 10^{-14} moles of sucrose in plant tissues. *Plant Physiol.* **60,** 379–383.

Chapter 7

1. Henriksson, J., Chi, M. M.-Y., Hintz, C S., Young, D. A., Kaiser, K. K, Salmons, S., and Lowry, O. H. (1986) Chronic stimulation of mammalian muscle: Changes in enzymes of six metabolic pathways. *Am. J. Physiol.* **251,** C614–C632.

2. Cole, B. R, Hays, A. E., Boylan, J. B., Burch, H. B., and Lowry, O. H. (1982) Distribution of enzymes of adenylate and guanylate metabolism in rat nephron. *Am. J. Physiol.* **243,** F349–F355.
3. Chi, M. M.-Y., Hintz, C. S., Coyle, E. F., Martin, W. H. III, Ivy, J. L, Nemeth, P. M., Holloszy, J. O., and Lowry, O. H. (1983) Effects of detraining on enzmes of energy metabolism in individual human muscle fibers. *Am. J. Physiol.* **244,** C276–C287.
4. Chan, A. W. K., Perry, S. G., Burch, H. B., Fagioli, S., Alvey, T. R., and Lowry, O. H. (1979) Distribution of two aminotransferases and D-amino acid oxidase within the nephron of young and adult rats. *J. Histochem. Cytochem.* **27,** 751–755.
5. Burch, H. B., Lowry, O. H., Kuhlman, A. M., Skerjance, J., Diamant, E. J., Lowry, S. R., and Von Dippe, P. (1963) Changes in patterns of enzymes of carbohydrate metabolism in the developing rat liver. *J. Biol. Chem.* **238,** 2267–2273.
6. Burch H. B., Choi, S., Dence, C. N., Alvey, T. R., Cole, B. R., and Lowry, O. H. (1980) Metabolic of effects of fructose loads in different parts of the rat nephron. *J. Biol. Chem.* **255,** 8239–8244.
7. Passonneau, J. V., Lust, D. W., and Crites, S. K. (1977) Studies on the GABAergic system in astrocytoma and neuroblastoma cells in culture. *Neurochem. Res.* **2,** 605–617.
8. Pusateri, M. E., Carter, J. G., Berger, S. J., and Lowry, O. H. (1984) Distribution of three enzymes of GABA metabolism in monkey retina. *J. Neurochem.* **42,** 1269–1273.
9. Chi, M. M.-Y., Manchester, J. K, Yang, V. C., Curato, A. D., Strickler, R. C., and Lowry, O. H. (1988) Contrast in levels of metabolic enzymes in human and mouse ova. *Biol. of Reprod.* **39,** 295–307.
10. Hintz, C. S., Turk, W. R., Cambon, N., Burch, H. B., Nemeth, P. M., and Lowry, O. H. (1985) A method for branched-chain amino acid aminotransfersse activities in microgram and nanogram tissue samples. *Anal. Biochem.* **146,** 418–422.
11. Burch, H. B., Cambon, N., and Lowry, O. H. (1985) Branched-chain amino acid aminotransferase along the rabbit and rat nephron. *Kidney International* **20,** 114–117.
12. Ichihara, A. and Koyama, A. (1966) Transaminase of branched chain amino acids. I. Branched chain amino acids α-ketoglutarate transaminase. *J. Biochem. (Tokyo)* **59,** 160–169.
13. Taylor, R. T. and Jenkins, W. T. (1966) Leucine aminotransferase. I. Colorimetric assays. *J. Biol. Chem.* **241,** 4391–4395.
14. Lowry, C. V., Kimmey, J. S., Felder, S., Chi, M. M.-Y., Kaiser, K. K., Passonneau, P. N., Kirk, K. A., and Lowry, O. H. (1978) Enzyme patterns in single human muscle fibers. *J. Biol. Chem.* **253,** 8269–8277.
15. Dietrich, W. D., Durham, D., Lowry, O. H., and Woolsey, T. A. (1981) Quantitative histochemical effects of whisker damage on single identified cortical barrels in the adult mouse. *J. Neurosci.* **1,** 929–935.
16. Lowry, O. H., Berger, S. J., Carter, J. G., Chi, M. M.-Y., Manchester, J. K., Knor, J., and Pusateri, M. E. (1983) Diversity of metabolic patterns in human brain tumor enzymes of energy metabolism and related metabolites and cofactors. *J. Neurochem.* **41,** 994–1010.

17. Lowry, O. H., Berger, S. J., Chi, M. M-Y., Carter, J. G., Blackshaw, A., and Outlaw, W. (1977) Diversity of metabolic patterns in human brain tumors. I. High energy phosphate compounds and basic composition. *J. Neurochem.* **29,** 959–977.
18. Hintz, C S., Lowry, C. V., Kaiser, K., McKee, D., and Lowry, O. H. (1980) Enzyme levels in individuat rat muscle fibers. *Am. J. Physiol.* **239,** C58–C65.
19. Bass, A., Brdiczka, D., Eyer, S., Hofer, S., and Pette, D. (1969) Metabolic differentiation of distinct muscle types at the level of enzymatic organization. *Eur. J. Biochem.* **10,** 198–206.
20. Burch, H. B., Narins, R. G., Chu, C., Fagioli, S., Choi, S., McCarthy, W., and Lowry, O. H. (1978) Distribution along the rat nephron of three enzymes of gluconeogenesis in acidosis and starvation. *Am J. Physiol.* **235,** F246–F253.
21. Chi, M. M.-Y., Hintz, C . S., Henriksson, J., Salmons, S., Hellendahl, R. P., Park, I. L, Nemeth, P. M., and Lowry, O. H. (1986) Chronic stimulation of mammalian muscle: Enzyme changes in individual fibers. *Am. J. Physiol.* **251,** C633–C642.
22. Burch, H. B., Bross, T. E., Brooks, C. A., Cole, B. R., and Lowry, O. H. (1984) The distribution of six enzymes of oxidative metabolism along the rat nephron. *J. Histochem. Cytochem.* **32,** 731–736.
23. Kato, T. and Lowry, O. H. (1973) Enzymes of energy converting systems in individual mammalian nerve cell bodies. *J. Neurochem.* **20,** 151–163.
24. Lowry, O. H., Roberts, N. R., and Lewis, C. (1956) The quantitive histochemistry of the retina. *J. Biol. Chem.* **220,** 879–892.
25. Lowry, O. H., Roberts, N. R., and Chang, M-L. W. (1956) The analysis of single cells. *J. Biol. Chem.* **222,** 97–107.
26. Kato, T. and Lowry, O. H. (1973) Distribution of enzymes between nucleus and cytoplasm of single nerve cell bodies. *J. Biol. Chem.* **248,** 2044–2048.
27. Curthoys, N. P. and Lowry, O. H. (1973) The distribution of glutaminase isoenzymes in the various structures of the nephron in normal, acidotic, and alkalotic rat kidney. *J. Biol. Chem.* **248,** 162–168.
28. Curthoys, N. P. and Kuhlenschmidt, T. (1975) Phosphate-independent glutaminase from rat kidney: partial purification and identity with gamma-glutamyltranspetidase. *J. Biol. Chem.* **250,** 2099–2105.
29. Lowry, O. H., Shulz, D. W., and Passonneau, J. V. (1964) Effects of adenylic acid on kinetics of muscle phosphorylase *a*. *J. Biol. Chem.* **253,** 8269–8277.
30. Hsieh, B., Chi, M. M.-Y., Knor, J., and Lowry, O. H. (1979) Enzymes of glycogen metabolism and related metabolites in preimplantation mouse embryos. *Dev. Biol.* **72,** 342–349.
31. Henry, C. G. and Lowry, O. H. (1985) Enzyme and metabolites of glycogen metabolism in canine cardiac Purkinje fibers. *Am. J. Physiol.* **248,** H599–H605.
32. Passonneau, J. V. and Rottenberg, D. A. (1973) An assessment of methods for measurement of glycogen synthetase activity including a new direct one-step assay. *Anal. Biochem.* **51,** 528–541.
33. Berger, S. J., DeVries, G. W., Carter, J. G., Schulz, D. W., Passonneau, P. N., Lowry, O. H., and Ferrendelli, J. A. (1980) The distribution of the components of the cyclic GMP cycle in retina. *J. Biol. Chem.* **255,** 3128–3135.

34. Cole, B. R., Hays, A. E., Boylan, J. B., Burch, H. B., and Lowry, O. H. (1982) Distribution of enzymes of adenylate and guanylate metabolism in rat nephron. *Am. J. Physiol.* **243,** F349–355.
35. Teutsch, H. F. and Lowry, O. H. (1982) Sex specific regional differences in hepatic glucokinase activity. *Biochem. Biophys. Res. Comm.* **106,** 533–538.
36. Seltzer, J. L. and McDougal, D. B., Jr. (1975) Enzyme levels in chick embryo heart and brain from 1–21 days of development. *Develop. Biol.* **42,** 95–105.
37. Burch, H. B., Kuhlman, A. M., Skerjance, J., and Lowry, O. H. (1971) Changes in patterns of enzymes of carbohydrate metabolism in the developing rat kidney. *Pediatrics* **47,** 199–296.
38. Holowach, J., Kauffman, F., Ikossi, M. G., Thomas, C., and McDougal, D. B. Jr. (1968) The effects of a thiamine antagonist, pyrithiamine, on levels of selected metabolic intermediates and on activities of thiamine-dependent enzymes in brain and liver. *J Neurochem.* **15,** 621–631.
39. Mourad, N., and Parks, R. E., Jr. (1966) Erythrocytic nucleoside diphosphokinase II. Isolation and kinetics. *J. Biol. Chem.* **241,** 271–278.
40. Goffeau, A., Pedersen, P. L., and Lehninger, A. L. (1967) The kinetics and inhibition of the adenosine diphosphate adenosine triphosphate exchange catalyzed by a purified mitochondrial nucleoside diphosphokinase. *J. Biol. Chem.* **242,** 1845–1853.
41. Lowry, O. H. and Passonneau, J. V. (1966) Kinetic evidence for multiple binding sites on phosphofructokinase. *J. Biol. Chem.* **241,** 2268–2279.
42. Kahana, E. E., Lowry, O. H., Schulz, D. W., Passonneau, J. V., and Crawford, E. J. (1961) The kinetics of phosphoglucoisomerase. *J. Biol. Chem.* **235,** 2178–2184.
43. Lowry, O. H., Roberts, N. R., Schulz, D. W., Clow, J. E., and Clark, J. R. (1961) Quantitative histochemistry of retina. II. Enzymes of glucose metabolism. *J. Biol. Chem.* **236,** 2813–2820.
44. Cornell, N. W., Leadbetter, M., and Veech R. L. (1979) *J. Biol. Chem.* **254,** 6627–6527.
45. Lowry, O. H. and Passonneau, J. V. (1964) The relationships between substrates and enzymes of glycolysis in brain. *J. Biol. Chem.* **239,** 31–42.
46. Pitts, F. N., Jr. and Quick, C. (1965) Brain succinic semialdehyde dehydrogenase. I. Assay and distribution. *J. Neurochem.* **12,** 893–900.
47. Middleton, B. (1973) The oxoacyl-coenzyme A thiolases of animal tissues. *Biochem. J.* **132,** 717–730.
48. Corder, C. N., Berger, M. L., and Lowry, O. H. (1974) Quantitative histochemistry of uridine diphosphoglucose-6-pyrophosphatase and uridine diphosphoglucose-pyrophosphorylase in developing rat kidney. *J. Histochem. Cytochem.* **22,** 1034–1038.

Chapter 9

1. Linderstrom-Lang, K., Holter, H., and Ohlsen, A. S. (1935) Studies on enzymatic histochemistry XIII. The distribution of enzymes in the stomach of pigs. *Compt.-Rend. Lab. Carlsberg, Serie Chim.* **20,** 66–125.
2. Glick, D. (1961,1963) *Quantitive Chemical Techniques of Histo- and Cytochemistry,* vols. 1 and 2 (Wiley [Interscience]), New York.

3. Cole, B. R., Boylan, J. G., Bross, T. E., Burch, H. B., and Lowry, O. H. (1988) Progressive enzyme changes within anatomically defined segments of rat nephron: Demonstration of a new technique. *J. Histochem. Cytochem.* **36,** 285–289.
4. Barbehenn, E. K., Wales, R. G., and Lowry, O. H. (1974) The explanation for the blockade of glycolysis in early mouse embryos. *Proc. Natl. Acad. Sci. USA* **71,** 1056–1060.
5. Chi, M. M.-Y., Manchester, J. K., Yang, V. C., Curato, A. D., Strickler, R. C., and Lowry, O. H. (1988) Contrast in levels of metabolic enzymes in human and mouse ova. *Biol. of Reprod.* **39,** 295–307
6. Outlaw, W. H., Jr. and Lowry, O. H. (1977) Organic acid and potassium accumulation in guard cells during stomatal opening. *Proc. Natl. Acad Sci. USA* **74,** 4434–4438.
7. Wenger, B. (1955) Personal communication.
8. Essén, B. Jansson, E., Henriksson, J., Taylor, A. W., and Saltin, B. (1975) Metabolic characteristics of fibre types in human skeletal muscle. *Acta Physiol. Scand.* **95,** 153–165.
9. Lowry, C. V., Kimmey, J. S., Felder, S., Chi, M. M.-Y., Kaiser, K. K., Passonneau, P. N., Kirk, K. A., and Lowry, O. H. (1978) Enzyme patterns in single human muscle fibers. *J. Biol. Chem.* **253,** 8269–8277.
10. Hintz, C. S., Lowry, C. V., Kaiser, K. K., McKee, D., and Lowry, O. H. Enzyme levels in individual rat muscle fibers. *Am. J. Physiol.* **239,** C58–C65.

Chapter 12

1. Lowry, O. H. (1963) The chemical study of single neurons. *Harvey Lect.* **58,** 1–19.
2. Matschinsky, F. M., Passonneau, J. V., and Lowry, O. H. (1968a) Quantitative histochemical analysis of glycolytic intermediates and cofactors with an oil well technique. *J. Histochem. Cytochem.* **16,** 29–39.
3. Matschinsky, F. M., Rutherford, C. R., and Ellerman, J. E. (1968) Accumulation of citrate in pancreatic islets of obese hyperglycemic mice. *Biochem. Biophys. Res. Commun.* **33,** 855–862.

Appendix

Good, N. E. (1962) Uncoupling of the Hill reaction from photophosphorylation by anions. *Arch. Biochem. Biophys.* **96,** 653–661.

Good, N. E., Winger, G. D., Winter, W., Connolly, T. N., Izawa, S., and Singh, R. M-M. (1966) Hydrogen ion buffers for biological research. *Biochemistry* **5,** 467–477.

Preface

Bergmeyer, H. V. (1970) in *Methods of Enzymatic Analysis,* 2nd ed. (Verlag Chemie, Weinheim).

390

Index

G